Dark Laboratory

Dark Laboratory

ON COLUMBUS, THE CARIBBEAN, AND THE ORIGINS OF THE CLIMATE CRISIS

Tao Leigh Goffe

DOUBLEDAY
New York

All rights reserved. Published in the United States by Doubleday,
a division of Penguin Random House LLC, New York, and distributed
in Canada by Penguin Random House Canada Limited, Toronto.

www.doubleday.com

DOUBLEDAY and the portrayal of an anchor with a dolphin are
registered trademarks of Penguin Random House LLC.

Page 343 constitutes an extension of the copyright page.

Front-of-jacket illustration: palm tree © Alhontess / iStock / Getty Images
Jacket design by Oliver Munday
Book design by Cassandra J. Pappas
Map copyright © 2025 by David Lindroth, Inc.

Library of Congress Cataloging-in-Publication Data
Name: Goffe, Tao Leigh, author.
Title: Dark laboratory : on Columbus, the Caribbean,
and the origins of the climate crisis / Tao Leigh Goffe.
Description: First edition. | New York : Doubleday, [2025] |
Includes bibliographical references and index.
Identifiers: LCCN 2024016657 (print) | LCCN 2024016658 (ebook) |
ISBN 9780385549912 (hardcover) | ISBN 9780385549929 (ebook)
Subjects: LCSH: Caribbean Area—Environmental conditions. |
Caribbean Area—Colonization. | Europe—Colonies—America—History. |
Imperialism—Environmental aspects—Caribbean Area. |
Climatic changes—Social aspects—Caribbean Area.
Classification: LCC GE160.C27 (print) | LCC GE160.C27 (ebook) |
DDC 304.2/8098611—dc23/eng20240904
LC record available at https://lccn.loc.gov/2024016657
LC ebook record available at https://lccn.loc.gov/2024016658

MANUFACTURED IN THE UNITED STATES OF AMERICA
1 3 5 7 9 10 8 6 4 2
First Edition

To the islands,
the coral reefs,
the blue-footed boobies,
the mountains made of guano,
the Calcutta mongoose,
the budding marijuana,
and the volcanoes
with which I have written this book.

Some of those that work forces,
Are the same that burn crosses.

—RAGE AGAINST THE MACHINE,
"Killing in the Name," 1992

We are dreaming all wrong. Art is otherwise, all that is
left, able to lift the grime and glitter caked under eye-
lids and halt, thereby, our crippled, crippling dreaming. Truth is otherwise. It risks all to be born, to be
unstoppably, irresistibly alive.

—TONI MORRISON,
The Foreigner's Home, 2018

Everyting flood out!

—BRIDGETTE "TUTTY GRAN ROSIE" BAILEY, 2013

Guanahani

BAHAMAS

CUBA

Cayman
Brac

CAYMAN
ISLANDS

JAMAICA
(Xaymaca)

Hispaniola

HAITI
(Ayiti)

DOMINICAN
REPUBLIC

Navassa
(Devil's Island)

CARIBBEAN   SEA

COLOMBIA

Montego Bay   Leeward
Maroons

Ocho
Rios   GoldenEye

Negril   Maroon Town

Cockpit
Country

Windward Maroons   Moore To

Kingston

Jamaica
(Xaymaca)

Maroon Territories

Bauxite Deposits/Mines

All-Inclusive Resorts

CARIBBEAN SEA

0   MILES

0   KM   5

The Caribbean

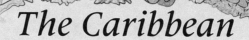

0 MILES 300

0 KM 300

—Montserrat

—Dominica
(Wai'tu kubuli)

—Martinique

—Barbados

St. Vincent and
the Grenadines

—Grenada

TRINIDAD
AND
TOBAGO

*ATLANTIC
OCEAN*

VENEZUELA

GUYANA

SURINAME

Contents

A Note on Terminology   *xv*

Introduction: Mountain Ballads   *xix*

A Map Through the Laboratory   *xxxv*

Book I

EDEN IS NOT LOST

Lo Ting and Mami Wata—Garden Interlude   3

1. **Island Laboratories**   7

   British Empire's Twin Experiments: Hong Kong and Jamaica   12

   In Search of Camouflage Vines   20

   Queen Nanny of the Maroons: Commander, Climate Warrior,
        Healer   25

   Mining the Red Earth: Jamaica's Global Bauxite Crisis   32

2. **Climate Crisis, Genesis 1492**   40

   Eden Isles: Black and Native Governance   44

   Columbus's Crisis   53

   Genres and Genealogies of Eden   58

Climate Denial, Racial Denial   62
What Is the Dark Laboratory?   67

3. **Natural History Museums, Shrines to Explorers**   74
   Enshrining Natural History   78
   Environmentalism and the Omission of Race   85
   Digitize the World: Endangered Archives, Endangered Species   87
   The Violence of Knowledge Production and University Museums   89

Book II
GEOLOGICAL TIME AND BLACK WITNESS

4. **Breathing Underwater**   95
   Coral Coloniality and Collectivity   104
   The Freedom of Black Underwater Worlds   106
   Corals and the Jagged Taxonomy of Race   110
   Darwin's Pacific Coral Pilgrimage   117
   Capturing Coral   123
   Regeneration Beyond the Future of Bleached Coral   127

5. **Guano Destinies**   132
   Shithole Countries: From Emerson to Trump   136
   A Diamond as Big as Haiti's Sovereign Debt   144
   Peru: Chinese and Indigenous Labor, Andean Agricultural Science   147
   "I Expect You to Die": James Bond's Guano Destiny   155
   Hawai'i: The Refuse and Refusal of Pacific Guano Economies   158

6. **Colonialism, the Birder's Companion Guide**   161
   Invisible Branches of the American Family Tree   168
   The Amazon: Birding as Colonial Practice   175
   Four and Twenty Blackbirds: Kerry James Marshall's Pecking
      Order   185
   White Flight, Black Mobility, White Fugitivity   192
   Jamaicenesis   196
   Birding Futures   198

Book III
LIFE IN THE GARDEN AFTER EDEN

7. **The Curious Case of the Calcutta Mongoose in Jamaica** 203
The Veil of Ignorance: A Collective Choreography of Mothering 211
Indian Natural History and the Panchatantra 215
"Massa Espeut's Ratta": Biopolitics and the Chimeras of Empire 216
Pedigree and White Flight 221
Sly Mongoose, the Predator 222
Indigenous Knowledge: Challenging "Invasive" Rhetoric 228

8. **Pedagogies of Smoke** 231
Ganja, the Ganges, and How Indians Brought Marijuana to Jamaica 235
The Glasshouse Effect: China, Scotland, the Caribbean 237
Genres of Gardens 240
Provision Grounds and Botánicas: Beyond the English
    Country Garden 244
Mutiny in the Garden: Twinning Tahiti and Jamaica 247
Commerce, Criminalization, Climate, and Legalization 249

9. **Affective Plate Tectonics** 254
Decentralized Island Futures: Climate Federation 258
The Pyroclastic Tragedy of Mount Pelée, Martinique, 1902 265
Black Irish Freedom and the Fire This Time 270
"Not in Our Time": The Volcano as Timekeeper 277
Ghost Chickens Come Home to Roost 280

Coda: Burial at Sea 289

Acknowledgments 301
Notes 309
Bibliography 323
Index 327

A Note on Terminology

When we surrender to language that limits our capacity to identify the structures of power that enforce racism, we have already lost half the battle. A colorblind approach prevents identifying anything at all about systemic inequality. I use Black with a capital *B* to denote references to Black people. I use a lowercase *b* in reference to the color black and abstract senses of blackness and anti-blackness. I use a lowercase *b* for "brown" because the term has a different political trajectory and is used by many disparate groups with varying and more diffuse meanings than the history of the Black Power movement. Relatedly, I use a lowercase *w* for "white" because while it is also an important racial category that should be spoken about more often, it does not have a coherent political history, community, or tradition, though white supremacy is a potent, divisive force.

Because there are many forms of slavery that vary as labor institutions across time and space, I use the specific terms "chattel slavery" and "racial slavery" to describe the European transatlantic trade of Africans who were enslaved. The distinction is important because historically (and in the present) there are multiple violent forms of slavery, but none have treated humans as property to be bought and sold at auction, stripping people of their basic personhood, in the way that the enslavement of Africans took place during the European colonization of the Americas, a dehumanizing and racializing process that took place from the sixteenth to the nineteenth centuries.

When referring to the system of temporary contract-enforced debt labor that coincided with and followed racial slavery in the nineteenth century, I use the term "racial indenture." This nomenclature is distinct from the "indentured servitude" of white Europeans from an earlier period. Muddying the terms erases the racializing nature of the labor scheme for people in the nineteenth century who were primarily Asian (Chinese and Indian) and were racialized by the process because of the proximity to racial slavery and Black people. There were also indentured laborers from Africa and Southeast Asia who were forcibly brought to the Caribbean. By referring to these people, who rebelled against inhumane labor conditions, as "workers" or "servants," we belittle the fraught nature of volition and consent under the mutilation of the whip on Caribbean plantations.

Race is a construct that is full of social meaning that translates as political power. I have carefully chosen the following terms not to enshrine false hierarchies, but to identify how people have historically identified themselves against the terms of colonial conquest. In regard to people indigenous to the Americas, I use the related terms "Amerindian"/"American Indian"/"Native"/"Native American"/ "Indian"/"Indigenous"/"First Nations," understanding that they are not synonymous in meaning. According to the regional context of what various Indigenous peoples in different countries have historically called themselves, I alternate between terms.

Relatedly, the overlap between the terms "Indian" (Columbus's misnomer used to describe Native Americans) and "South Asian" is pertinent to the premise of this book. In the Caribbean context, I will sometimes use "East Indian," "Indo-Caribbean," or "Hindustani" interchangeably and according to regional context.

I use the terms "enslaved African/African/Afro-," which are not interchangeable, to describe different Black identities, and avoid using the word "slave" or "Negro" or any of their derivations or translations in other European languages.

I use the term "Chinese" in favor of "Sino" based on regional preference. Sino tends to have a foreign-policy connotation in the Western world that can at times flatten Chinese heterogeneity. I avoid the term "coolie" because of the racial injury that it carries when spoken in varying national contexts to refer to Indian, Chinese, and Southeast Asian subsistence laborers and people.

I do not use terms such as "half" to describe people or other racial fractions because the eugenicist logic of blood quantum, used by the U.S. government to police Native peoples, is so casually spread throughout our everyday language. Relatedly, ahistorical terms such as "biracial" are not useful to me or the arguments in this book. For similar reasons, the term "mixed-race" is not useful to me or the arguments put forth in this book (though I recognize these are commonplace terms that others identify with). People of mixed heritage have existed for generations and centuries without such designations that can often serve to splinter possible political coalitions. Beginning to disavow these mathematical terms is a step toward rejecting these norms of valuation and categorization of human life on a pseudoscientific genetic scale we know often to be Eurocentric or anti-black.

Introduction
Mountain Ballads

Mountains hold the echoes of history. The vibrations and shock waves of the climate crisis are written in stone, absorbed over the course of geologic time dating back more than four billion years. These mountain ranges were once submerged underwater. If we measure the span of existence by the recent rock record, it tells a layered climate history of precious materials stolen from the earth—mountains of coral, gold, bauxite, and guano—sedimented. Time avalanches with the heaviness of histories of labor exploitation as we reckon with the overwhelming and inevitable ecological crises of the twenty-first century. Within each chapter of climate history exists a labor history, and we draw upon the energy of those who have gone before us. Those forced to extract from the land continue to be deemed disposable for the price of so-called technological progress. The racial regime determined who was forced to extract ores from the earth, and it continues to this day with unnamed millions mining cobalt to source lithium-ion batteries to satisfy the demand for the rechargeable batteries of our electric cars and smartphones.[1]

This is our current perverse geological reality of powering so-called green technologies and clean energy solutions. Standing at the precipice of climate collapse, we feel the increasing pressure for the planet to implode with each breath we take. Yet, as a global community, we continuously fail to address the origin of the problem. Without economic and historical analyses of the origins of the climate crisis, how can we expect to understand its sedimented layers?

While it is easy to picture plumes of smoke as the primary output of global warming, we must ask questions about the global economy that preceded our dependence on fossil fuels and what remains unseen. The brutal order of plantations organized the world before there were smokestacks and factories. The economies and ideas of plantation slavery have irreparably scarred the natural environment. Before idyllic pastures, mass deforestation was necessary to clear the way for farming. Plantation owners and overseers mutilated the flesh of those they held captive under the whip. With each tree felled, carbon was released into the atmosphere. Sacred branches, hundreds of years old, that had witnessed the first European colonizers were razed in an instant. Gone with this botanical life were the multitude of medicines and materials critical to Indigenous life and traditions. Vacating the land of vast and complex biomes and ecosystems, seventeenth-century monocrop agriculture transformed the planet as new agricultural practices stripped the soil of nutrients. While precious metals and rocks—gold, silver, and quarries of marble—had long been prized throughout the human history of mining, coal and oil have transformed how the world breathes. Agricultural and mining industries asphyxiate the future, the smoke emitted from fossil fuels burning our lungs as it propels a cycle of greed and disposability. Today, wind currents carry debris from faraway forest fires that irritate our airways as we inhale a sky turned a burnt shade of orange. The air is hazy and thick with the ashes of those who have been deemed disposable.

High in the mountains where the air is fresh, the land holds the memory of existential hope amid ecological catastrophes. Mountains carry messages that have been communicated across time and space. The land recalls things that we cannot. Across Black poetic traditions, mountains have held significant political, religious, and environmental meaning. "Go, tell it on the mountain." James Baldwin returned to the lyrics of the African American spiritual for his 1953 book.[2] "Over the hills and everywhere." Initially, Baldwin was going to name his Jim Crow narrative *Crying Holy;* the sound of the Black Church in these mountain lyrics was Baldwin's choice for his urgent articulation.

During the nineteenth century, the lines reverberated from African Americans as populations passed secret instructions about how to travel and communicate safely across mountain ranges to freedom. The message of freedom from the Christian Testament of the Bible was amplified and reinterpreted alongside the prophecy of Black hope and liberation. The good news of the Underground Railroad spread across the landscape of Appalachia and the Caribbean and the Atlantic seascape. It stretched from the Bahamas to Mexico to Nova Scotia.

When, in 1967, Martin Luther King Jr. famously asked, "Where do we go from here?," he could have easily been asking a climate question.[3] From the mountaintop there is a scale of global imagination and collaboration needed for salvation from the climate crisis. King made this speech the day before he was assassinated, perhaps prophesying that he would not make it to the Promised Land, but that God had shown him—as he had shown Moses—a view from the mountaintop. Protest songs echo across mountains as climate movements grow with the same demands for racial justice and land sovereignty.

Freedom for everyone requires a confrontation with the capitalist greed upon which Western society was founded. That sin of chattel slavery was cultivated on plantations across the United States, the Caribbean, and Latin America for centuries. Stolen land and stolen life extend across the hemisphere from sea to shining sea, and so too do the freedom trails that Black and Native peoples built during the time of enslavement out of the wilderness from Vermont to Haiti to Argentina. Black runaways had long carved clandestine routes out of the forests to freedom alongside the paths of Native peoples established thousands of years earlier. Both Black and Native peoples looked to the stars, reading constellations toward routes of freedom.[4] For centuries, nations of Afro-Indigenous peoples found sanctuary in swamps and bogs. From the Seminoles of Florida to the Garifuna of Belize, people echoed declarations of Black Native sovereignty in parallel and in unison. Over the centuries, survival has become an art form that is also an ecological philosophy; Black and Indigenous peoples understand what it means to survive multiple time lines of apocalypse. The genocide of colonial encounter is ongoing, visible in the uneven effects of climate crisis on people of color.[5]

By the year 2050, it is estimated that one billion people worldwide will be displaced due to natural disasters. Within the next twenty years, the majority of the world's biodiversity will have been lost due to the endangerment of key ecosystems, including coral reefs, mangroves, and mountains.[6] Tsunamis, active volcanoes, tornadoes, earthquakes, and unrelenting heat waves abound. While I am not a climate scientist, as a professor of literature and history, I have evolved my lesson plans to face the cascade of global ecological crisis. It is the reality and the responsibility of the individual to face this crisis and our role within it. The weather is not simply changing, as climate deniers would like us to believe. We are experiencing the consequences of a centuries-long cycle of exploitation of people of color, whom European colonial powers have forced to extract resources from the earth. The archival research I undertake on the history of racial slavery feels detached without reckoning with the climate crisis as an ongoing colonial crisis.

Poring through ledgers in Caribbean archives, I return each year not knowing what hurricane season will destroy of the delicate yellowing papers, the records the British, French, Danish, Dutch, and Spanish left behind of their exploits. Academic researchers seem to care more for the humidity and climate control of such documents than they do for human life on these islands, the descendants who have survived these regimes. To conduct research in a vacuum is a cold, unethical, and extractive approach. While my career began in such dusty record offices and official state archives in tropical countries, it soon led me to the outdoors and the importance of reclaiming environmental histories for people of color. I began to hear the birdsong that we are losing. I knew these soundtracks had been there over millennia and that we were in danger of losing them note by note just as the birds lose their habitats. The research methods I learned spending a decade in Ivy League classrooms at Princeton and Yale—close reading and other forms of textual analysis—were important, but they no longer served me. Why set about the mission of trying to extract some hidden meaning that only those specifically trained and deemed sufficiently erudite can interpret?

Instead, I learned to let go, to surrender to my surroundings. I became adaptive and iterative in my approach, which led me to become the founder of my own initiative, the Dark Laboratory. My lab is a space for research on climate, race, and technology, and, more important, it is a philosophy. We at the lab understand that climate crisis cannot be solved without solving racial crisis. The two are inseparable. With over thirty members worldwide, from Helsinki to Hilo to Portland to New York, we collaborate on a wide range of creative storytelling projects centering on Black and Indigenous ecologies.

Once I realized that European colonial archives are evidence lockers full of crimes against humanity, I began to stop arguing the case in court, as it were. As it often occurs with genocide, there is simply *too much* evidence. Those who have benefited from the crimes of colonialism refuse to be convinced. The weight of transatlantic history and its impact on our current climate crisis is overwhelming. The evidence is so proudly and meticulously preserved. It becomes liberating then to evaluate not what lies enclosed within the walls of colonial architectures, but to begin to comprehend the significance of what is outdoors. What was stolen from the natural environment and can still be salvaged. Ways of life, scientific traditions, philosophies of environmental coexistence. The magnitude of what was stolen in the past is only a small fraction of what can be restored in the future. There are unborn worlds of scientific possibility from multiple traditions with answers, strategies, and solutions for tackling the climate crisis. Denial of the colonial condition, denial of the origins of the crisis, limits our imagination and how to live after. Denial clouds our memory of the thriving world markets that functioned before the advent of modern capitalism.

In my second year of graduate school, I enrolled in a course taught by former U.K. prime minister Tony Blair called "Faith and Globalization." After the photo ops were done, I realized that class taught me how at odds my presence as a human being was with the curriculum. It was befitting that this was a cross-listed Yale business school and divinity school course. Future CEOs and reverends sat side by side, and then there was me. The final exam was to create a business plan for a religious startup company. Many churches are, after all, business enterprises. During class discussion, when asked to answer the question of when globalization began, I raised my hand and said, "Fourteen ninety-two," noting the Christian expedition undertaken in the name of Spain. Others, including the teaching assistant, were content to celebrate the 1970s as the genesis of globalization. They saw it as a booming period of economic and technological growth that brought the four corners of the earth together like never before. But what about my ancestors who arrived here hundreds of years ago? I asked. I thought of the violence of the modern world wrought by Christopher Columbus's arrival in the Caribbean carrying the cross. Faith and globalization. Africa, Asia, the Middle East, and Europe were all joined across the chain of islands where Indigenous peoples were dispossessed and murdered.

This is my inheritance.

I gathered the courage to ask: Was this not globalization? The TA moved on. After class, when everyone else had gone, he apologized to me for being incorrect with his 1970s definition. He had never heard my definition of globalization or modernity beginning in 1492. I am sure Professor Tony Blair was not ready to address this time line of global modernity and the British Empire's role in it. Often absent from class, he was on his own crusade in the Middle East, where he was part of a special envoy of so-called peace mediators. (Though he admitted to us students that he could not accurately draw a map of Palestine and Israel.) Since I took that course and grappled with its contradictions, the urgency of the communities I am accountable to has motivated my research on stolen land

and stolen life. We cannot afford to wait for the change of peace-keepers; these populations have already withstood centuries of violence, and the pace of peer-reviewed publications behind academic paywalls is too slow. We are operating on a different timescale of reckoning with the long and violent impact of globalization.

In the mountains, the soundtrack of nature provides answers to what sustained peoples who have managed to maintain their sovereignty for centuries, such as Maroons across the hemisphere and Native tribes. History textbooks cannot convey the visceral nature of what is at stake. And so, this is an attempt to write neither a revisionist history nor a science textbook. Instead, it brings together historical evidence and scientific methods to ask urgent, philosophical questions about whether we want to live differently. While books are currently being banned in Florida, the planet's survival depends on us writing history together and strategizing beyond national and disciplinary boundaries. A call-and-response between the sciences and the humanities, between arts and technology, is more critical than ever. I hope *Dark Laboratory* shows the necessity of interdisciplinary methods for our collective global survival beyond the climate crisis.

This book asks questions about how deeply committed we are to the cannibalism of capitalist economic growth. The United States and the West consume the largest share of corporate profits, emit the most carbon, and control the environmentalist conversation. Conducting research abroad has shown me how other worlds and economies that don't degrade the natural environment are more than possible; they are thriving. Why do we refuse to heed the successes of green and sustainable island nation communities?

While hiking in the mountains of the Caribbean and Tahiti's tropical rainforests, these philosophies became even more apparent to me. There are other ways of breathing, of being, that require reciprocity and tuning in to nature. At dusk, I made field recordings of what sounded to me like jungle symphonies of toads, birds, and beetles, noting the similar timbre from the paradise of the Pacific to the Eden of Atlantic islands. The rising volume of the chattering birds, patient buzzing cicadas, the fluttering wings of bat colonies, and the mating call of tree frogs—each sound wave resonates through me. Once you tune in, you will not be able to unhear it. Your ancestors have heard these soundtracks for generations, and so have mine.

Mountains are political geographies of boundary-making and -blurring. Dissonances radiate. Rain showers pour, filtering nature's sounds and playing different notes than were possible before. The water absorbs and refracts the energy. The sound on thatched roofs feels different from that which reflects off corrugated zinc rooftops. Through the drops of water, staccato claps of thunder bellow as the lightning intensifies. The earth drinks with thirst. Whether urban or rural, our surroundings are loud with a rich cacophony, ecologies living symbiotically, parasitically, and in forms of relation we don't yet have the language to understand. Ancestral ballads are the ambient sounds that surrounded your forebears and mine. But many of nature's sounds will never be heard again. With each extinction event, the irrevocable silence grows heavier. The earth's condition is terminal and has been for some time, and therefore so is ours.

From peak to peak, encrypted messages are sent from the hollers of Appalachia to the hollows of the Blue and John Crow Mountains in Jamaica to Mauna a Wākea in Hawai'i to Tai Ping Shan in Hong Kong to Arthur's Seat in Scotland. Mountain ballads hold these promises to protect the climate that has nurtured us. The musical notes are layered with lore and with the protest of sacred chants repeated against mountaintop removal in Kentucky, the construction of the Dakota Access Pipeline at Standing Rock, bauxite extraction to produce alumina in Jamaican Maroon territory, and the construction of the NASA Thirty Meter Telescope on Hawai'i's tallest mountain. Mountains are biodiverse biomes under threat. They are sacred sites of worship and mourning. Businesses raze consecrated burial mounds of Native peoples to build the slopes of luxury golf courses watered by automatic sprinklers daily to maintain the pristine illusion of pastoral beauty, suburbia, and rolling hills such as that at Shinnecock Hills Golf Club on Long Island.

Due to isolation and lack of technological infrastructure, mountain dwellers are among the world's most at-risk and impoverished communities, protected by (and protecting) valleys and foothills.[7] From the lowlands to the highlands, they know how "to live on little," to echo bell hooks's Affrilachian elegy for her native Kentucky:

> *we a marooned*
> *mountain people*
> *backwoods souls*
> *we know to live on little.*[8]

So many have had no choice but to keep living *after*, immediately preparing for the next disaster. The soil reverberates the song of sovereignty by those who have been the protectors for millennia. The poetry of the future is just this, to *know to live on little*. To fight for the climate beyond crisis, we must learn this song. We must not waste our precious nonrenewable natural resources. The global environmental struggle is a chain of call-and-response over the generations.

Profiteering empires and transnational corporations have devised systems to silence the songs of dissent that echo from mountaintops. In 1883, the British banned African drumming in Trinidad in response to the Canboulay riots. The British also controlled and criminalized drumming in West Africa. In 1793, drums were banned in South Carolina during a rebellion by enslaved Africans. Yet defiant anthems travel from mountain range to range: freedom chants, forbidden drumbeats, coded spirituals, and now the battle cries of climate activists protecting the lands and the seas.

Knowing precisely where to stand in the valley to receive and send these messages is a science.

Acoustemology describes a sonic way of knowing and being in the world.[9] This is the method we need to face the crisis, tuning in to what is written in the mountains, inscribed in the layers of the geologic record.

✳

Freedom was found in ecologies beyond the contiguous United States. Just as Alpine yodelers in Switzerland sang mountain reverberations as a folk tradition, many others sent mountain messages as an ancient form of climate storytelling. Across the mountains of the Americas, similar advanced sonic communications systems were used to signal the arrival of conquistadors. The climate crisis is not a foregone conclusion of dystopia because there are ecological lessons yet to be decoded in each echo from the past. There is a chance for ecological renewal only if we commit to divesting from the economies designed to kill us.

"Go tell it on the mountain, to set my people free," the Wailers, featuring Bob Marley and Peter Tosh, sang in their 1970 rendition of the classic song.[10] Soundtracks of Black music have inspired many others to amplify the demands of justice for all. There is no justice for anyone without ecological repair for everyone. The reality of Black pain does not travel as fast as the popularity of Black music. The global recognition of Black genres of the Americas—jazz, reggae, hip-hop—reliably delivers the encoded message of rebellion to those far and wide. Black struggle inspired the soundtrack of the Dalit Panthers in India and Aotearoa's (New Zealand's) Māori people, whose battles are inherently climate struggles too, for land, sea, and climate sovereignty.[11] Both looked to Oakland in the 1970s for a model to articulate how the militancy of the Black Power movement translated. The Black Panthers themselves were multinational, with members from the Caribbean. These kindred anti-colonial philosophies connect across geographies; anthems of freeing the land are paired with anthems of freeing the people.

Wherever oppressed people are in the world, silence can never be fully enforced. In guerrilla formations, small wars and warriors become one with the natural environment, relying on the natural setting to get free or to retain sovereignty. The Vietnamese guerrilla force depended on the tropical brush from which to execute stratagems against the incursion of invading Americans.[12] Eritrean freedom fighters depended on the topography of their native mountainous terrain in the war for independence against Ethiopia.[13]

Nepal was never successfully colonized by European powers because of the country's mountains and the artistry of Indigenous warfare shaped by the climate. In 1865, the British renamed Qomo-langma (ཇོ་མོ་གླང་མ), meaning "Goddess Mother of the World"—Mount Everest—after the surveyor general of British India. Even George Everest rejected the proposed renaming because he realized the syllables could not be pronounced by Indigenous peoples. But this, of course, was the imperial vocabulary of dispossession. Every year Europeans and North Americans trek to the site and when they leave, they leave behind waste—plastics, rubbish, and literal tons of human feces. Depending on the guidance and geological knowl-edge of sherpas, tourists will never know the mountain's name in Nepali, which is Sagarmatha. She is the Goddess of the Sky, and when the land tremors vacationers will ignore the relief support that is needed.[14] The mountainous nation of forty million people suffers from natural disasters, including avalanches, drought, and wildfires. Climate guerrilla activism is the battle cry that makes the possible sovereign futures of these interconnected geographies tangible.

Learning from what our ancestors ate teaches us lessons about botanical co-adaptation. Which ecologies have sustained us? Learn-ing from the materials from which they built their houses also pro-vides information about sustainability. Agriculture and architecture need not exist only in service of colonial expansion. We have been adapted by the earth as much as we have adapted it. Thus, I reject the academic analytic appellation of the so-called Anthropocene that was coined in the 1980s and popularized in the year 2000 because it is jargon, which means it can only fail *us*.[15] The term, using the root *anthro*, or human, reveals a self-centered approach. It is invested in the myth of human supremacy. The term "Anthropocene" com-mits the anthropocentricism it claims to critique. That humans have destroyed the natural environment is no surprise. The debates of the Anthropocene often fail to distribute the blame where it is due, on so-called developed nations. To use the word "Anthropocene" is to universally blame all human societies.[16] Meanwhile, the specific-ity of the calamity wrought by European colonialism is unmistak-able. Furthermore, the Western scientific method and traditions of

Enlightenment knowledge production have relied on centuries of discrediting Indigenous, African, Pacific, Middle Eastern, and Asian schools of thought.

To address the climate crisis, we must question our assumptions, go beyond the predominate Western climate science approach, and acknowledge who has been excluded from the category of scientist to the detriment of the planet's survival. Structural inequalities, including racism, sexism, homophobia, transphobia, classism, and ableism, have limited the pool of scientists and thus scientific approaches and solutions in the climate crisis. In truth, technological innovation is hindered by the refusal to acknowledge the ills of scientific racism and that eugenics is not science. If we do not grapple with *how* it could have been accepted as science, we are doomed to repeat the same mistakes. Eugenicist and white supremacist principles are embedded in climate science—not to mention that they form the foundation of natural history as a discipline. Racism has severely limited who has been able to pursue science as a career. Yet many scientists believe that matters of diversity are not relevant to the sciences, because everything is determined by merit and empiricism. This is far from true.

Peoples from the Andes to the Himalayas understand the stakes of our shared relation intimately. Indigenous peoples across the world protect a large percentage of the planet's biodiversity. Over millennia they have adapted genetically to life at a range of altitudes. Mountain communities often subsist from all aspects of the land, from the peak to the valley. These societies have developed modes of camouflage, hiking, foraging, and practicing techniques of terraced cultivation. Nepali sherpas and the people of Appalachia live worlds apart, but mountain climbing is second nature to them both. Mountain dwellers can hear the vibrations of the source of fresh water within the aquifer. From freshwater ghats and ravines, mountain peoples know where the runaway stream joins the saltwater sea. From the ridge to the reef, they know how to navigate safely at low tide to avoid coral bollards.

Regardless of who caused the climate and racial crisis, our fate will be the same, but the global consequences are uneven. In the global south, is there even time to mourn before the next cata-strophic climate event? The share of carbon emissions that devel-oped nations are responsible for expands as these nations sanction developing nations. But what would our climate future look like if small nations, if islands, had been allowed to lead and were given the financial resources to do so at COP 30 (the major yearly UN cli-mate conference) in 2025?

Because I have spent most of my life living in cities and suburbs, I hadn't realized how much mountains had shaped me. Maybe the peaks and dales closest to you have shaped you too. That my high school mascot was a mountaineer seemed inconsequential at the time. Bearded and clad in a coonskin cap, he was the familiar colo-nial, colonizing white figure of American lore. Davy Crockett or Daniel Boone. When I was invited to the bell hooks archive at Berea College in 2023, I learned their mascot was a mountaineer too. Those fur hats, I later learned, were co-opted from Native Ameri-can traditional clothing in Appalachia by white U.S. frontiersmen. It was not until Native tribes across the region started to win back the rights to their sacred peaks in 2020 that I began to reckon with New Jersey's Native history, present, and future. Alongside the chants of Black Lives Matter, in U.S. courtrooms three million acres of Okla-homa land were granted back to Muscogee Nation in July 2020. The legal battles fought to regain Native sovereignty through the Land Back movement for many Indian nations are ongoing.[17] And, over time, I came to appreciate the mountain reservations that sur-rounded my New Jersey childhood neighborhood, which are named for clans of the Lenni-Lenape. The voice of the NJ Transit transit lines calls out the names of stops: Piscataway, Rahway, Metuchen, Paramus, Ramapo, Watchung. But these names are missing from the school curriculum despite the Indigenous presence hidden in plain sight, a colonial strategy of erasing the Native future much like mascots.

The Ramapough Lunaape Turtle Clan in the mountains of Ring-wood, New Jersey, is one of the numerous Native communities in immediate proximity to toxic dumps and wastelands. In 1980 the U.S.

government established the designation of Superfund sites, under the Environmental Protection Agency (EPA). They are defined as contaminated due to the poisoning of hazardous dumping, and many communities of color find themselves with this designation, and without restitution.[18] The correlation between Superfund sites and environmental racism for historically Black and Indigenous settlements is deeply disturbing. Located on the clan's land at the northern border of New York and New Jersey, the Ford Motor Company is responsible for pouring paint sludge down mineshafts in the woodland territory in the 1970s. Chronic debilitating health outcomes, including increased prevalence of cancers, are part of racial and climate injustices, especially for Afro-Indigenous peoples who live in these areas today. Environmental racism manifests in medical racism. By stark contrast, the New Jersey Botanical Garden enjoys 96 acres of gardens and 1,000 surrounding woodlands of the Ramapo Mountains nearby in the same township, Ringwood. On their unceded territories, the Ramapough Lunaape Turtle Clan established the Munsee Three Sisters Medicinal Farms for botanical healing, CBD, and food sovereignty in 2019, which we will explore in further detail in chapter 8.

*

Shifting from New Jersey's mountains to the Caribbean and my ancestral mountain ballads: for a long time, I could not hear these songs. Being British and becoming American, my ancestral homelands were abstract and romantic ideas. It was not until I began my career as a professor that my research brought me to the Caribbean and to China. Though my heritage is not indigenous to the Americas, being a Black person in and of the Americas requires reckoning with what indigeneity means. Learning about Chinese indigeneity has taught me a lot and, I believe, has lessons for the world on nomadism and land sovereignty. Simply due to the realities faced by my African ancestors—abduction, deracination, and violence— there is so much I will never know. Mountain dwellers in Native Jamaica—which was known as Xaymaca or "the land of wood and water"—met Christopher Columbus in 1494. He would not have survived without the shelter and provisions offered to him and his men for over six months by the Arawak.

A more contemporary popular name for the island is Jamrock. My mother was born on the Rock when it was still a British colony. A massive mountain surrounded by the Caribbean Sea, the British army could never fully conquer it because of its karstic limestone terrain. The mountain goats Europeans brought, however, understood and adapted to the altitude. Swarms of endemic fruit bats understood this as a colony in flight, seeking refuge in ancient caverns where Maroons planned their ambushes. I didn't know my ancestors had been buried in those tropical mountains, where the generations who came before me had labored and died on plantations under the tyranny of sugar production. Three hundred and forty-five of Jamaica's mountains were named by the British. I have learned to listen to the vibrations that shaped the limestone caves where Maroons and Amerindians plotted survival against the Spanish in the seventeenth century. What are the true names of the mountains where Afro-Indigenous alliances were forged? The British described the Maroons as having a wild mountain empire in Jamaica. Masked are the original Koramantee and Yorùbá names of the Gold Coast, but the mountains will never forget these echoes.

I hadn't realized that where one of my grandfathers was raised in Hong Kong near Victoria Peak (Tai Ping Shan) was too a mountain island with its own Indigenous ballads of contested Chinese sovereignty. Hakka women sang these songs to one another as they tilled the mountain soil of South China in the fifth century CE. One imagines their songs traveling to Jamaica across the currents of the Black Pacific during the nineteenth century. My people are buried in those mountains too. The family of one my grandmothers also came from the southeastern coast of China, and its mountainous borderland terrain. I found my way here serendipitously in 2017 and learned what it means to be Black in China.[19]

The British brought the Chinese to Jamaica as agricultural laborers in the nineteenth century; they were not enslaved but not free.

The rocky, fertile terrain of the Caribbean would prove to be familiar to the indebted Chinese laborers, installed by the British to toil on sugarcane plantations for low wages under the whip. Chinese laborers, like Black laborers before them, protested the tortures of the plantation. Soon I learned about the mirror territories of British colonial desire in the West African woodlands of Atakora, going back centuries ago, my African ancestry. I hope one day to visit this part of the continent so that I can be immersed in the music of the Gold Coast's vibrant ecologies. I have learned from dear friends who are biologists that some of the names of birds in West Africa are the same as those spoken in the Caribbean. It gives me hope that somehow the language was not lost in the face of centuries of colonialism. The woodlands of West Africa and the Caribbean are mirror territories of refuge. What might the Akwapim mountain range stretching from Togo to Ghana to Benin be able to tell me about the intergenerational duty to protect the land? Across the globe, my kin, it turns out, were mountain people who recognized familiar terrain in unfamiliar countries.

Even if we can no longer touch the soil our ancestors intimately knew, there are ways to reconnect with ancestral ballads by listening to the land as a storyteller. The songs are not lost, though the dislocation of diaspora makes them feel far. To be part of a diaspora is to have the memory of a dish we have never tasted. That dish feeds our future. The plot of European colonialism has always been one of ecological degradation and of forgetting. The key to undoing the violence of this erosion lies in restoring the land and the living vespertine rhythms of plants, animals, and geologies. Intergenerational knowledge can be activated, but it is also critical that we learn to protect territories to which we are not native.

In *Dark Laboratory*, we will touch on Chinese geomancy, Yorùbá philosophies, and Hawaiian principles such as Aloha ʻĀina as ways to approach the land. The art of arranging life in harmony with the natural elements is at the core of Chinese geomancy. The scale of Yorùbá temporality reckons with the near future and the far future in forms that are nonlinear such that time is not a problem to be solved. In Hawaiian, Aloha ʻĀina means "love of the land" and the interconnectedness of life. Our shared planetary future depends on tuning in to all possible ancestral knowledges, which comprise global traditions and sciences.

In opposition to the land, the colonial approach has been one of razing and dynamite, eroding Indigenous relationships to geology. Take Mount Rushmore: fictive nationalist genealogies are inscribed into sacred mountains. From 1927 to 1941, when the American architect and sculptor Gutzon Borglum etched the faces of the founding fathers into the landscape, his unabashed white supremacist mission and ideology and close affiliations with the Ku Klux Klan were known.[20] Regardless of the vandalism masked as a patriotic national project, the Lakota still know this granite rock formation as the sacred summit Tunkasila Sakpe Paha. For them, the Arapaho, and the Cheyenne, this is the center of the universe. I find hope in that there are an estimated 108,706 independent mountains unnamed. Perhaps these unreachable, undiscovered mountaintops can remain free from removal and mineral extraction. If there are yet geological zones unmined and without the vandalism of carving supremacist myths, there is hope for environmental repair and justice for Indigenous communities.

A MAP THROUGH THE LABORATORY

Each chapter of this book follows a path of hope by tracing a different nonhuman ecology to chart strategies for how to live after climate crisis. There are climate ballads and lyrics throughout the chapters that we need to tune in to as a planet. Parochialism is killing us, so this book offers global strategies from the vantage of what islands teach us about crisis. Blue-footed boobies, coral reefs, Calcutta mongooses, and buds of marijuana are storytellers that lead us through the dark laboratory of possibility. Beyond the borders of the nation-state, each ecosystem shows us new orientations that tackle ecological crises.

The book is divided into three sections with three chapters each.

The first section, "Eden Is Not Lost," explores the premise of the Caribbean as the origin of the climate crisis after Christopher Columbus's arrival there in 1492. It looks at island case studies in Jamaica, Hong Kong, and Dominica based on my travels over the years. On these islands, we contend with Maroon, Asian, and Indigenous philosophies of time and science. It posits the ways in which many Western museums are mausoleums entombing artifacts from these societies. The second section, "Geological Time and Black Witness," looks at the delicate balance of island ecosystems from the sea to the sky through the interrelation of corals, guano, and the birds. It focuses on regeneration and rebellion in the Pacific and Atlantic worlds. Could it be that guano is the climate solution that could save coral reefs threatened by ocean acidification? The final part, "Life in the Garden after Eden," celebrates ecologies thriving against the odds by highlighting the evolution of zoological and botanical island transplants. Looking to strategies of mothering and healing in the animal and plant worlds, it centers economies and circuits of care across islands. For all the impending natural catastrophes to come, when will the people most impacted be at the helm of emergency climate-crisis planning? These nine chapters ask what it means to survive beyond the lab of European colonial experimentation.

We begin in chapter 1 with Maroon ecologies and the contemporary geological crisis of bauxite mining in Jamaica. This chapter pairs extractive British colonialism in Jamaica with its parallels on the island of Hong Kong. I visit with Maroon leaders to learn more about the ongoing battle for climate justice and Black sovereignty. Chapter 2 travels to 1492 and the ripple effects of Christopher Columbus's arrival in the Caribbean. What of the Chinese admiral Zheng He, who is said to have circumnavigated the world before Columbus? Each island of the Caribbean archipelago is a parallel universe of cross-pollination, mineral extraction, amalgamation, transplantation, and animal husbandry. I visit Kalinago Nation to understand more about the Indigenous Caribbean experience and futurism. The third chapter traces the history of natural history by looking to museums and their institutional role in conservationism and environmentalism. What, I ask, does it serve these institutions to omit matters of race in the twenty-first century?

Turning to island bedrocks in chapter 4, corals become the protagonist as the very foundations of certain types of islands. We dive deep to learn underwater lessons with these endangered organisms about how to adapt to rising sea levels and temperatures. The imperiled invertebrate animals have seen a decrease of 90 percent across the Caribbean in the past thirty years. Corals, though, have an incredible power to self-regenerate when damaged. Listening underwater to the symphony of biodiversity from snapping shrimp to leatherback turtles, it becomes apparent how the reef is a shelter for multiple life-forms. I ask how ancient corals bear witness to the cargo ships of the transatlantic trade of enslaved Africans. I imagine the sharp, stony outcroppings of reefs that tore into wooden hulls, sinking ships in the early Americas. A home to sharks, plankton, and clownfish, the coral reef is an architecture that shows us the power of rebirth.

Chapter 5 brings us to the era of guano trade in the nineteenth century—and how the sedimented bird droppings used as fertilizer form the missing link between the Agricultural Age and the Industrial Revolution. The tragic history of the guano trade shows how collective labor movements of refusal formed across Pacific and Caribbean islands. Coalitions of hundreds of thousands of African American, Pacific Islander, and Asian guano diggers echo in rebellion against colonial overseers in this chapter. Some leaped to their deaths in the mines where they labored, becoming part of the biomatter of the island's bedrock, bodies decomposed in the shallow graves.

Chapter 6 turns to the birds who produce the guano to examine how bird-watching developed as a colonial hobby related to racial thinking and categorization. Through binoculars, we trace white flight and fear of Black emancipation in the nineteenth century by reading the compendium *Birds of America* by John James Audubon. One of those who fled the Haitian Revolution, the famous naturalist Audubon illustrates the American pecking order of race through what we can piece together of his biography. In the present moment, the chapter looks to the contested grounds of leisure and bird-watching in Central Park to underscore painful histories of surveillance for Black people.

Chapter 7 follows the Calcutta mongoose, a foe to many ground-nesting birds in the Caribbean. The weasel-like animal arrived by ship from India to Jamaica in 1871 alongside Indian indentured laborers. One of the first documented "invasive species" case studies, the mongoose—who was imported by plantation owners—teaches us about what it means to be a predator in the plantation ecosystem. Hopping across islands, the mongoose with its furtive choreography forms a blueprint for mammalian survival through collective mothering. Mongooses demonstrate maternal and reproductive acts of intergenerational care, birthing and raising their young according to the moon. Mongoose survival continues to defy scientists and government officials alike, stymying their plans for animal control.

In the eighth chapter, we trace another Indian transplant to Jamaica: the buds of cannabis that sprouted across the Americas, eventually forming a shadow economy. The sacred stowaway was cultivated by indentured Indian laborers on the same merchant ships that arrived in the Caribbean. Between the spiritual worlds of Hinduism and the emergence of Rastafari herbalism in the 1930s, worlds of botanical healing and cultivation thrived beyond the plantation. From the Caribbean botánicas of New York City and Brixton to the pristine royal hothouse botanic gardens at Kew, different forms of horticultural knowledge shape the contested grounds of diaspora.

The ninth and final chapter offers new frameworks for futurist policy led by island nations. I call for an "affective plate tectonics" by returning us beneath the ocean's surface to the thermal vents of the island of Montserrat, a British territory in the Caribbean. The volcano that erupted in 1995 is set to explode soon, part of a thirty-year cycle. What has been learned since then about the friction and fault lines of race and nation? Our collective future depends on developing federations that do not depend on the governance of the UN or NATO. We urgently need nations to connect across the divisive fault lines of race and nation. The earth's crust is a dynamic system, a conveyor belt of instability. If we can learn to feel *for* and *with* the unnatural disasters of nations not our own, a potential affective plate tectonics for climate grief beyond national boundaries can be formed. I discuss how coalitional struggles for Palestinian freedom have existed in the Caribbean. Dispossessed peoples in disparate geographies face the same crises of climate in the global battle for sovereignty against state-sanctioned genocide.

Layer by layer, the chapters of *Dark Laboratory* form the geological strata of an island. Survival is a way of life that is an art form in the Caribbean, where increasing severity in natural disasters shapes everyday realities. An intergenerational refrain of repair is audible once we know where to listen in the mountains. The twenty-first century requires new genres of climate storytelling to encompass the multiplicity and nonlinearity of salvation if it is possible. Beyond the epic that ends in doom and extinction, there are shanties, ragas, ring shouts, lullabies of hope. Strategies of regeneration and repair are part of the rebellion rallying across the globe, a call-and-response with the natural world. We require unprecedented collective environmental policy and action beyond the boundaries of nation-states for the twenty-first century.

We must refuse to betray future generations, especially because we have been forsaken by so many before us. Technocrats have convinced us the only option is continued extracting, geoengineering, and polluting. We need new stories, new technologies, and new forms of nature writing. The armchair approach of the old boys' club has failed the earth. Nature documentaries, filmed from the eye of God with classical music scores, have failed us and detached us. Transcendentalism has failed the planet in its abstract romanticism when what we need is action led by the global majority, which is to say Africa, Asia, the Caribbean, the Middle East, the Pacific, and Latin America. We must tune in to the world's mountain ballads to remember what so many have forgotten—that we are not separable from the natural environment.

BOOK I

Eden Is Not Lost

Lo Ting and Mami Wata

Garden Interlude

What other way is there than the sea? Lo Ting (盧亭) and Mami Wata are two of our guides through this book. From China and from West Africa, they met in the garden of the Caribbean, where religious practices came together. Eden is full of such gods who traveled to the Americas with the rituals and beliefs of stolen people and those forcibly brought to those shores. Their saltwater children were born, those between international port cities. He, Lo Ting, is a hybrid, a fish-man native to Hong Kong's fishing villages. It is said in Chinese folklore, he arose from the water centuries ago. He was resurrected as a political figure by dissident artists, beyond ancient myth, to rally claims for Hong Kong sovereignty in the 1990s. Mami Wata is a West African mermaid deity of the waters, a guardian of those forced to cross the oceans. Or her name is maybe Watramama or River Mumma. Kin to Yemayá, she is amphibious and mermaid-like, a snake charmer. Perhaps a manatee? She became the mother whose children are fish. Born of Lo Ting and Mami Wata, they were called impure, because they were of the saltwater, in Chinese grammar not fully Chinese, born in foreign waters and of diluted bloodlines. My lineage is of salt water.

In 1838, Britain's colonial labor experiments between harbors connected

the Black Pacific and Chinese Atlantic port cities Kingston and Hong Kong.
Both are vertical modern cities and today both are flooding. Lo Ting sobs
briny tears because he knows that with over sixty surrounding islands,
Hong Kong is drowning. Hungry ghosts form a loud chorus asking to be
venerated in secret corners and underwater caves. The lore of Lo Ting, the
"Thing with Scales," shapes the underwater. The South China Sea islands
are in geopolitical and climate crisis. Blank A4 sheets deny the piercing
gaze of the surveillance state.

Victoria Harbor was submerged again in 2023, with Super Typhoon
Saola's floods sweeping away infrastructure. Hot and humid days brew as
cyclones threaten the city, already channeling the dissent of over seven million residents. What other way is there than the sea? It is said that Lo Ting
"fled to the islands and lives wildly there, eats mussels and uses shells to
build walls." For generations, Hong Kong's Tanka people have built houses
called pang uk (棚屋) that are engineered on stilts over the water. These
traditional Chinese architectures are the blueprint for the future, for the rising tide of a vertical metropolis. With the rising tide of the climate crisis,
will houses across the globe need stilts to walk across the waters?

In the Caribbean, ghosts called jumbies walk on stilts, having crossed
the Middle Passage from West Africa. What other way is there than the
sea? Caribbean fishermen will not learn how to swim because they believe
it is bad luck. They revere the waves as a way of life and livelihood.

Kingston's Tutty Gran Rosie is another one of our guides through the dark
laboratory. She bears witness. Like the late New Orleans YouTuber and
Bounce rapper Messy Mya, she is back by popular demand. In Jamaica and
beyond she rose to fame because she was inconsolable. A Black woman,
a mother of nine, Rosie went viral in 2013 with the exasperated refrain
"Everyting flood out." Cocking her head back, Bridgette Bailey became
dubbed Tutty Gran Rosie because she demanded 30,000 Jamaican dollars
($192 USD) in reparations to replace her belongings washed away by flood-
ing. Who will pay the reparations of climate crisis? Echoing the despair
of Katrina in New Orleans, Tutty Gran Rosie's home on Sunlight Street
will flood again. She wants back her flat-screen TV, laptop, phone, desk
lamp, an inventory of what was swept away. "Everyting flood out." She
demanded a name-brand car and name-brand furniture. She screamed

with the ire of all people, Black and poor, swept aside by urban floodwaters by state-sanctioned neglect. The babies would get meningitis from the "shit water" of the overflowing gullies, Rosie warned in the viral video. The world laughed, ignoring her despair and demands for restitution, her ringing of the alarm of the failure of infrastructure of a bridge poorly built to handle climate crisis. Such is the case in many Black impoverished neighborhoods. It was a failure of the Jamaican government to meet the accelerating environmental crises. She is still waiting for her thirty grand. Her exclamation, "We need justice!" echoes.

1

Island Laboratories

M y bright yellow Wellington boots were not at all the right shoes. I stared down at the vertical drop of the forested hillside, my hand folded tightly into Colonel's. We were hiking on his property in Jamaica's Blue and John Crow Mountains. Chief Wallace Sterling of the Windward Maroons is known by the honorific "Colonel" and is the descendant of a long line of Black warriors who have maintained their people's independence for over 350 years on the island nation. Sterling is the longest serving chief, having been elected the leader of Moore Town (Windward Maroons) in 1995.[1]

An important climate history about sovereignty is enfolded in Maroon history. Among the African people abducted from the Gold Coast were Akan and Fon generals and prisoners of war taken aboard cargo ships to the Americas. Colonel's forebears plotted their rebellion at sea and were ready when they landed. They devised ways to fight from the moment they were shackled. They never stopped fighting; no enslaved person did. As much as the Caribbean is made of islands of European experimentation and exploitation, the region is also full of hidden laboratories for Afro-Indigenous sovereignty. In many ways the warfare between West Africa and

Europe is ongoing today—despite the abolition of the transatlantic slave trade in 1807—because these Maroon and Amerindian communities still exist off the grid.[2]

Maroon warriors include peoples fragmented across the hemispheres, from Virginia to Suriname to Haiti to French Guiana. These communities have long protected the natural environment as part of their martial philosophy. *Cimarrón,* a Spanish word derived from the Arawak language, signifies the renegade descendants of the escaped people who had been enslaved from the seventeenth and eighteenth centuries. Maroon autonomy has depended on a philosophy of unity with the wilderness. The Maroons' identity as Black people who defied the British, the Dutch, the French, the Spanish, and the Danish has also been core to their survival and to the retention of African cultural and spiritual practices. The fate of the Maroons depends on land sovereignty. Where they live tend to be the most biodiverse spaces of wildlife refuge. The European colonial plot to conquer African and Indigenous nations was a twin campaign that also hinged on the depletion of tropical soil, subsoil, and bedrock.

As I looked downward along the mountainside, the sound of the stream rushing below us in the wilderness of Moore Town calmed me. I thought of the duality of how the Maroons are perceived: both as traitors to the Jamaican people and as valiant, silent warriors. It's a troubling contradiction, and one of many in my lineage. Today the Jamaican government is betraying the island while Maroons fight to protect the natural environment from the land grab of European mining. They fight to save the mountain aquifers, from which the island derives 40 percent of its drinking water, from contamination.[3]

For those like me, descended from Africans who were enslaved, escaped, and manumitted, the contradictions of Black sovereignty are disturbing and liberating. Blackness is a lineage more complicated than the binary of free or enslaved. It is not a monolith. There are numerous shades of unfreedom. Maroon genealogies are heterogeneous pedigrees that include collusion and bloodlines with Native peoples. Capture and return of runaways to plantations in

Jamaica were the terms of the British 1739 treaty. This clause outlines the requirements of betrayal:

> Ninthly, That if any negroes shall hereafter run away from their
> masters or owners, and shall fall into Captain Cudjoe's hands, they
> shall immediately be sent back to the chief magistrate of the next
> parish where they are taken; and these that bring them are to be
> satisfied for their trouble, as the legislature shall appoint. [The
> assembly granted a premium of thirty shillings for each fugitive
> slave returned to his owner by the Maroons, besides expenses.][4]

The British army sued for peace, fearing they would lose the whole
colony of Jamaica to the Maroons in a protracted war. But only the
mountains know the truth of the rival allegiances and what was
required for the Maroons to retain Black sovereignty.

We do not yet have the full vocabulary to describe the ethics
of these alliances because we rely so much on European languages
(which is to say colonial languages such as Spanish and English) to
articulate radical coalitions. European lexicons are not only inadequate but are also antithetical to Black, Indigenous, or Asian liberation. At Dark Lab, one of our ongoing initiatives is the composition
of a collective *Decolonial Glossary*. The concept is to demilitarize our
language before we can decolonize our imaginations. Some of the
terms we will ask contributors to define may be guttural sounds,
kissing teeth, common gestures, or specific diasporic registers of
humor.

So many other languages and literacies other than Eurocentric
ones exist. The language they told us was broken carries the answer
to a broken present. In 1739, the price of freedom was forged under
and beyond colonial sanctions of the British Crown in the secret
headquarters of the limestone mountains. How could this negotiation be adequately translated from the Koramantee language and
English? To answer this question, I hear the messages of the abeng,
a Maroon heraldic musical instrument made from a cow's horn and
used to communicate across far distances. Such sovereign reverberations of the battle for land rights and sovereignty echo along the

spine of the Americas. Across the hemisphere, these communities remain unconquered.

It was rainy season, and I was glad to be wearing my yellow boots, but I teetered on the muddy mountainside. While I have spent the past decade as a researcher traveling across the Caribbean taking part in field research and archival trips, I have never found it easy to acclimate to being so far from New York City.

As a Black person who is also of Chinese descent, I feel the pressure that looms over every climate debate in the Caribbean: the distrust of the Chinese. I believe the contested island geographies—Jamaica and Hong Kong—reveal the key to unlocking strategies that might help to solve this crisis. So here I trace Maroon and Chinese histories in relation to the questions of climate and sovereignty after British Empire. Part of my family history traces back to the borderlands of China and Hong Kong. The Chinese presence in the Antilles spans more than two hundred years, back to at least 1804.

Today, in Jamaica, Chinese-backed loans provide funding for infrastructure to fight climate crisis, but these arrangements also fuel widespread suspicion. What motivates the favorable rates? There is outrage over the extractive practices and the ecological degradation that has occurred. Many Jamaicans fear being indebted to China. There is no shortage of headlines from Western outlets such as the *Economist* and the *New York Times* that stoke the anti-Asian rhetoric about China in Latin America, the Caribbean, and Africa. I hear the Yellow Peril sentiments of a century prior echoing throughout these articles. Few denounce the United States, Canada, Russia, and European nations as loudly for their continued role in the Caribbean's climate degradation.[5]

Where the colonial West has abandoned and neglected the region, China has offered relief in the form of concrete financial investment over the past twenty-five years. The Jamaican prime minister welcomed an investment of $300 million JSD (almost $2 million USD) by Chinese telecommunications corporation Huawei into the capital city of the country in 2022.[6] The presence of a 9,000-square-foot office in the heart of uptown Kingston on Hope

Road worries the United States most of all. However, to critique the extractive capitalist policies of China without fully reckoning with the ongoing role of the United States, Europe, and Canada is to remain complicit in false climate innocence and amnesia. Canada is home to 75 percent of global mining companies, which are head-quartered there.[7] Toronto is the heart of many colonial extractive prospecting projects and lists 60 percent of these corporations on the Toronto Stock Exchange (TSX).

China's presence in the Caribbean long predates the twenty-first century; the earliest mass labor migration to Jamaica took place in 1854. Before the voyages of indentured laborers, we must consider the seafaring of Chinese admiral Zheng He, who is said to have arrived in the Americas in 1421. Though often dismissed as coun-terfactual history by Western historians, how does Zheng He alter the possible definitions of colonial desire? China's colonialisms are multiple, and their legacies are evident across Southeast Asia and do not manifest with the same extractivist design of European settler colonialism.

As a small minority community, the Chinese and people of Chinese descent have been a target of frustration across the Carib-bean. Loyalties and allegiances are questioned because of their prominence in the merchant and entrepreneurial trades. Many non-Chinese people collapse the difference between centralized Chinese state power and decentralized diaspora formations. Separating these actors, the cycles of migration, the Cultural Revolution, and these spheres of influence is critical to grappling with the Chinese role in Caribbean climate crises. Migrants—chiefly subsistence farmers and small retail shopkeepers who migrated in the nineteenth and twentieth centuries—often subscribe to an uneasy sense of Chinese identity outside of China.

Maroon and Chinese histories flow in my bloodlines, with rival philosophies, territories, and dialects. Many minority groups, in-cluding manumitted and free people of color, find themselves at these crossroads in Jamaican history. Have they betrayed the project of Black freedom? Such fractures are the design of the British. How do we envision autonomy after colonialism? The contemporary Jamaican government has been complicit in selling the future of the island to the highest foreign bidder. The exploitation of the islands by British, Canadian, and American absentee colonialism is so normalized that it no longer seems to merit interrogation, while the Chinese are figured as a visible and present target. Canadian banks, such as Scotiabank and Royal Bank of Canada (on which Caribbean nations depend) are guilty of a financial practice called de-risking since 2019. Shuttering banks across the region, they are destabilizing the banking infrastructure of average citizens because it is no longer lucrative. Canada sees more opportunity in Latin America, where there is strong population growth, as opposed to what they consider to be stagnant about the Caribbean. Abandonment is a form of imperial economic control. More financial dependencies are created by the sudden vacuum left in the economy.

Perversely, since racial slavery is the majority history of the region, we lack details about the interior lives of enslaved Africans. What I glean I source from plantation records that document rum and sugar production from those plots of land where generations of my family toiled. Even if they did not escape from plantations to form runaway communities like the Maroons, I know they participated in everyday forms of maroonage or rebellion. Self-determination is an everyday practice of survival through major and minor acts of refusing the colonial reality.

BRITISH EMPIRE'S TWIN EXPERIMENTS: HONG KONG AND JAMAICA

Before he knew Hong Kong, my mother's father knew Jamaica, where he was born in the parish of St. Elizabeth. In 1934, at the age of seven, my grandfather boarded a ship in Kingston Harbour with his little brother. Scared and alone, these boys, born to different Black mothers and the same Chinese father, were part of a small category of exceptions. These children traveled to China to *become* Chinese with the intention of returning to Jamaica years later, fluent in Chinese ways and customs and the Cantonese language. They were able to circumvent British race-based exclusionary laws targeting the Chinese in Jamaica. Just as the United States became a gate-keeping nation during the Chinese Exclusion Era from 1882 to 1942, across the hemisphere nations formed their identity by excluding Chinese immigrants. Westerners depicted them as foreign invaders who would encourage a criminal element in society, and they were denounced as importing heathen religious practices.

In the 1920s and '30s, these "saltwater children" were born of China's port cities, from Shanghai to Hong Kong.[8] The Chinese grammar of the saltwater denoted a racial dilution. As dual citizenship holders, these children had the mobility to travel back and forth. There is a history written of the sons and daughters of Chinese merchants who were also of white parentage.[9] But the stories of the Black saltwater children are missing.

Born a colonial subject, my grandfather traveled between the two mountain islands of Hong Kong and Jamaica. He and his sib-

Map of Hong Kong and the South China Sea

lings, it was believed, would never be "pure" enough to be deemed as full-formed people, imbued with humanity, in China. Other derogatory Chinese names used to describe children of mixed heritage included Half Brain (bàn nǎo shi), Mud Duck (ní yā), and Eleven O'Clock children.[10] Ducks made of mud cannot fly. Eleven o'clock children were not a full twelve o'clock, not complete, not pure. How did my grandfather receive these Chinese slurs and taunts? Did he find a sense of refuge diving in and out of the saltwater caves and grottos as a teenager? Did he hear Lo Ting's Hong Kong mountain songs? Was Mami Wata there too as a protector across the waters of his Black Pacific childhood?

Gah-san is the Hakka ritual of tomb sweeping for the dead. When I learned that the characters mean "climbing mountains" (because traditionally our ancestors were buried in mountain crypts), the moniker made a lot of sense. Who will sweep the tombs of generations lost to the climate crisis? On the Chinese national festival for the dead, Qingming, observed every April, I wonder what my grandfather, who was schooled under a colonial British Hong Kong curriculum, knew of natural history and the geological origin stories of the occupied islands. As a youngster, he would have known that no matter how perfectly he spoke Cantonese, others perceived him as alien. Did he worship the memory of his African ancestors while he lived in Hong Kong's New Territories?

As one who had also risen out of the seas of migration as a child transformed by transatlantic currents, I felt eyes peering at me when I arrived in China in a way I have never felt in London or New York City. Walking the streets in Shanghai, Hong Kong, and Beijing in 2016 with my family, I knew the glares were reserved for Black people in China. White people I met there described an altogether different gaze. While both may be met with awe and fascination, the hierarchy of racial difference between Black and white is stark.

A friend, who is Nigerian American and a professor, and I walked on a busy street in the rain for an hour trying to hail a cab in Shanghai. Finally, a cabdriver who called us Obama picked us up. It was in that city that I first heard the word "nigger" used toward me, and

The author's grandfather and his cousin,
Hong Kong, ca. 1940s

in that moment on the subway I knew what my grandfather's Black boyhood in Hong Kong during the 1930s would have been. I stopped wondering.

My mother found a pile of small black-and-white photographs among her father's last effects, in the basement of her family house in Queens. These yellowing photographs from the 1940s are the only clues I have to my Black Pacific inheritance. In random snapshots of my grandfather in Hong Kong, the composition and the landscape make me wonder if he took these photographs with a timer. Did he own the camera? When I knew him in the '90s, he was always the one silently and dutifully holding the camcorder to record family memories. In the photograph above, the mountain ranges form a backdrop shadowy with the contrast of the geologic texture of the New Territories. Two adolescents, my grandfather and his cousin (perhaps during World War II), etch themselves into the geology of the South China Sea Islands chain, occupied by the Japanese and the British in the 1940s. The two young cousins gaze off into the distance, posing with a wistful sense of purpose, perhaps looking across the Pacific toward the island of their birth, Jamaica.

I imagine my lineage as descended from the mythic Hong Konger fish-humanoid creature that arose from the sea: Lo Ting. I also see River Mumma or Riva Maid in the mountains of Clarendon in Jamaica, where my mother spent part of her early childhood. As a youngster she vaguely remembers hearing the streaming water near Crawle River rushing at Arthur's Seat, a remote mountain village. When I ask, most Jamaicans have told me they have never heard of Arthur's Seat. Scottish people, though, know it well, as a peak in Edinburgh. The Jamaican namesake is too small to exist on the colonial map or Google Maps.

The verticality of the two colonial capital cities, Kingston and Hong Kong, tells a story of British sweat and domination. The built and natural environment racially delineate and segregate. The peninsula of Kowloon, the woodlands of the jutting protrusion of the mainland, would welcome my grandfather as a young boy. In the forests of the New Territories, the borderlands of Hong Kong and China, he would find his father's home. In Fanling Lau (粉嶺樓) we were lucky to find his house in 2016 because it is one of the few that the Hong Kong Antiquities Advisory Board protected from redevelopment as a historical landmark. A classic of 1890s architecture, my grandfather's childhood home is a walled property designed for clans, and thus preserved from the grasp of hyper-industrialization.

According to colonial-era laws like the Small House Policy of 1972, any male in the New Territories of Hong Kong who can trace his male lineage back to Indigenous villagers has the right to a plot of land where he can build a three-story domicile.[11] What rights are guest dwellers entitled to as the seas rise? What if their heirs do not conform to this binary notion of gender? While we debate these questions, Hong Kong's government is spending $80 billion USD ($624 billion HKD) to build artificial islands off Lantau near Kau Yi Chau to, as they say, alleviate the housing crisis.[12] The housing crisis is also a climate crisis.

My ancestral city on my maternal grandfather's and grandmother's side of the family goes back millennia and sits just across the Hong Kong–Shenzhen border in the foothills of Wutong Mountain, the tallest mountain in Shenzhen and the source of the Shenzhen River. These foothills are a rare scenic locale known as the lung of the city. The Chinese countryside is vast, and this part of it holds many ecological lessons about what it means to be a Special Economic Zone in communist China. The small city (a population of 310,000 in 1980) has ballooned to an estimated 14 million today. When former Chinese chairman Deng Xiaoping enacted these changes, there were a few potential economic models of growth possible for developing nations. Among them was that of the Caribbean-born economist Sir Arthur Lewis, a Nobel Prize winner in 1979. Under his model, a developing economy can propagate a capitalist sector, which can be incentivized to eventually eclipse the subsistence economy. What would they have said to each other, the chairman and the knight, had they met? Global knowledge and economic exchanges from the Caribbean to China have been taking place for centuries.

Near the border of Luohu and Yantian is my landmark for ancestral wayfinding, should I ever want to make my way back. Maybe I won't. The town is too small to be found on Google Maps, like Arthur's Seat in Jamaica. Thanks to Geoffrey, a kind friend who took me to the village in 2017, I found my way there. It just so happened that, by chance, a friend of his had just retired there. When we first met the year before in Hong Kong, the mystery of the ancestral village was one he set about helping me uncover. I was doing research in Shanghai and was glad to receive Geoffrey's message on WeChat. When we reached the village, my distant cousins welcomed me without question, though I do not speak Putonghua. Together we all went to a restaurant for a feast to celebrate the ancestral homecoming around a banquet table and ate river fish and delicate, fragrant Hakka preparations. Bustling cities sprouted overnight in Shenzhen; in some cases, subsistence farmers were handsomely compensated. In hidden ways, these foothills are still governed by a zodiac of animals correlated with the stars. Capitalist time governs factories as pollutants are expelled into the air and leach into the soil. Every day, millions read the forecast for the Air Quality Index (AQI) in China, now a permanent meteorological fixture. You don't want to leave the house without knowing how hard it will be to breathe. The direction of the winds shows how a geodynamic trans-national climate policy beyond nation-states is needed for the flows of oceanic and air currents.

Today the Chinese government funds infrastructure projects that carve highways out of Caribbean mountains. A four-lane cross-island motorway nicknamed the Beijing Highway cost $730 million JMD ($4.7 million USD). Most Jamaicans will not use the North-South Highway because of the $32 JMD ($5 USD) toll for a one-way trip, which is not easy to afford for many Jamaicans.[13] Roads long neglected by climate crisis impacts have been secured against the increasing severity of hurricane season. On the road from Norman Manley Airport into the capital city of Kingston, huge boulders help to break the storm surges and prevent the flooding of the highway to the Palisadoes to enter Kingston. This forms a daily reminder to Jamaicans of Chinese aid in climate infrastructure.

Victoria Harbour and Kingston Harbour, both named for British monarchy, echo each other. These are territories whose sovereignty is deeply contested after layers of British occupation. Both Maroons and Hong Kongers wrestle with the validity of the ninety-nine-year lease, as part of what undersigns their authority is already compromised by the corruption of the British Crown. Maroons know not to trust the *obroni* landlord. The *obroni* is a ghost to Maroons and to Ghanaians. The Chinese also name outsiders as ghosts: *gwái*. Yet the non-Maroon Afro-Jamaican is also referred to as *obroni*, the Government of Jamaica. Hong Kongers understand too that they cannot trust their future sovereignty to foreign devils.

Both Jamaicans and those from Hong Kong understand the betrayals of colonialism and the necessity to challenge colonial historiography. The First Opium War (1839–1842) overlaps with the time line of emancipation in the British West Indies (1834–1838), and both shape political currents and futures. When time was up in Hong Kong in 1997, rural Shenzhen was ready for an economic boom of hypercapitalist development. Declared a special autonomous zone because it abuts Hong Kong, Shenzhen also exists in a state of exception, but there are consequences. The entire region suffers air pollution from steel production. Shenzhen's Industrial

Revolution shapes the climate future of the planet. The price paid for the velocity of production of electronic goods is dear for the world. When the old model of a cell phone is disposed to trade in for the next, the ores extracted can never be replenished.

Racialization shifts across oceans; my grandfather probably did not become Black until he arrived in Hong Kong. Jamaican theorist Stuart Hall wrote in his memoir of the importance of the year 1938 for the Caribbean and blackness. Until that year, he did not identify as Black. Labor rebellions swept the region, ushering the first Black becoming. Not brown, not yellow, not red, not mixed. Black became a powerful and politicized category, and though my grandfather missed this moment in Jamaica, I like to imagine that he was inducted into this modern and transnational sense of blackness in Hong Kong. *Partus sequitur ventrem*: "That which is born follows the womb." He was born to a Black mother, but that which follows the womb was not the order and, in this case, he was sent away. Yet my grandfather's existence, as a person of both Hong Kong and Jamaica, translated as a certain type of mobility. He left his mother and father in Kingston to be raised by his Chinese stepmother in Hong Kong's New Territories, leased by the British in 1898. Mirror territories, both aqueous and mountainous, the landscape of British extractive capitalism exhibits the same vertical design over the natural environment. The vertical axis of colonial landscape architecture shows how the British always scouted perches strategically to look down on the territory.

Hakka (客家) origin stories always begin with an emphasis on our Han Chinese identity, showing how precarious it is to be a minority in China. The mythos of migrating from Northern China to the South illustrates the contours of ethnonationalism and what it means to be a minority in China today. *Hakka* or *Keija* translates to "guest families," and, unsurprisingly, nomadic peoples have always represented some threat to those anchored to specific land. It is their relationship to climate that we must interrogate. Whether Native Americans, the Bedouin of al-Naqab (Negev), or Roma people, those who travel are characterized by Europeans as uncivilized by nature of their nomadic lives. Those who opt to live beyond the colonial grid pose a threat to society because they prove it is possible to live beyond its structure, defying technologies of surveillance and traditional conventions of landownership.

Other nations determine the future of Jamaica, though it declared its independence in 1962. Not yet entirely sovereign, it remains a part of Britain's Commonwealth. Debates swirl in Jamaica about whether to follow the defiant path of Barbados to achieve the status of a republic. The list of demands for reparations also mounts as local groups rally for what is owed by Charles III and his descendants, whether they are considered Black or not. Yet, without accounting for the climate future, there can be no hope for sovereignty or racial justice.

IN SEARCH OF CAMOUFLAGE VINES

Returning to the mountainside of Colonel's property in Portland, he spotted the cacoon leaves and reached for the pods. Every plant is believed by Maroons to have its own distinct wavelength. The green garments, full of musical notes, had held them in military camouflage safety for generations. They also call them ambush. Green vines have long been a strategy of retreat into nature. Like many African traditions that have survived across the islands, camouflage represents the sacred guerrilla relationship with ecology as a hiding place.

While one of my grandfathers grew up in Hong Kong, the other had grown up in Portland, Jamaica, in the 1930s. I did not have the chance to visit the parish till 2022. Over the years I had traveled to twelve of the other municipalities on the island, but this one stood out because of the windward history of Moore Town and the gravesite of Queen (or Grandie) Nanny. I was on the island to debut an artwork I created on the history of Queen Nanny, to be exhibited at the Institute of Jamaica in Kingston, the nation's natural history museum. It was powerful to see Queen Nanny's final resting place, Bump Grave, a national monument. Colonel explained to my friends and me how Queen Nanny's philosophy was an ecological one, such that it is easy to see why she has been taken up as a climate warrior today.

Colonial powers derisively called the Maroons "bush people" and "Bush Negroes" (an official Dutch term), to malign their communities as primitive, but Colonel's people depended on and embraced these vines for safety and as a mark of tradition. Climate has long been part of their strategy of survival using guerrilla warfare. Camouflage has meant 350 years of Black sovereignty and sanctuary in Jamaica. Alas, the numbers of Maroons living on designated and protected lands in Jamaica has dwindled with each new generation as people are pressured to integrate and assimilate into the broader society.

According to Linnaean nomenclature, the cacoon vine is called *Entada rheedii*. The species is named for Hendrik van Rheede, a colonial administrator of the Dutch East India Company born in 1636 and the author of the naturalist book *Hortus Malabaricus*. Could Rheede have heard the wavelength of the ambush vine the way Maroons did? The famous Afro-Jamaican musicologist Olive Lewin, who spent extensive time learning from the Maroons, notes each plant's distinct sonic wavelength in her writings.[14] Rheede did not spend time with the Maroons and could not have known the secret power of the African dream herb, also known as the snuff box, or the sea bean, even as he attempted to classify it. Queen Nanny of the Maroons knew its power instinctively. As an herbalist, she remembered that when the pods arrived they could be used to induce vivid dreams in West African traditional medicine. She is said to be the first Obeah practitioner, and the British outlawed the practice after learning of its power. As a practice, dreaming is core to Maroon science and its African-derived cosmology. Visitations take place in dreams. Traveling by water, the seeds of the snuff box sea bean take a few years before the scarified plant embryos can germinate. The dreams and visions are a vital conduit to the Gold Coast and otherworldly planes of being, an intimate part of Maroon initiations, pedagogy, and midnight rituals.

Prophecies that were delivered to Grandie Nanny in visions from West Africa helped her nourish her troops by drawing on nature's bounty, thus defeating the British army in the 1720s. She planted three pumpkin seeds given to her in a dream. A gateway for communing with the spirit world, this plant is one of many forms of African botanical life and horticultural knowledge that thrives in the Americas. It is a staple vegetable of "provision grounds," which are plots of land farmed by African people when they were granted time away from plantation labor. Chapter 8 will detail more about the significance of this vegetation. Older women Maroons lured British troops into cleared plots of land like this and would then surround and overpower them in a coordinated ambush. Did Grandie Nanny prophesy the climate ruin that has befallen the Caribbean today?

Peering high up into the rainforest canopy, I knew that even though I was not a climber I could not get lost. Colonel's profound, simple words have echoed with me long after. What does it mean to be lost, anyway? There is power in not being located on the bureaucratic map. Numerous distinct African-derived musical traditions can be heard across Jamaica, including Ettu, Goombeh, Kumina, Nago, Tambo, Revivalism, and Rastafari. Being absent from the colonial census might have ensured the survival of these songs across the Western Hemisphere. It is easy to lose sight of the heterogeneity of Maroons across different countries, but decentralized fragmentation has been at the core of their survival. The Aluku, Kwinti, Mātāwai, Ndyuka, Paamaka, and Saamaka have maintained their lands and traditions in Suriname and French Guiana. In many cases the government, or powers that represent the state, flattens the heterogeneity of Indigenous Amerindian nations like Kalina, Lokono, Trio, and Wayana, if it recognizes them at all.[15] Often they are placed in the category "Other." The discovery of new oil, in 2015, imperils Maroon and Indigenous governance in places like the Guyanas. Guyanese president Irfaan Ali, who is of Muslim Indo-Caribbean ancestry, has been an outspoken voice against Western colonialism, but falls short in answering how the environment is being protected and where Indigenous representation in climate policy can be found today.

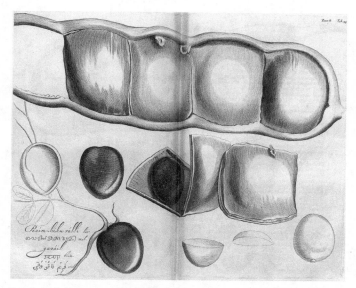

Botanical drawing of *Entada rheedii Hortus, Indicus Malabaricus*

I am reminded often about how Colonel Sterling taught me to realign my feet to the hinterland terrain of my forebears. Heel first: my expensive yellow English boots probably prepared me just as the British army was ill-prepared for the Jamaican mountainsides in the eighteenth century. The Maroons and the mountainous terrain defeated them. By the nineteenth century, Wellington boots had become a staple shoe of the British aristocracy, named for the duke. The Hessian boots were a modification of the eighteenth-century model.

The Jamaican Maroons maintained their sovereignty in hushed Koramantee whispers such that the Ghanaian language remained preserved in a time capsule. When he was growing up, my father knew that he would never be privy to this language, though it was a part of his heritage. Those who were not true-born Maroons were not to even hear these clandestine intonations. After Jamaica's independence from Great Britain, many Maroons eventually took the path toward assimilation into greater Jamaican society and the diaspora because of the socioeconomic pressures for a liveli-

Colonel Alexander G. Fyfe and six Maroon soldiers
in cacoon camouflage, 1865

hood. The promise of social mobility after 1962 was a major pull of
Maroons away from the protected territories.

Anxious and conflicted thoughts raced through my mind as
Colonel and I continued to descend the woodland slope. His two-
year-old puppy gleefully yipped between us, thinking this the perfect place to play as we hiked. The dog understood Colonel's words,
"You cannot get lost out here." Before long, Colonel found what
we were searching for, retrieving the plant called antidote. The seed
pods are often found on or near water and can survive immersion in
seawater. The plant's medicinal and military uses included cleaning
wounds in battle, treating burns, and healing jaundice in children.
The bark of the tree is also medicinal and is used to treat diarrhea,
dysentery, and parasitic infections.

Learning this, it occurred to me that the concept of being lost
is a very Western preoccupation. One of the great thinkers of the
Caribbean diaspora, Audre Lorde, wrote, "When we view living, in
the european [sic] mode, only as a problem to be solved, we then
rely solely upon our ideas to make us free, for these were what the

white fathers told us were precious." Lorde, who was born in New York to Carriacouan and Bajan parents, also writes of the alternative philosophical orientation of the Black mothers. I interpret Queen Nanny to be one of them, though she had no direct descendants. Lorde's elocution gives us a strategy for the climate crisis. The white fathers say, "I think, therefore I am," and the Black mothers say, "I feel, therefore I know I am free."[16] Black feminism has been tasked, time and again, with articulating the world's grief. Still, we manage to create a poetics out of that which wishes to destroy us and the planet. How else will we able to live in "the after"? We must reassess what a problem is. Living is not a problem, as Lorde reminds us. I would add that dying is not a problem either. Decomposing is essential to the natural order and cycle of life. Living *at the expense of others* is a problem. The Eurocentric existentialism that Lorde warned of creates the problem, invents being lost as a concept to sell a map. The U.S. Department of Defense engineered the Global Positioning System in 1973. As I could only have learned in the mountain villages of my ancestors in Jamaica and China, satellites cannot penetrate everywhere, and that is a good thing.

＊

Colonel's words reverberated through me: "You cannot get lost out here."

＊

QUEEN NANNY OF THE MAROONS:
COMMANDER, CLIMATE WARRIOR, HEALER

Queen Nanny's name echoes across the mountain ranges of the archipelago because she refused to sign the eighteenth-century British Treaty. She refused to betray the future. Through her we can center Maroon climate science. Black and Native peoples have long been the climate scientists, philosophers, doctors, griots, shepherds, healers, and horticulturalists of the Western Hemisphere. Grandie Nanny's descendants—human and nonhuman—are many, though she did not mother biological children. Before a famous battle, she

planted three pumpkin seeds given to her in a dream and was able
to provide food for her army. Herbalism and botanical cultivation
were essential to the adaptive warfare she practiced. British history
continues to discredit Maroon history and defame Grandie Nanny
as if she never existed.

The triumvirate of leaders—Nanny and her brothers Quao in
the east and Captain Cudjoe in the west—could not get lost in the
Jamaican wilderness. They knew the British planters, militia, and
army alike would see the mountainous terrain as a problem to solve.
They would get lost. The Maroons adapted to landscape quickly and
learned from the Arawak, with whom they intermarried.

"You cannot get lost out here."

When we arrived at the Institute of Jamaica for the exhibition opening, Hans Sloane's *Voyage to the Islands Madera, Barbados, Nieves,
S. Christophers and Jamaica* (1707) sat open in a display case. This book
is the basis of the British Museum, which was founded in Sloane's
honor. The British physician was the personal doctor of one of the
governors of Jamaica. By serendipitous coincidence, the 284-page
folio happened to rest on the page depicting the antidote cacoon, a
different type of cacoon from Colonel's vine.[17] We had encountered
this mysterious pod on his property too. Sloane depicts the islands
of the Caribbean as laboratories for medical science. In his volume,
Sloane documents the scientific knowledge produced by Black and
Indigenous peoples of Jamaica. He often dismisses these practitioners, writing that "the *Negros and the Indians* use to Bath themselves
in fair water every day, as often as conveniently they can." Sloane
disregards their expertise, but, interestingly, he notes similarities
in methods of medical treatment between the East Indies (Southeast Asia) and West Indies. There is a similarity in the perspective
remarked on enslaved African people. With this landmark book,
Sloane began a tradition of natural history writing about the Caribbean that we see continued by Irish physician Patrick Browne's volume *Civil and Natural History of Jamaica* in 1756, which also notes the
peculiar antidote cacoon and its healing properties.[18]

My 2022 painting *Queen Nannies,* which we had arrived to install, is still displayed at the Institute of Jamaica as part of a climate art exhibition on Grandie Nanny. I wanted to depict her as a Pan-African and Pan-Caribbean figure, so I used the traditional tartan textile pattern of Dominica paired with that of Jamaica. I also included collaged elements of Black Guadeloupean women photographed at Ellis Island in 1911 to gesture to the Atlantic circuits of Caribbean migration. These women were domestics, women paid to care. They carried with them a knowledge of West African and Caribbean botanicals that healed many and derives from the legacy of their foremothers like Queen Nanny. For the backdrop of my painting, the red bauxite earth gestures to current climate debates on the extractive economy and what Maroons have done to protect it from mining. I sutured three canvas panels with gold thread to bring these elements together. A QR code at the natural history museum activates the sonic component of the painting I designed. The soundtrack I produced is an assemblage of sounds of nature I

Tao Leigh Goffe, *Queen Nannies,* Institute of Jamaica, 2022

have recorded over the years. I chose to mix these with dancehall
riddims: Jamaican instrumentals that are shared and remixed. Titled
"Kill Them with the No!," my piece remixes the Filthy Riddim created in 1998 by Jamaican producer Danny Browne. I chose to re-
interpret a popular Mr. Vegas song from the late nineties, "Heads
High," in the spirit of Queen Nanny because of its refrain of "No."
To me, it amplifies the radical defiance of saying no in a very Jamaican and feminist register, even if the original song is not feminist in
intent. It has become an anthem to me. The Maroons of Jamaica
have been saying "No!" as a refrain for centuries.

I recalled an experience of my own years earlier in the rival
Maroon territory across the island from Nanny Town in Cockpit
Country to the west. I had learned from the patter of thousands of
bats swooping and diving in the dark. They knew exactly what to
avoid and evade. To be in the Windsor Great Cave at that moment
was to tune in to the soundtrack of the rebellious mountain territory. The bats showed us the way to their roost through echoloca-
tion, demonstrating that there are multiple paths home.

Just as science writer Ed Yong explains in *An Immense World*,
there is much we can learn from the *Umwelt* of the range of animal sense perceptions.[19] The German word describes how each animal has its unique way of perceiving the world. Yong uses the term
because it does not quite translate to existing English phrases for
proprioception. Assuming that human forms of cognition are the
only ways of knowing is the first misstep. Not all cultures subscribe
to this anthropocentric form of science in their traditions of natural history. Yong expands on sensorial theories by zoologists Jakob
von Uexküll and Thomas A. Sebeok to show how animal communication is only a fraction of the animal cognitive experience. It is
this sort of imagination of expression and relation beyond binary
modes of epistemology that Maroons have likely long understood
about bat proprioception across the Western Hemisphere. Communing with the rhythm of the Windsor Cave bats as one example
of a rainforest ecology, I learned there are many sacred lessons in
navigating the dark wilderness without sight. The flight path of the
bats is an important pollinator pathway of night-blooming flow-

ers such as those of the sacred silk cotton trees. The British militias never would have had the inclination to comprehend this in the eighteenth century as they batted away mosquitoes for fear of malaria.

The Jamaican philosopher Sylvia Wynter also touches on the proprioception of bats in her meditations on Caribbean epistemology titled "Towards the Sociogenic Principle: Fanon, Identity, the Puzzle of Conscious Experience, and What It Is Like to Be 'Black.'" Drawing on U.S. philosopher Thomas Nagel's "What Is It Like to Be a Bat?" on phenomenology, Wynter speculated on the bat's subjective conscious experience to analyze the colonial condition.[20] In her analysis of race and its formation, she turns to Frantz Fanon's *Black Skin, White Masks* to grapple with alienation. What is the *Umwelt* of the Black experience? Relatedly, her concept of indigenization has been key to negotiating the complex process of how Black people have become *of* the Americas over centuries of separation from West Africa. What have these Jamaican bats witnessed of these processes? Did Nagel know that Jamaicans call fruit-eating bats "rat bats," as opposed to bats which are moths? Wynter, of course, knows these vernacular distinctions; maybe she even knows the chill of the air from the cathedral-like caves where the cold seeps out of the limestone caverns in Cockpit Country.

The karst limestone formations that shaped those caves had deposits of guano at least two meters high. The strata suggest the accretion over millennia. Guano is considered a sacred substance across the Americas to many Amerindian tribes. Telling a story of shelter and ceremony, the caves deep within Cockpit Country are sacred architectures of resistance carved by water that flows out of the limestone hills. Petroglyphs estimated to be 1,500 years old tell a story too. Plantation runaways had seen these cave paintings over the centuries when they found shelter in the same caves. Mining threatens to collapse these Jamaican caves, which would disappear into the mountain. These rock formations should be protected from dynamite, bulldozers, and scrapers. Miners often find this ore in flat deposits beneath the earth's surface and extract it using the Bayer process, in which sodium hydroxide, or the caustic substance lye, is instrumental. Fragments of the island itself are being mined and exported away. As of October 2022, the first shipment of 36,000 metric tons of Jamaican limestone was being exported to Savannah, Georgia. While the Jamaican government sees this as a victory for the diversification of the Jamaican economy, Lydford Mining Company is literally selling the ground Jamaicans stand on to the highest imperial bidder.[21]

There are a reported twelve species of bats that live in the Windsor Great Cave. It should not come as a surprise that, though Cockpit Country is Maroon territory, the land outside the main entrance to the cave is owned by the World Wildlife Fund. It was sold to them by Dame Miriam Rothschild and had been owned before her by Willy Donald-Hill as part of the Windsor Estate. When I met an Afro-Jamaican worker named Sugarbelly—who is technically the manager at the Windsor Great House—the significance of his name and subordinate position was not lost on me. His boss, a white docent and conservationist from Arizona, made a joke about what she called "blue-black people." People with skin so black they were blue, she said. The white American conservationist also warned us about the Chinese and the nefarious impact they were having in mining bauxite from Cockpit Country. She conveniently omitted the role of British, Canadian, American, and Russian corporations in establishing the Jamaican bauxite industry.

Guano had been mined from the caves in the 1930s before it became publicized for tourism. Jamaica's limestone caves form an intricate underground network. I arrived here in search of guano after reading *Dr. No*, Ian Fleming's 1958 James Bond novel about the precious substance. He must have known of these caves and the guano deposits in them, and his was a militaristic imagination that stemmed from his time as a British naval officer during World War II. Men like him were not only permitted but encouraged to be intellectually promiscuous, though they lacked formal university training. Attending Eton was enough to qualify a person as an authority about everything. Why shouldn't we imagine the renegade Maroon and Arawak midnight meetings held in these caves over the centuries? The bats and freshwater crabs that inhabit Jamaica's caves are witnesses to this history of the island itself.

A groundbreaking study conducted in 2021 on Jamaican bat guano sedimented over 4,300 years traces the Industrial Revolution through chemical composition.[22] Prior methods relied on the analysis of lake sedimentation. Climate change can be tracked through the accretion of droppings from long before Christopher Columbus arrived on Xaymaca. From the fourteenth century onward, the carbon, sulfur, and nitrogen composition shifted drastically. Agricultural shifts can be identified as well; for example, the use of nitrogen-based fertilizers, sugarcane, and fungicides. From the 1760s, increases in cadmium, mercury, lead, and zinc become apparent in the excrement. The shifting diet of the bats demonstrates what it means for actors within the natural world to digest the toxins that industry produces. The fruits, a substantial part of the bats' diet, were exposed to different metals and pollutants, which were ingested. These shifts can be tracked through isotopes that even show the history of when leaded gasoline was introduced on the island.

To find our way back to the estate and wooden barracks, our sleeping quarters, we only needed to follow the flight of the green-glowing peenie wallies lighting the path. The flying beetles form a trail in the way that writer Aimee Nezhukumatathil powerfully describes as the bibliography of a firefly, speaking of the beetle's cousins in Mississippi.[23] The sound of the Greater Caribbean echoes in the American South and in Jamaica, connecting geographies, economies, ecologies—each luminous flicker a citation of woodland knowledge. Synchronous lessons of nature such as these disappear as extractivist capitalism accelerates. Maintaining Maroon sovereignty is a part of global climate policy. Three hundred and fifty years of self-determination did not require external permission; though they signed the treaty with the British, it is not the governance of the United Kingdom that granted their sovereignty, not truly. Even if the Jamaican government refuses to give Chief Richard Currie of the Accompong Maroons his day in court, it cannot renounce the Maroons' rights to the sovereignty of their ancestral lands.

MINING THE RED EARTH:
JAMAICA'S GLOBAL BAUXITE CRISIS

While Portland—in the east of Jamaica, where Colonel reigns over
Moore Town—is known for its striking shades of sapphire and
cobalt blue water, in the west there are rich shades of red where
the earth is the color of blood. Bauxite deposits exist here in the
interior rainforest of Cockpit Country where the Leeward Maroons
(Accompong) have lived for over 350 years. Unmined, the earth
is plentiful with stores of ferric oxide tinting the ruddy oxidizing
hinterlands. The Accompong Maroons have been living on this
1,500-acre spot and have left it unmolested. Many bauxite mines
elsewhere across the island have been depleted, and the Maroon
territory is what remains untouched. The same chemical component (*hematite*) that makes blood red gives the Jamaican mountain
soil its rusty pigment. From Miocene volcanic ash deposits sedimented over time, bauxite is a sedimentary rock that has a high
aluminum content. Found near limestone, it is made up of ancient
coral compressed under the weight of mud and sand. Bauxite is
the compressed and compacted remnants of the last climate crisis, and now it is used to craft the missiles of destruction of our
next crisis.

Jamaicans who live near bauxite mining sites are plagued by red
dust. The microparticles are like tiny daggers, abrasive dust residents inhale daily with no other choice. No matter how much people who live in these mountain communities try to vacuum, mop,
and clean their homes, the red dust film of bauxite particulates
invades their pores and scratches their skin.[24] There are increasing reports of pulmonary defects that lead to a slow death from
carcinogenic exposure, an exposure that has gone unregulated by
the Jamaican government. Contaminated waters form red lakes
where plants and fish are poisoned. Since its discovery in the 1950s,
there has been only one medical study in Jamaica on the impact
of bauxite on the body. Why should the eroding of mountainsides and the shorelines of small islands in the Caribbean Sea matter to the rest of the world? The Caribbean is a bellwether for the

climate crisis because of what Columbus's arrival inaugurated for globalization.

British geologists first noted the presence of bauxite in Jamaica in 1869, but it was not until the Canadian and U.S. corporations Alcan, Kaiser, and Reynolds arrived there in the '50s to survey what they called the "red ferruginous earth" that the trade began. The interior continues to be carved up and sold to the highest international bidder. First mined in 1952, the ore from which aluminum is made transformed the British colonial island's economy, just a decade before Jamaican independence was achieved in 1962. Jamaica had become the world's leading exporter of bauxite. Depleting the mountains of this precious ruddy ore, Russia and China now also divide the Jamaican earth and its future. From aluminum foil to house siding to construction tools to parts for engineering commercial airplanes and rockets, the global demand for metals has only increased post–World War II.

Even with the best intentions in the 1970s, democratic socialist prime minister Michael Manley miscalculated the environmental bargaining cost of the island's freedom. The Socialist and capitalist approaches that originated during the Cold War and that completely ignore climate have failed us. Communist forms of government did not at the time offer an alternative to the extractive economy of pollution. The race for economic and political dominance took the environmental toll of industrialism.

At 4,000 square miles, Jamaica is the largest island nation in the Commonwealth Caribbean, presenting a microcosm for the climate crisis from "ridge to reef." This is one of the many slogans that the National Environment and Planning Agency of Jamaica used from 2001 to 2005. The island is a laboratory of ecological and economic history, as are all the Caribbean islands and coastal areas. Colonial powers will not allow the earth to heal or the people to be sovereign. Against the odds of ongoing exploitation of natural resources and citizens, we can chart survival through a series of case studies after apocalyptic events.

If Maroon elders such as Colonel were at the policy table, how would our planet's climate fate shift? What would it mean for them to have a constitutional voice in the Caribbean? Is a table even the appropriate furniture for the climate strategies we require? Our dreams require other architectures.

Maroon history becomes complicated in regard to their purported role in suppressing the revolt of Takyi, an African who was enslaved in the parish of St. Mary, against the British in the 1760s.[25] Takyi was gunned down and decapitated, with his head placed in the town center as a clear message. Maroons also played a role as a militia in suppressing the Morant Bay Rebellion in 1865, capturing Paul Bogle and turning him over to the British authorities.[26] Much of this distrust will never be resolved; however, taken as a historiographical matter, colonial history as told by the overlords will always be designed to sow seeds of doubt.

On May 6, 2022, not long after the apology to the Jamaican people, Chief Currie sued the Jamaican government for crimes against climate. In his thirties, the chief represents a much younger and more defiant generation of Maroon future leadership. In some instances where the three other leaders of the Secretariat—the Scott's Hall, Charles Town, and Moore Town settlements—have met at the bargaining table with the government of Jamaica, Currie has refused. He has taken to Facebook videos and GoFundMe to amplify his message to protect Cockpit Country. He maintains that the 1,500 acres designated should exist "forever hereafter in a perfect state of freedom and liberty," as the British stipulated. So far, the GoFundMe has earned $24,000 USD out of a goal of $70,000 USD. Supporters worldwide continue to write notes on the site. Some simply say, "Maroon descendent." Several of the online supporters and funders thank Chief Currie for his climate advocacy.

According to Currie, threats have been made against his life to subvert his reign. He says that planes fly low across the Accompong territory at two in the morning. He interprets these tactics as intimidation, but Currie also notes that the small sovereign nation has garnered attention globally beyond the Caribbean. Currie lists the new media outlets that have reported on the crisis of sovereignty. Perversely, their freedom depends on a piece of paper from 1738. Technically, Jamaican Maroon sovereignty depends on the British keeping their word and their Commonwealth. If Jamaica becomes a republic, what sort of legal claims can be argued about the boundaries of the land? Chief Currie's representative, attorney-at-law Charles Ganga-Singh, and legal team members Isat Buchanan and Alessandra Labeach will have to devise alternative juridical ways to challenge the Jamaican government and the National Heritage Trust. In disregarding the claims and failing to take the charges seriously, the government disrupts the rule of law on the island nation.

Without many updates, the last report was that the Jamaican government was making all efforts to have the Maroon lawsuit thrown out. The rationale for the delay is unclear. The hearing continues to be postponed. The bureaucratic, legal, and procedural forms of denial show what is at stake globally when it comes to Indigenous peoples challenging other nations. Currie has sought diplomacy, statecraft, and international recognition of Maroon sovereignty from the Nigerian high commissioner, Yvette Clarke of the U.S. House of Representatives, and Asif Ahmad, the former British high commissioner to Jamaica. He has said that "when no one is speaking to us locally, we go speak for ourselves" at the United Nations. The Annual Maroon Treaty Celebration is a global platform to declare friendship with the U.K., which is the basis of their freedom and defined boundaries. Chief Currie understands the optics of this battle unlike any Maroon leader before, deploying social media as a tool with the hashtag #SoundDiAbeng, signaling the clarion call for continued Afro-Indigenous sovereignty and climate justice. He says, "Not until the last tree has been cut, last fish has been caught, and the last river poisoned, will #Jamaicans realize that we cannot eat Money!" Embracing technology, the platforms become a digital abeng for the Leeward Maroons of Accompong under Chief Currie's leadership.

Later that year, Chief Sterling of Moore Town accused Chief Currie of a plot of upheaval. Sterling, who had gone unchallenged for years, ran against an eighteen-year-old woman named Lamorra "Hope" Dillon, vying for the position of leadership of Moore Town.[27] Photographed draped in cacoon vines, the young woman attempted to rally support for her candidacy in a more youthful style of leadership. Not since Nanny had there been a woman leader, and she evoked the image of Queen Nanny. While Dillon garnered only thirty-six votes, the signaling of the camouflage vines was clear.

The disputed boundary determines where bauxite and lithium mining are permitted. With written newspaper and TV coverage by Al Jazeera and Vice News, global attention has reached Accompong. In the podcast episode "The Fight for Jamaica's Red Gold," journalists emphasize how the climate battle is not over. Three hundred years on, how do we measure the validity of the treaty based on Britain's colonial reign? Land grants, whether in Mannahatta or Xaymaca, are land grabs, upon which the treaties and terms of surrender are dubious.[28] Notions of property or land ownership are not universal and so should not be taken for granted. How would Maroon historiography and Maroon jurisprudence orient these matters differently? The hue of dried blood, the protected soil, is fugitive land stewarded by generations of clandestine custodians. For nearly four centuries, the undisturbed life-giving volcanic soil, fertile and rich with iron, has nourished generations of Maroons. They have not mined it, only farmed the land.

Bauxite and limestone form a chiaroscuro of geological landscape, as the ruddy Jamaican earth contrasts with the white craggy karstic limestone environment. Scientists estimate that 65 percent of the island, measured by weight, is limestone. The pristine white sand of Caribbean beaches is often a fiction, too, imported from other parts of the region that have been deforested and mined.[29] Sandals Resorts and other luxury all-inclusive resorts are among the largest proponents of bluewashing, or a hollow commitment to responsible climate practices that protect the ocean. These resorts have been the strongest protectors of the Caribbean oceans only because of profit margins. Depending on the resort, a weeklong stay in Montego Bay or Negril can reach more than $15,000 USD. Across the island, genealogies are etched in the sand, the bauxite, the limestone, coastal and cliff erosion. Privatized beaches made up of imported white sand show us the unequal distribution of the climate crisis. Resorts operated by European corporations have denied Jamaican people of color entry. The erosion of the shorelines of the public urban beach Hellshire shows how colorism divides access to nature.[30]

*

Manhattan, an island, is the site of my laboratory. (New York City is sometimes called a Caribbean city of the North because of the large communities of people from the more than forty islands that make up the Caribbean archipelago.) Geographically, Long Island is part of the same landmass and island that includes the boroughs of Queens and Brooklyn. It was not until 1898 that the boroughs were incorporated to become the urban conglomeration of New York City as we know it. Brooklyn had been a different city altogether from New York, which shows us the shifting constructs of land and water in a global metropolis. These dividing lines have affected biodiversity, especially across urban wetlands. Cold Spring Harbor, located on Long Island's North Shore, was the home of the Eugenics Records Office from 1910 to 1939. It was founded by Charles Davenport, the director of Cold Spring Harbor Laboratory, in 1904. The laboratory was the headquarters of race science, and American eugenics experimentation deeply connected to the Caribbean and other islands.

Cold Spring Harbor Laboratory was the breeding ground for the poisonous thought that would eventually morph into Nazi Germany's death camps. A locus of contradictions of modern science and pseudoscience, it is also where discoveries about the crystallography of DNA were made by Rosalind Franklin before James Watson and Francis Crick took credit for the double-helix theory. Watson's later legacy as a eugenicist shows us to heed the empiricism of scientific discovery. Racist thought is never far when it comes to the theorizing of genetics, although the science shows there is no biological basis for "race." The Americans at Cold Spring Harbor modeled their studies from the eugenics movement that was born in Victorian Britain.

As a biologist, Davenport was deeply inspired by Francis Galton's biometric methods for interpreting evolution—the theory of

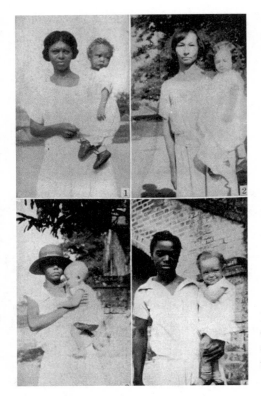

"Fair-skinned babies from dark mothers." Davenport and Steggerda, *Race Crossing in Jamaica.*

Galton's cousin, Charles Darwin. Davenport was also inspired by Karl Pearson, the British mathematician, which leads us to how biostatistics are at the core of the race science theorization. Species thinking applied to humans is predated by a hundred years by scientific racism, the nomenclature of classification that Swedish taxonomist Carl Linnaeus developed. The taxonomic impulse is the taxidermy impulse. Though it may seem intuitive to scientists that human beings seek to categorize based on a desire for pattern recognition, it is a far leap to the pathological logic of eugenics, a genocidal science. Davenport remained an ardent supporter of Nazi Germany until he died in 1944 at the age of seventy-seven.

In a striking arrangement of four photographs in his 1929 book *Race Crossing in Jamaica,* a quadrant of four unnamed Jamaican women hold their toddlers. The portraits remind me of how people like my grandparents who were born in the 1920s and 1930s were pathologized in colonial Jamaica. Davenport and his assistant Morris Steggerda—who was a physical anthropologist by training—described what they believed to be a genetic phenomenon: "fair-skinned babies from dark mothers."[31] Each mother protectively holds her baby and glares at the lens with a defiant gaze. Davenport and Steggerda noted these and other racial anomalies in their book. The relentless alienating logic of the plantation cuts deep and extends to the 1920s with endless attempts to classify people by genetics. *Race Crossing in Jamaica* offers a textbook formulation of American pseudoscience and how the Caribbean has long been under the microscope of racial experimentation. These islands were fetishized under the myth of perfect isolation as hermetic laboratory settings, when in fact islands and archipelagoes are dynamic systems of exchange and trade. The Caribbean was a laboratory for the U.S. colonial racial imagination. Adding to the colonial design of Europeans projected across the region, the occupations are multiple and layered, all ongoing and compounding. These histories connecting Hong Kong to Jamaica to New York to Britain show that no island is deserted; rather, it is a set of relations, ideas, and ecosystems that sometimes converges as family history.

2

✳

Climate Crisis, Genesis 1492

What is the origin story of the climate crisis?

✳

It is difficult to say, but Eden is one answer. Presented by the Hebrew Bible as a utopia, it began as a failed, impossible project, and God's first punishment for humanity. Eden is a fantasy of innocence disrupted by a tree of knowledge. When Europeans arrived with the cross in the Caribbean, they could not help but see Eden. Who, then, is Adam, and who is Eve? Were the Natives, who were described as childlike, the innocents, or were the innocents the Europeans? Today Europe feigns naivete about the origins of the climate crisis, as if the Industrial Revolution was the price of progress. The green fantasy of clean energy is often a violent one, with bloody consequences and collateral damage. Cascading, world-ending events from the microscopic to the geopolitical level have disproportionately impacted the life expectancy of Black and Indigenous peoples. Although our current ecological annihilation was initiated in 1492, so many peoples, plants, animals, and other living organisms have survived beyond that crisis.

If Eden is but one origin story, there are multiple others of hope in *Dark Laboratory*. What of the other gods present in the Garden of the Caribbean? Traditions were transplanted to the Caribbean with the arrival of Hindus, Christians, Jews, Muslims, Buddhists, Taoists, and secular people. Columbus could not have conceived of this flourishing when he planted the flag for Spain on Guanahaní, the present-day Bahamas. While captors erased histories and names, wrenching families apart to subjugate people, they could not fully erase the lore of gods and orishas not forgotten. Native peoples like the Lucayans were enslaved and forced to dig, mining from the earth to extract resources for Spain. Those who held the whip could not, however, steal everything, not the primal memory of ancient lullabies. If we embrace a nonlinear understanding of time, there are multiple origins and apocalypses. Among the current crises is one of political memory. The cycle of forgetting is dangerous. It is so easy to lose everything community-led struggles have fought for if we don't study what it means to organize politically. Amnesia and hubris lead to reinventing movements that already exist. Once we view the struggle as intergenerational, there is hope because past blueprints of resistance show us the problems are not new and the battle against complete European conquest of the Americas, Africa, and Asia is ongoing.

The origin story of the Hebrew Bible—the book of Genesis—designates that God's first punishment was itself a climate crisis. Banished to the barren earth, Adam and Eve were the first climate exiles. Since then, there are those who have been trying to find a way back to the innocence of the garden by spilling blood. Eden is a story constantly being told and retold through the mapping of a colonial desire that sacrificed preexisting worlds. Erasure took place in the name of Eden on the three ships of the apocalypse. The innocence of virgin land requires annihilation, and climate innocence requires a modern amnesia that vacates the landscape. Forgetting the apocalyptic consequences of 1492 is part of the plot of colonialism. The European quest for Eden has pushed the planet to environmental ruin. The capitalist mission has been one of extraction and greed. Yet race, labor, and colonialism are often conspicuously missing from conservationist debates. The climate crisis requires a theory that addresses how non-Western cosmologies were erased in order to lay claim to the land, and how Christianity's rise annihilated other belief systems. Erasing multiple creation stories—each a way to relate to the environment—paved the way for the capture of and extraction from the land.

Each summer is the hottest summer on record; the global forgetting about the last climate tragedy intensifies. How are we to process the cascade of levees breaking during Hurricane Katrina in 2005? The melting of the tarmac on British airport runways? California wildfires setting woodlands ablaze? Flooding New York City basements and subway systems? Devastating fires in Hawai'i? Tragedies accumulate toward the earth's exhaustion as the grievances mount. How can unwritten climate policies save us without acknowledgment of the violence of the Edenic origin story? To the world's collective detriment, scientific racism is embedded within Western natural science, including climate science. Omitting race while it remains a material reality is both racist and unscientific. Western science has been not only complicit in racism and colonialism but it has also been a core tool to enforce it. This crisis is ours, and we must own it because it possesses us. Our ecological fate is shared, but we are all dying at different rates due to climate risks according to race, nation, class, and gender. All will not be lost if societies could collectively confront these risks at the United Nations Climate Change Conference COP 29 in 2024. Would they be willing to confront the Edenic myth of climate and how its origins are deeply tied to alibis of the fifteenth century? We are still paying the ecological price for the justifications used by Crown and Cross to claim the Caribbean (and the rest of the Western Hemisphere).

Carbon is merely one current metric for greed, and the world must reckon with Christopher Columbus's colonial footprint of dispossession and displacement. The earth's axis shifted permanently, with dire environmental consequences, when Columbus went searching for Indian spices and ended up south of what is now the United States. In an island chain that the Europeans christened the Caribbean, Columbus planted the colonial fantasy of Eden. He also planted sugarcane. Columbus's navigation set the foundation for a globalizing world powered by the hyperdrive of monocrop plantation agriculture. The opposing logic of the monocrop, and perhaps its antidote, is to give from a source within. A Black feminist ecological ethic approaches this mode of being and doing: To undo the colonial regime, multinational corporations cannot continue to incentivize processes that lead to ecological degradation. At many global climate conventions, negotiations are made while the world is held at ransom by the interest of transnational corporations and fossil fuel lobbyists.

The lost Italian sailor Columbus never set foot on what is now U.S. soil, nor was that his intention. Yet on the second Monday of October every year, the young nation officially commemorates the federal holiday in his name. Columbus died without knowing he had not reached Asia.[1] Many Americans who celebrate Columbus Day will never know Columbus did not set foot on the U.S. mainland. Many Native American communities celebrate Indigenous Peoples' Day in protest, and many across the Americas reject the commemoration of Columbus because he was a colonizer. Yet even among those who reject Columbus as a national hero, very few look south beyond the contiguous United States to where he landed.

The Caribbean isn't often recognized as ground zero of colonial conquest, but its islands form a dark laboratory of colonial desires and experiments. As such, this island chain is the epicenter of the modern globalized world. Many can only imagine the Amerindian peoples Columbus encountered in abstract terms. He described Native peoples in his journals as being childlike, anointing them erroneously as Indians.

Edenic myths of U.S. exceptionalism erase the Caribbean arc. The bureaucratic map of colonialism has effectively erased Indigenous reservations across the isles, proof that Columbus's conquest has been incomplete. Contemporary Amerindian communities suffer neglect from the state but have also managed to survive and retain their cultures because the state has ignored them. More than a thousand small islands have been dubbed "America's backyard": a time-share that the U.S. bought from Europe. The islands became dependencies, new markets for U.S. exports. The myth of "virgin islands" hides the ongoing stakes of Cold War military ploys. Each island is a chess piece, but island climates can teach us much more than what exists in field guidebooks. Each summer, the planet approaches the wet-bulb temperature, or the atmospheric conditions at which human life is no longer possible because it's too hot. At what temperature does asphalt melt entirely?

EDEN ISLES:
BLACK AND NATIVE GOVERNANCE

The annihilation of many Indigenous societies and ecological worlds commenced across the hemisphere when Columbus first saw the mangrove-covered island of Guanahaní, one of the islands of the present-day Bahamas. The 700 coral islands of the Bahamas in the North Atlantic Ocean remain the unceded territory of the Lucayan people (*Lukku-Cairi*), a branch of the Taíno (Arawak). The Genoese navigator was not innocent, and the violence in the name *Caribbean* invokes indigeneity while simultaneously erasing it. Columbus took 500 Taíno people with him to Spain, of whom roughly 200 died on the voyage.[2]

"Carib" is a name that implies cannibalism. Distinct from the Taíno or Arawak, this rival society is the mistaken namesake of this archipelago. The so-called Caribs form the tribe who today call themselves the Kalinago, who are most prominently represented on the island of Dominica, often mistaken for the Spanish Caribbean country of the Dominican Republic. British law legally established the Kalinago reserve in the Eastern Caribbean nation Dominica in 1903, after George III allotted it to the U.K. in 1763. The island's name owes to the French settlement before the British occupation. The Kalinago tribe rejects the brutality of "Carib" and its origins, and, like many Indigenous nations and Maroon settlements, it has continuously maintained its sovereignty for over four centuries since the European invasion. The island has maintained this autonomy via communal landownership and systems of regenerative agriculture. Although the living conditions in the territory are fraught with many external pressures, with an estimated 3,000 members of the Kalinago tribe out of the broader total of 70,000 Dominicans, the Native tribe maintains its own governance, and periodically votes on village councils and for the position of chief.

Near the end of the Caribbean island chain, Dominica has been a haven for Maroons and Native people from other islands as they retreated from European colonial incursions. Tall is her body, and even on the plane to the island, American and European tourists do not know how to pronounce her name—DOM-ih-NIQUE-uh. Their tongues want to lead them to say the name of the more familiar island tourist destination of the Dominican Republic. In 1978, when Dominica achieved independence from Great Britain, the 3,700-acre mountainous district was protected under the Carib Reserve Act.[3] This protected area suffers from deforestation and land erosion. Due to the rotation of tenant land usage, intensive rains on mountainous slopes, high-velocity winds from the Atlantic Ocean, and thin topsoil, agriculture must be hardy and adaptive.[4] The tribe has remained steadfast in protecting communal landownership, though the Dominican government pushes for privatization and modernization. There are deeper philosophical and legal questions at play regarding the land usage and private property, questions that transcend European colonial notions of land tenure. Can one ever legitimately own the soil, the bedrock, the wilderness? If settlement were not a possibility, how would we live in relation to the land? Could we abolish territorial attitudes? How are architecture and infrastructure used to make claims to the land?

Many of the territory's residents are among the poorest on the island. Yet at Secret Bay, tourist resort rooms are booked at $1,000 USD per night. The Kalinago Territory was ravaged by Hurricane Maria in 2017, like the rest of the island. For repairs, the tribe had to depend on themselves, as did all the people of the Eastern Caribbean island who went without electricity, phones, or internet for several months. The Secret Bay facility, though, has a private bunker, should a hurricane like Maria touch land again. And it will. What plan will prepare the residents of Dominica for the next inevitability? The tourists have a built-in hurricane contingency plan that ends in evacuation from the island. The plan does not include offering shelter to the Kalinago, an hour and a half away by car. They will have to stay. I ask you: Will the Black hotel staff and housekeepers stay in a segregated part of the bunker?

During my visits to Kalinago Territory, I have learned about Native survival, governance, and futurist architectures.[5] The boundaries of the reserve are tightly guarded against outsiders who are not expressly invited, or whose clearance has not been granted by tribe members. Birthright and Kalinago parentage determine membership and residency in the territory. Despite the catastrophic aftermath of Columbus's landing, the plot of land still exists for many Black and Native peoples as a pact. Dominica—the Eden Isle—became an island of refuge, where Maroons and Indigenous peoples canoed from other Caribbean Islands. The Native communities protect the territory, and it protects them. The urgent question of the climate crisis today is, who will protect the land and sea from us? Climate sovereignty is impossible without giving the land rights back to Black and Indigenous communities.

Dominica's mountainous climate most likely protected it from European colonization. The French arrived late, in 1715, and the British supplanted them in 1763. My purpose in visiting the Amerindian reserve on this island was to learn about local efforts and artistic practices to support Dark Lab. My organization sees the many islands across the planet as spaces of colonial experimentation. At Dark Lab, which emerged for me in the aftermath of the Black Lives Matter global protests and the Covid-19 pandemic, we explore how race, climate, and technology are entwined.

From Black women's farming cooperatives on the island, I have learned how to follow the phases of the moon's calendar of cultivation. This calendar is a guide that designates when to cut the grass, when to cut one's hair, and when to harvest from the carambola tree. Kalinago architects and engineers follow the moon, too, choosing certain types of palm fronds to thatch roofs. Elders across Dominica in Black and Indigenous communities came together after the decimation of Hurricane Maria. The collective of women farmers told me about the cycles of waxing and waning that determine the rising tides of the rivers flowing to the Atlantic Ocean. Maria's sudden landfall at 9:15 p.m. on September 18 in 2017 overwhelmed rivers, flooding graveyards and unearthing the long buried. The storm claimed a total of 3,059 lives across the isles.

The Caribbean has long been under the microscope of the West. As Caribbean theorist Édouard Glissant reminds us, the West is a project, not a place.[6] Current research methods are failing us because the project of colonial expansion is immense and totalizing. Anthropologists stationed on the Eden Isle take morbid notes that they offer as climate theory before a postmortem of deaths from the storm is even complete. Scientific researchers from the U.S. and Western Europe parachute in to create algorithms from the sperm whale songs off the shores without regard for the people.[7] They will never hear the frequencies of the tragic and valiant Maroon parables. In 1940, a white American woman named Hester Merwin arrived with a pencil in hand to draw "Carib Indian portraits" in the district of Salybia.[8] So many acts of preservation are colonial projects. Now part of the National Anthropological Archives at the Smithsonian Institution in Washington, D.C., the drawings present us with the ethics of entombment through artistic practice. Hyper-visibility of this sort is no better than the omissions of such people written off the map and out of history. Spoken for, or spoken out of existence, it is as if these peoples are a primitive ancestor or a specter.

In my travels, I met collectives of grandmothers and great-aunts who have other plans.[9] Reading and writing their own almanacs, these women are teaching the next generation of young people approaching their twenties. Together, these Caribbean elders and youth devise intergenerational methods of future climate justice that transcend government intervention. They speak about secure drinking water and post-hurricane sustainable farming practices. Younger and older generations are listening deeply and are responding courageously in a necessary dialogue. They demand to know where the island's rubbish goes after it has left the dump, why plastic is not sorted from regular trash. The youth refuse to accept that recycling is pointless here. The Caribbean may be a small place, but it is tragically among the world's largest contributors to plastic waste pollution because of a lack of recycling facilities.[10] Trinidad and Tobago in the Eastern Caribbean is reportedly the largest contributor of plastic waste per capita in the world, with 1.5 kilograms of waste daily.[11] Per capita, the island of Saint Lucia produces four times more plastic waste than China.[12] And yet more trash from the West is dumped here. Existential environmental issues of a scale such as this weigh heavily on the Caribbean.

Caribbean seagrasses are some of the largest vessels of carbon in the world's oceans.[13] They remove as much carbon as fifteen hectares of rainforest emissions. Photosynthesis is a vibrant cycle of genesis happening constantly just below sea level and aboveground. Regenerative crop rotation cycles on land and sea are part of the sacred knowledge systems devised over centuries by Black and Indigenous peoples. They function alongside West African collective banking systems of loan rotation such as *sou sou* and Caribbean Chinese *hui* communal lending.[14] These forms of social organization and economics predated Columbus and have persisted long after him. Long sustained by homegrown intellectual traditions of financial, architectural, and agricultural engineering, these communities have optimized climate sustainability and repair. For centuries, Caribbean communities have pooled money together, trusting each partner until it is time to access the collective pot.

In 1997, the World Trade Organization's ruling against the European Union's special trade arrangements with the Caribbean (versus the U.S. and Central America) ruined the banana trade for many Caribbean economies. How could this not transform the climate? Dominica has still not recovered from these actions, which former U.S. president Bill Clinton set in motion. Jamaica, too, has never recovered. While the monoculture of bananas was never a sustainable model under the British rule, the replacement of such industries with tourism is no better for the environment. Protective tariffs that had been guaranteed to Africa, the Caribbean, and the Pacific by the British were repealed and many small island nations could no longer compete with Central America's banana republics. These countries suffered too, becoming dependent on one crop and the foreign capital and influence on elections. The devastating winds of Maria in 2017 led many people in Dominica back to the ways of the land through small back garden plots, where they grew provisions and tropical fruits for themselves, not for a Sainsbury's or Tesco in London. A Dominican tour guide said to me, "The land is so fertile in this Eden, if you throw a lime seed on the ground during the rainy season a tree will grow in a matter of days." The dry seasons are getting longer, and each second is an opportunity for the next brushfire to ignite.

To whom is Eden lost? The Caribbean is a small place that continues to be forgotten. But in Dominica's capital city of Roseau, we discover that Eden was never lost. The roof of the public library was blown away due to the Category 5 hurricane. Where does Dominica's archival knowledge live? White French thinker Jacques Derrida wrote that political power is derived by those who control the constitution of the archive and its interpretation.[15] Those who control the writing and interpretation of science also control this power. In the laboratory of the Caribbean, who are the technicians and who are the test subjects?

Dominica is an island comprised of five volcanoes from the Holocene era. Neither the French nor the British could tame it. People who live in Dominica know it is paradise but must fight to remember this against the pull of brain drain. The force of the colonial meritocracy beckons the best and brightest on the island to the United Kingdom and the U.S., which dangle opportunities for further education.

On a scouting expedition to Dominica in 1905, U.S. president Theodore Roosevelt, who was a naturalist, arrived hoping to extract the mountainous forest's lumber. The U.S. failed as pitifully in their mission as Columbus, the British, and the French had. The mountains refused. Afro-Indigenous people found refuge here. They will never surrender. Although Dominica was named for Sunday, or the last day of Columbus's final expedition in 1493, the island's true name in Kalinago reverberates across the spine of dormant volcanic mountain peaks: Wai'tu kubuli, or "tall is her body."[16] The

Waitukubuli National Trail is a treasure of the Nature Island, winding through coastal rainforests. Igneous rocks form 115-mile rugged
paths for extreme-sports tourists between multiple live volcanoes.

The all-inclusive resorts and package deals of Punta Cana, affordable for middle-class travelers, are not to be found on the Nature
Island. Accommodations range from outdoorsy options and cabins
to a small number of high-end luxury rentals. Wai'tu kubuli is a
bit of a secret even to those who live on other Caribbean islands.
It is both protected and neglected by its small runway. When I flew
in during hurricane season, we could not land on the mountain
because of an intense storm, and so we circled the tiny mountaintop runway several times. As of 2023, the minister of tourism is
pushing to extend the runway for larger passenger aircraft. The captain made the call that we would land on sunnier Antigua and the
airline put us up for a day, trapping us at an all-inclusive resort with
bottomless rum punch bowls.

Out of sight, out of mind. Global aid and relief efforts for Hurricane Maria often forgot the island, which has become common
for majority-Black countries like this. It was estimated that 98 percent of the country's infrastructure was left in ruins; the hurricane
destroyed bridges, roads, and more. Nevertheless, island communities have been quick to help one another, though they barely have
enough themselves. The Chinese government has offered Dominica
infrastructural aid and loans. The only hospital on the island (the
Dominica-China Friendship Hospital) was built by the Chinese government in 2019. Since the 1980s, a network of small retail Chinese
enterprises has also unfolded across the island, and includes variety
stores, or retailers that sell general merchandise such as hardware,
convenience items, clothing, and groceries.

When I spoke to a fellow nature-lover in Dominica, he told me
that he had never experienced racism on the island. He speculated
that perhaps racism was an American problem, because he had wonderful friends in Canada and the U.K. who were white and treated
him, a Black man, like an equal. In the next sentence, he recounted
the traumatic events of Hurricane Maria and the lack of funding
and aid Dominica received. I wondered why this man, a resident

of a former British colony that has been independent only since 1978, failed to connect the dots: racism is the system that ensures Black people across the globe will die before others when extreme weather events such as hurricanes occur. For parallel reasons, Indigenous peoples, including those who are also of African ancestry, are dying before their time.

An analysis of white Caribbean identity reveals the perversities of how race works across the islands. The author Jean Rhys is the Eve of this Eden. Rhys lived a life as an insider-outsider and something of a fallen woman in England; she was born and raised in the Caribbean and struggled at length with alcoholism. At the age of seventy-six, she scripted herself into the twentieth-century literary canon by penning an Edenic prequel, the novel *Wide Sargasso Sea*. Rhys is among the most renowned modernist authors of the twentieth century. She crafted her white Caribbean girlhood into fiction, and made a trope of the Jamaican madwoman in the attic who burns down the stately British house. Interweaving her birthplace with the landscape of opulent English manors, Rhys introduced the literary world to the lush paradox of the Caribbean as Eden.[17] *Wide Sargasso Sea* is a climate parable, and it begins with the Great House ablaze. Rhys's writing highlights how the architectural grandeur of eighteenth-century England was underwritten by the terrors of the Caribbean plantation. The abandoned wealth and scaffolding of empire was fated to burn in Rhys's novel. From ashes as fertilizer, nature reclaims its dominion as the island becomes overgrown.

Currently, Jean Rhys's Sargasso Sea blooms red, and it is far from poetic. It is prophetic, though: yet another consequence of the climate crisis is that sargassum overgrowth imperils the delicate balance of many marine ecologies in the North Atlantic. Seaweed blades and stipes block out the sun, preventing the photosynthetic processes that plankton and many other life-forms depend on. The strong stench of red weed currently extends from the North to the South Atlantic. It has become toxic to those who live on coastal lands.

The rancid seaweed smell greeted me in 2022 when I visited Rhys's home island. Our tour guide, a member of the Kalinago

Nation, pointed outward to the sea from the cliff on which the reservation stands. He explained to me and two German researchers that the Sargasso could be seen from satellites in space and was blooming out of control. Entrepreneurs in Mexico collect tons of red sargassum for luxury hotels, converting it into bricks for sustainable architecture for the global housing crisis.[18] With or without government sanction, new homegrown industries will form to convert refuse into the building blocks of a new climate-resilient society. At Dark Lab we seek just these sorts of inventive civic modes of rebuilding. Community-led efforts motivate us to respond creatively to conditions of scarcity. Decentralized forms of governance are necessary as we adapt to the everyday nature of climate precarity.

Here in the Caribbean, I learned many lessons about how nature's equilibrium has been disturbed since Columbus's incursion. Being in Dominica, I was struck by how true to life *Wide Sargasso Sea* remains. The novel structures itself around island life, and yet even Rhys's house, which should be a historic landmark, sits in deteriorating ruins. I stayed in a secluded rainforest mountain cabin in Trafalgar Falls. One of the largest and most vertical waterfalls in the Caribbean, it is a twenty-minute drive from the capital city of Roseau. Since I never learned to drive at home in Manhattan, I was all but stranded in the rainforest, and I did not expect to be living in a wooden barrack exposed to the elements. The British person who invited me to visit Dominica had not explained the island's lack of tourist infrastructure, which was made available only to the wealthiest visitors. I learned to ignite the gas stove manually so that I could make jook, a Chinese dish also known as congee, with three simple ingredients: rice, salt, and water.

My first lesson on the island was from *Solenoposis azteca*. The fiery giant ants bit the soles of my feet in a daily parade on the rough cement floor. The insects taught me where not to step. They taught me the collective choreography of proper burial while I was mourning the toll of the Covid pandemic. The ants reliably carried out a programmed ritual of cannibalistic decomposition of their kin. I watched with amazement as, triggered by the release of hormones upon death, the living ants became coordinated pallbearers.

So many of the world's funerals had been delayed or denied in 2020 and 2021. How do we find the collective global strength to carry the dead claimed by climate catastrophe? Through coordinated direct action. Marching dutifully single file toward the ant carcass I had squashed with my toes, they surrounded the insect's remains. I watched them carry it away, and I knew the decomposing body would nourish the colony.

Nature offers many such models from which we could be inspired to use biomimicry for collective care, protest, and action. By honoring cycles of life and death, we might realize how every funeral is of course also an excuse for a feast. Termites marching across the rotting logs of the rainforest taught me the necessity of cycles of decomposition for future growth. Avoiding the burrows of tarantulas and the flight path of tarantula hawks, I learned to be attentive to their subterranean infrastructure. The tropical house geckos or Afro-American house geckos (*Hemidactylus mabouia*) soon became my trusty companions as I grew used to their daily routines, and they, mine. We listened to rolling thunder from the bathroom window together and felt the warmth of the sun as magnificent rainbows appeared after the midday mountain storms subsided. The hummingbirds visited me to drink nectar and rainwater by the heliconia. By the third day, I gave up clinging to my conical tattered mosquito net and shooing away reptiles. I realized I was a guest in their house in this tropical rainforest, and I was exhausted from having exerted so much energy to keep the critters out. The giant cannonball trees (*Couroupita guianensis*), some standing 115 feet tall, had witnessed Columbus, and they had survived both him and Hurricane Maria. In these small details of the island, the finite scale of colonial time became apparent to me. Caribbean time, on the other hand, is infinite and nonlinear, extending back so much earlier than 1492.

COLUMBUS'S CRISIS

The myth of Eden erases Amerindian peoples who continue to live in the protected Kalinago Territory. Not to mention Black, Indigenous, and Afro-Indigenous communities living in St. Vincent and the Grenadines, Guyana, Suriname, and French Guiana. The expedition Columbus waged in the name of Spanish dominion was a form of biological warfare due to the diseases Indigenous peoples of the Americas became exposed to. In the twenty-first century, we must ask ourselves what the definition of war is so that we might grapple with the magnitude of the climate crisis, a war we cannot quite see. Ferdinand and Isabella designated Columbus as the Admiral of the High Seas, giving him his mandate of capture and dispossession.

When he arrived in Hispaniola, Columbus likened the island to biblical Sheba or Saba. In Santo Domingo, on the eastern half of the island, the Columbus Tree stands as a monument because it is said that he moored his caravel there. Part of Columbus's remains are buried on the island per his request, while the other remains were interred in Spain. Is the Columbus Tree an arboreal witness? What about the trees that became impediments to Columbus and his men? The ones that communicated with one another in complex mycelium networks underground, and that are still living, holding memories of the origins of the climate crisis in 1492?

In the American South, the concept of a Witness Tree came to prominence to grapple with the haunting legacy of the Civil War and lynching. White supremacy defiled the boughs as a Jim Crow gallows that became an American pastime: hangings during which white families picnicked. Other witnesses, documented in the Library of Congress, are southern magnolias, willow oaks, cherry trees, and elms, living specimens hundreds of years old that are listed as part of the Witness Tree Protection Program, established in 2006. This state-sponsored registry at once memorializes and endangers the trees by drawing public attention to their existence. Some of these trees are marked because they are said to have witnessed the March on Washington. Others were present for the 1995 Oklahoma City bombing perpetrated by Timothy McVeigh. The climate politics of national tragedy and witness by American trees will always be vexed.

Many such ecological witnesses of the natural world have resisted European colonialism across the broader Americas. One of the doctors present on Columbus's voyages wrote of the thicket of mangroves on the coast of Guanahaní when the *Nina*, the *Pinta*, and the *Santa Maria* first ran aground in 1492. The trees were so thick, he noted, that a rabbit could jump across them. It was not long before the naturally waterproofed wood of mangroves became an endangered commodity. The roots of the plant exude a waxy substance called suberin, which is how mangroves thrive in saltwater, removing salt at their roots. Coveted by the Spanish Crown, the special wood was extracted by Amerindians at the demand of Ferdinand and Isabella, who mandated a yearly tax of mangroves for shipbuilding. What do the remaining mangroves witness and remember of 1492? Important agents of carbon sequestration, the tangled roots of mangroves also act as seawalls protecting coastal communities from tropical storms. Mangroves are strangled daily by single-use plastics and other nonbiodegradable rubbish that gets caught in their roots. Thanks to their deterioration, the coastal ecology has undergone a total reduction of one-third in the Caribbean over the past thirty years.

In 1498, upon arriving on the South American mainland, Columbus declared his belief that he had reached the Earthly Paradise. His environmental extractions for Ferdinand and Isabella led to the calamitous loss of the sanctity of nature in the Caribbean. By now, Columbus is a myth himself. Although the fidelity of Bartolomé de las Casas's translation of his direct correspondence, *Diario,* has been disputed, his writings provide an opportunity to better understand our own current crisis.[19] According to de las Casas, Columbus recounted seeing the Lukku-cairi on the dawn of October 12, 1492:

They go about naked as they were born, the women also . . . everyone appeared to be under thirty years of age, well-proportioned and good looking. The hair of some was thick and long like the tail of a horse, in some it was short and brought forward over the eyebrows, some wearing it long and never cutting it. Some, again, are painted, and the hue of their skin is similar in colour to the people of the Canaries—neither black nor white.[20]

He compared them to other islanders and immediately scripted them into an epidermal taxonomy that quickly became racial. His emphasis on their lack of garments echoes the tale of Eden but also becomes a justification for seeing them as primitive, an attempt at validating his quest in the name of Catholicism.

Columbus described Amerindian architecture and canoe technology, which gives us clues about Native infrastructure and arboreal life. "They came to the ship with dugouts that are made from the trunk of one tree, like a long boat, and all of one piece, and worked marvelously in the fashion of the land, and so big that in some of them 40 and 45 men came."[21] The capacity of the boats indexes the age of the tree. The silk cotton tree (*Ceiba pentandra*) was harvested to construct long canoes used to navigate interisland transport across the arc and coastlands. Sprouting on average 80 feet, the tree can grow up to 240 feet high. The tree was sacred to the Arawak in Jamaica and exalted in other parts of the Caribbean, Mexico, and Central America. It is known as the kapok tree in southeast Asia, the Akpu-ogwu to the Ìgbò, and Rimi in Hausa. The silk cotton tree is also known as the àràbà tree in Yorùbá, and its large, twisted roots are portals to Orun, the home of the ancestors. It is known as the duppy tree because it is believed, as part of Obeah, that spirits of the deceased collectively inhabit the bottom under the towering trees. Obeah is a healing tradition practiced across the region and is adapted from West African–derived beliefs about the natural and spiritual world. Some believe mermaids make their homes in blue houses under the cotton trees. In Trinidad, one of these trees is known as the Devil's Castle because it blooms at night and bats visit it. In some West African beliefs, including Ifa, trees are understood to be ancestors. The Lukku-cairi believed in *zemis,* spirits of each individual element of nature. Across the Bahamas some of these images and artifacts have been preserved, revealing another epistemology-revering nature. There were other gods in the garden, and among them were the practices and rituals of the original inhabitants of the islands. Caves were sacred spaces of origin to the Lukku-cairi, and they were also burial crypts.

Justice of any kind, including climate justice, cannot exist without the autonomy of Black and Native communities across the globe. The terms of surrender have not been settled by any valid treaty. Afro-Indigenous governance, it occurred to me on the remote island of Dominica, is a reality inseparable from any hope of solving the climate crisis. The Native protest calls in the United States for Land Back and the Pacific cries in Aotearoa for Oceans Back are rallies in unison for anti-colonial sovereignty as climate solutions. True decolonization requires the phenomenon of a violent upheaval, as Afro-Caribbean psychiatrist and philosopher Frantz Fanon heeds.[22] For those of us living as part of the diaspora like me, part of Land Back is returning to the lands of our ancestors and kin. Sometimes these are not our ancestral lands because we are intergenerationally displaced multiple times.

I am not of Indigenous South American heritage, but I have found long-lost cousins in the Chinese banquet halls of Suriname—the former Dutch colony—who are. My story there was guided by the Black, Asian, and Indigenous bird-watchers I have met in the Amazon rainforest of Suriname, and the Maroons who have led me in Jamaica's Blue Mountains. And then there are those who found me—the hungry ghosts of ancestors and the unfamiliar duppies loudly haunting island graveyards and European colonial archives alike. They demand an audience, so you will have to meet them, too. These unexpected guides, along with the nonhuman storytellers, are "deliberately hidden, but not forgotten." This is the way Toni Morrison describes the power of haunting, and I am indebted to each of these layered presences and to her. The cast of characters of this book demands an audience. Morrison writes of her mode of storytelling, "My preference was the demolition of the lobby altogether." She refused what she called the "seductive safe harbor; the line and demarcation between . . . them and us."[23] Through climate as a lens, we can all rescript the storytelling of ecological crisis and what happens next. Only by becoming invested in how the story of climate will unfold can we invent and put into action what is required to reverse this crisis.

GENRES AND GENEALOGIES OF EDEN

A path carved by philosophers, statesmen economists, and novelists before me, Sylvia Wynter, Eric Williams, and Jamaica Kincaid, to name a few examples, confronted the entangled relationship between stolen life and future land rights for Black people.[24] I have remembered William Makepeace Thackeray's colonial critique of bourgeois London's merchant economy in *Vanity Fair* (1848), written as a serial tale entirely undergirded by the sugarcane plantations in St. Kitts. As a child reading about Miss Swartz, whom Thackeray "affectionately" describes as the "wooly-headed mulatto," and the heiress at the British girls' finishing school, I recognized the texture of my unruly hair. Wool that broke brushes and combs trying to tame it. I was a problem. To the Victorian audience, Miss Swartz was a figure of fun, heiress or not. I recognized Black ancestors of mine who arrived in Britain from the West Indies in the 1800s to be educated as lawyers and doctors. Long before the HMT *Windrush* disembarked at the Tilbury Docks in 1948, I also understood why the Afro-Trinidadian anti-capitalist revolutionary figure C.L.R. James declared *Vanity Fair* a favorite. British empire is a vanity fair, a place of ostentation or empty, idle amusement, and frivolity. The underside of colonial commerce is a climate history of entangled relations, half siblings, and invisible branches. The accumulation of wealth has been a systematic process of environmental impoverishment and ecological disinheritance for the global majority.

When I teach William Shakespeare's play *The Tempest* as a professor, I introduce it to my students as a 1611 climate critique. The winds of the tropical storm of destiny ushered in the violence of globalization during the Elizabethan era. These currents pulled my ancestors in wooden ships crafted from trees departing West Africa and East Asia to Eden. I pair the play with Aimé Césaire's Martinican rejoinder, *A Tempest*. Much like the changeling child stolen from India to serve the faerie princess in *A Midsummer Night's Dream,* I have questioned just how inevitable the direction of the trade winds was that filled the white, pregnant sails of merchant ships in the nineteenth century. I pair that classic with W. E. B. Du Bois's novel *Dark Princess,* inspired in part by Shakespeare's play. Du Bois takes us to the post-plantation soil and cultivation of Virginia's farms to sit with the meaning of broken promises enslavers made to African Americans: forty acres and a mule. Again, climate is a radical part of the plot and plotting of future reparations in such works of literature. Land and compensation are essential to reparative climate justice.

The Caribbean as Eden is formed by interpretations of the Hebrew Bible that have long shaped how the Antilles have been seen as refuge from religious persecution. Some of the earliest Jewish settlers to the Americas sought refuge in the Caribbean, establishing some of the first synagogues in the hemisphere (1655) when they escaped the bloody Catholic terror of the Spanish Inquisition. Many non-Christian cosmologies and traditions continue to thrive in secret corners of the Caribbean alongside Catholicism and Protestantism: Winti, Vodou, Hinduism, Taoism, to name a few. Each reveals a climate philosophy of potential repair. What ceremony is there for those who die due to the impact of the climate crisis? In what formation will the pallbearers gather when the cemeteries are being flooded because the rivers are overflowing? The duppies of the plantations still haunt these grounds.

After their exile from Eden, Adam and Eve were climate refugees banished by God to the barren earth. The Bible is full of climate changes: floods, plagues, pestilence. Does this naturalize the response to major environmental shifts? Reading the holy ancient Indian scripture the Ramayana in Trinidad, was Hanuman—Hindu monkey god of strength and determination—vengeful? He traveled across the *kala pani,* or what translates from Hindi as the dark waters, in ships with indentured Indian laborers. Hanuman was a guardian for the pregnant Calcutta mongoose, whom we will meet in chapter 7. In Anansi's garden, what did the griot portend of Yemaya's vengeance? The Yorùbá orisha of the ocean evolves into the form of River Mumma in Jamaica, where she is a mermaid-like figure known to inhabit the island's streams. In Haiti, we meet the form in the water deity Agwé, an aquatic guardian of enslaved Africans during the Middle Passage. What can these cosmic beings tell us about philosophies of climate globally?

How can we expect meaningful climate policy to address the current ecological crisis—let alone repair it—if we are not prepared as a planet to agree on its origins? *Dark Laboratory* tells the story of potential strategies and futurist technologies of repair informed by Black, Indigenous, and Asian science and what could come after. From the threat of glacier melts to the acidification of the oceans to coral bleaching to mangrove devastation to disastrous floods to severe hurricanes to increasing volcanic eruptions, the crisis is accelerating as capitalism goes into hyperdrive reinventing itself.

What's past is prologue, and so often it is conveniently forgotten that the prelude to industrialization was the plantation economy. Columbus inaugurated ecological devastation by clearing productive farmland and bringing sugarcane to the Caribbean in 1493. Segmenting the earth into plots was a tactic to dispossess Native communities of their sovereignty. From the first theft and transport of West African people from the Gold Coast, the transatlantic trade of enslaved Africans has been an original sin against the natural world. While the geographical location of the Garden of Eden is speculated to be at the intersection of the Tigris and the Euphrates Rivers, the fifteenth-century European tradition mapped Eden onto the Caribbean Isles.[25] Eden is a metonym for the urgent ethical question of the origin of our contemporary climate crisis. If we expect a sustainable future, we must ask who *we* includes—and excludes. A future without race does not equal a future without racism. It equals genocide. The origin story unfolds through botanical and zoological experimentation from the Agricultural Age beyond the Industrial Revolution.

My island lineage offers me the vantage of *island thinking* as a model necessary for grappling with capitalist ruination. I imagine a horizon of climate repair that will require sovereignty for dispossessed peoples. The cycles of reinvention in Hong Kong and Jamaica give me hope because they are dynamic spaces of artistic and political renewal. Island stories are far from ones of isolation. Insularity is the current political crisis; our shared fates are determined by the recent Westphalian construct of the nation-state and racial tensions. Before the islands of Jamaica and Hong Kong, the coastal villages of West Africa and Southeast China were my ports of origin. Coastal communities live attuned to the tide and the phases of the moon. The climate crisis is not news to them. The global color line is a tectonic fault line with a widening chasm between those who have suffered the most from the severe and immediate consequences of natural disasters.

From the perennial flooding in the Bangladeshi plains to the breaking of the levees of Louisiana during Katrina to the magnitude-five eruption of submarine volcano Hunga Tonga-Hunga Ha'apai in Tonga, the debt the West owes is collapsing the rest of the world. Seven years after the Tongan island rose out of the sea, it collapsed into a volcanic eruption. Islands are not forever, and the shock waves were felt across the globe. Within the same neighborhood, color and class differentiate the effects of environmental racism, worsening as the denial of greenhouse gases—carbon dioxide, methane, sulfur dioxide—grows deeper.[26] Regardless of politics, the denial will eventually asphyxiate us all. Racism structures the climate crisis because it was a part of its origins. The belief that some races are more expendable for the progress of others is a white supremacist tenet that shows up in environmentalist thought.

What is climate science without climate ethics? Without the

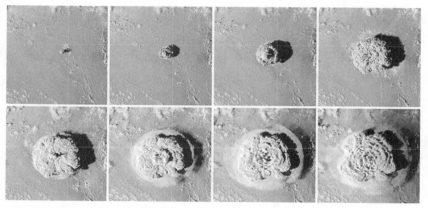

A sequence of images showing the magnitude-five eruption of the submarine volcano Hunga Tonga-Hunga Ha'apai in January 2022

global majority leading the ecological conversation, the plot for continued extraction and extinction is clear. The ethical principles and
scientific methods that led us to this colossal crisis will not be the
ones to save our planet. Repair will require challenging the assumed
principles of the scientific canon as much as the core of the humanities. The true crisis of the humanities is one of contradictions at
the foundation of the West. Universities are defunding humanities departments while they bolster and expand STEM fields. The
empiricism of Western science continually reveals its hypocrisies.
It fashions alibis and exceptions and ignores glaring evidence of
crimes against humanity and the natural environment.

An interrogation of "natural history" leads us to see that it is
not a coincidence that many of these scientific principles and disciplines of science emerged in the Victorian era. This was also the era
of the abolition of the transatlantic slave trade, with emancipation
following across the Western Hemisphere. Race science and eugenics have never been far from these advances in scientific knowledge
and engineering. In the 1880s, plantation soil had been exhausted,
as had the people abducted and forced to till it, so now an alibi was
necessary to enable a return to the colonial project of mining in
Africa. The project of colonial expansion is an extractive one that
turned to minerals. In the Caribbean, in the American South, and in
South America, plantations were left abandoned. Walter Rodney, an
Afro-Guyanese radical intellectual of the Black Power movement,
phrased it as the underdevelopment of Africa. Just before he was
assassinated at the age of thirty-eight in 1980, Rodney perceptively
understood the economic plot as an ecological plot connecting the
Caribbean to the African continent.[27]

CLIMATE DENIAL, RACIAL DENIAL

Edenic innocence requires the denial of these calamities. Forced to
cultivate the land of the Americas, Africans and Asians became *of*
this hemisphere when their blood was spilled to labor on its plantations. Africans were enslaved and Asians were not. Each found
inventive ways to adapt to a terrain not native to them. They became

people of color together as the climate changed them. Vacationers with package deals to all-inclusive resorts in the Caribbean seek Eden and do not see the poetry of the red earth and the African agrarian tradition. They perpetuate the myth of "virgin land" and penetration, not caring that these islands had first seen the dawn of an extractive regime in the fifteenth century. Luxury tourism fuels beach erosion and the demand for sand mining. The Caribbean is both a hub of globalization, tax shelters, and offshore banking and the logical endpoint for the fate and financing of the climate crisis. This is not only because oil has recently been found in Guyana. It should be no surprise that cryptocurrency clowns and conquistadors want to buy and settle on Caribbean islands as digital nomads. At tech conferences in Europe, these self-important venture capitalists have confessed to me that they seek a discovery like the thrill of the conquistador's first step onto the gangplank bound for America, as bad as they know this was for the Native peoples.

They pay upward of $50,000 USD for passports to buy vacation homes and claim tax benefits, whether or not they will get a tan or step foot on the island. They have allegiances to no nation but financial capital. The heat generated by high-performance computing data centers and crypto mines must continuously be cooled and mitigated to avoid downtime, draining vast amounts of electricity from fragile grids. Across the region's islands and other low-lying coastal territories, the state of emergency is unfortunately unexceptional. Climate survival across the archipelago is an everyday practice of living beyond multiple colonial time lines of apocalypse.

With decades of *climate denial* since the greenhouse effect was acknowledged in 1988 (though it had been hypothesized in 1859), our shared futures have become dangerously politicized. Since then, there have been cycles of denial. Reports from the 1950s and even *National Geographic* magazine show that the climate crisis is not new. In the archives of Kew Gardens, I read records of colonial officials warning of the climate changing due to rapid deforestation in British India. The denial of the origin story of the climate crisis *is* the crisis. There are many who deny race is a factor. There are many who deny race exists at all. Climate inequality is one of the ways we see this violence manifest. We must connect the dots between the brutal system of chattel slavery and the degradation of the natural environment. As partisan politics slows any chance for climate repair through policy, climate scientists and climate philosophers can help us understand how environmental racism is a part of our current ecological crisis. To remain willfully blind to race is to enforce racist modes that lead to the premature death of racialized people.

It bears repeating that the scientific principles that led us into the crisis will not be the same science to redeem us. While many accept the fact of climate change, they fall deeper into denial about its origins as racial in nature. The etymology of the crisis demonstrates how invested capitalists' interests are in the marketing of this debate for short-term gains. The vocabulary of "change" consistently fails us. That our climate is ever-changing is inevitable; climate crisis is not. Fatalism and long-termism justify continued corporate greed transnationally. Meanwhile, nonhuman and human lives both suffer at the bargaining table of stillborn climate negotiation. We must liberate ourselves from the interests of lobbyists because the planet's future is being auctioned off to the highest bidder. Not interrogating the assumptions of linear technological progress has led the world to this environmental catastrophe. Empire is an act of world-building, and its calamity can be easily reproduced if the lessons of the past are not heeded. The worlds Europeans built depended on making the lives of some disposable.

Billions are spent on research and development each financial quarter in service of more and more advanced machineries of extraction and geoengineering. Amid the abandonment and ruin left by deindustrialization, many hope for reindustrialization as a solution. How could it be? While it should be apparent that more mining and pollution cannot be a sustainable answer, investments in geoengineering are often touted as the carbon-neutral solution. Economic geology as an academic discipline came to prominence in the 1970s, funded by gas and oil companies. It's currently reinventing itself, and it's easy to forget the fossils in fossil fuels, and how mining our material and decomposed past is foreclosing the future. The collective death of the planet is being incentivized while today's prospectors and robber barons look for the "new oil." They will need to invent new philanthropy to greenwash it. Why isn't transnational decarbonization being incentivized? How seriously are geothermal solutions, wind turbines, and solar power banking being incentivized by governments? The future is being gambled away without the consent of the global majority. The mines grow deeper, and transnational corporations put them to use as landfills. The wealthiest nations bet on the future of our climate every single day on the Dow and the Nasdaq. The top 1 percent of impoverished nations also gamble the future away on speculative profit-driven punts.

The best geologists are philosophers, too, faced with the existential magnitude of how to tell the story of origins by reading the layers of sedimentation over billions of years. Western science is but one of numerous scientific methods. It would be unscientific not to consider and draw on all the existing scientific methods—Black, Asian, Pacific, Middle Eastern, and Indigenous scientific traditions. Theories and epistemologies, ways of knowing, offer nonlinear alternatives that were retained in cosmologies. Myth is born when meaning takes form, and grafting Eden onto the Caribbean is part of the genesis of greenwashing. Climate science models are used to project backward and forward, predicting future annihilation based on past obliteration.

*

In Houtouwan, former fishing villages on Shengshan Island, the earth is overgrown. Ghosts dance here, and nature is the revenant. Ruinate houses are covered in thriving vines, camouflaged perhaps—guerrilla architecture. Where the sun shines, photosynthesis reclaims inch by inch, exponentially. Whether humans will be the protagonist of the next century is the crucial question when it comes to climate crisis. The ruination stems from the fact that for too long we have been the main characters in a tragedy we scripted.

Like the ghost islands off China's coast, Caribbean hinterlands, too, are overgrown. These plots are designated as land people have chosen to reclaim: "capture land," in Jamaican parlance. Reggae singer Chronixx sings a ballad of Black landownership in celebration of those deemed squatters by the state. The 2014 song celebrates the reclamation of the Great Houses of the Antilles overgrown.[28] Reckoning with Columbus's crimes, he provides an alternative Rastafari historiography. In Chronixx's lyrics, we hear the historical cast of fifteenth-century colonizers, from his mention of King Ferdinand to the naming of Columbus as a thief. A bard of the climate and Black sovereignty crisis, the singer challenges the Eden alibi and derides the genocide of Native peoples of the Caribbean. As the singer notes in these lines, Columbus's wrong turn led to Indigenous annihilation. Thus, Chronixx poetically articulates Black claims to indigeneity in the Americas.

For Black people, capturing land back was a revolutionary act in the post-emancipation landscape and continues to be. Squatting is a political act of dissidence with climate implications. The colonial land grab is at odds, of course, with Black sovereignty and the claiming of land rights. Immediately after emancipation, Britain sent in Scottish militias in an attempt to seize the captured land back from newly emancipated Africans. European colonialism has deployed this design of substitution and the recruitment of mercenaries time and time again. As white enslavers treated and abandoned their grand plantation houses, nature reclaimed and reset the violent architecture of sugar wealth. New patterns of overgrowth, from China to the Caribbean, claim sovereignty over the land and seas.

When profiteers and privateers always prosper, it should be clear who instigated the infighting. We must ask, Which historians have profited from conflict by writing textbooks that uphold nationalist myths? There is no choice but to be disloyal to European, U.S., and Canadian empires. What has the cost of imperialism been for the natural environment? The Eden alibi must be interrogated to contend with the origin stories of armchair theorists. Men of science in England's Royal Society operated a system of credentialing touting the role of the explorer. The Caribbean and other tropical isles were their laboratories. Economic and labor experiments took place here that were racial experiments. These experiments mirrored those of the citizen scientist, the amateur necrophiliac collector of taxidermy, or gentleman farmer, who simply needed the endorsement of three fellows in the Linnean Society to become a member of what continues to be a literal old boys' club. A hobbyist legacy of extractive practices by fraternity brothers continues to this day. Should we not interrogate this intellectual tradition, which is the very foundation of natural history? The mandate of discovery was a justification for ecological degradation. The next alibi was the advancement of scientific knowledge, when really the dominion of the British Crown was the ideal advanced. Climate crusaders are often climate paternalists—not saviors but rather the horsemen of the apocalypse seeking martyrdom to save the world by lionizing themselves.

Diving beneath the ocean's surface, there are underwater worlds, aqueous territories of multiple thriving life-forms. Be wary of corporate environmental social and governance (ESG) initiatives led by those secretly invested in cryptocurrency, accelerating financialization, decentralized or otherwise. How clean can "clean energy" ever be? Corporate initiatives drain energy and reduce liability, much as most procedural diversity, equity, and inclusion (DEI) initiatives do to many people of color. Often uncompensated labor is required under DEI from the person who names the problem. Why should it be up to the impoverished nations to save the world from climate ruin? Aren't they belabored enough as they fight to survive, with nations like Tuvalu and the Maldives struggling as they sink below sea level?

WHAT IS THE DARK LABORATORY?

The Caribbean has long been a dark laboratory to European powers. In a Black feminist register, I am reimagining the terms of experimentation by redefining research. The Dark Laboratory is an attempt to create a space of refuge and possibility, against the daily grind of what universities demand of us. The lab is a space where we emphasize the "labor" at the root of that word. Labor histories, labor politics, all these matters seem to disappear in the scientific space of the lab. However, hierarchies govern laboratories just like any other social space. Labs have been places of discovery and death. Labs are hermetically sealed. Is consent even possible with such parameters?

When 2020 arrived, my instinct was to do something generative. The pandemic amplified the reality that some institutions needed to fall in the face of so many other destructive forces.[29] I founded the Dark Laboratory to respond to the rallying call of Black Lives Matter and Land Back for Native sovereignty as two deeply interconnected movements. I invited humanists, coders, lawyers, and biotechnologists to form a collective with me and challenge what is known as the canon. Together, we analyze the shape-shifting nature of colonialism and its role in the climate crisis. Using creative technology tools such as VR headsets, 3D architectural modeling, drone filming, and DJ'ing software, we are rewriting the plot of climate and colonialism.

Having been a member of a working group during my PhD schooling, I had been deeply influenced by Black British cultural studies scholar and sociologist Stuart Hall. At the Centre for Contemporary Cultural Studies at Birmingham University in the 1980s, his methods of collective research took the shape of working groups and then shifted to the Open University. He embraced media technology using television as a platform to teach Britain.[30] I welcomed choreographers working on robotics, lawyers who have been advocates for LGBT asylum cases, and software engineers. The advisory board represents the vision of what it would mean for Black and Native leaders to be in the boardroom. The selection was based on a syllabus of collective reading that forms a constellation of Afro-Indigenous philosophy. Even with this defiant statement, white allies looked at this board with amazement, and with far lesser accomplishments requested to be board members.

In the lab, we center race, climate, and creative technology across many different institutions. Our methods are fragmented, decentralized, and site-specific. We have found land grant institutions and universities to be especially useful for methods of self-critique that we can employ to follow the paper trail of Native dispossession and African enslavement. Universities across the nation are slowly beginning to own up to the source of their endowments. At Dark Lab, we encourage models of research for reckoning with stolen land and life. We turn to design and curation as modes of repair and activation. We apply for interinstitutional grants, and some of our projects involve creative ventures such as filmmaking on coastal erosion and folklore-inspired video game design.

In 2020, "BIPOC" instantaneously became a buzzword to differentiate the unevenness of the experience of people of color in the United States and Canada. Other countries latched on, though their own demographics would call for other vocabularies and reckoning with blackness and indigeneity. Indigenous Britons have different concerns than Indigenous peoples of the Americas, India, or Africa. All are relevant to the land grab of colonial desire. Together at Dark Lab, we ask intellectually urgent and ethical questions about the Western scientific method and interrogate which parties have benefited from the exclusion of people of color. From Hilo to Harlem to Helsinki, members collectively consider: Who are the test subjects in the lab of colonialism? How do scientific ethics shift when the researcher becomes the subject of research?

The Dark Laboratory is a space of creative possibility that requires collaboration beyond traditional institutions. There is a genealogy of darkness as power in the writings of W. E. B. Du Bois and in William Shakespeare's Dark Lady. We play in the dark, understanding that the Western imagined notion of "darkness" has nothing to do with the reality of our African lives or those of people of color. I named the lab after Toni Morrison's 1992 book of literary criticism *Playing in the Dark*. Her atelier model of teaching at Princeton University, which she imagined in the French style of workshopping, moved me.[31] I wanted to see what the parameters of a laboratory as a frame could offer us. Now we have the physical space to put this imagination in action with offices at 695 Park Avenue on Manhattan's Upper East Side. My pedagogy as a professor is led by the lessons of Toni Morrison. I was one of her last students at Princeton before she retired, and I remember dearly what she taught us about the dark. We read tales of Grendel and of Belgium's genocide in Congo. She warned us that evil would appear in many seductive forms to us, signs of the impending apocalypse. Wearing a top hat and a coat with tails, riding in a four-horse driven carriage. Would we be equipped with the tools to recognize evil in its many forms and fight it? Reading historical texts and novels together in her class called "The Foreigner's Home" was a way to decipher the sheer brutality that human beings are capable of. It has provided a guide for me in creating my lab.

Videoconferencing technologies opened new possibilities for us to begin this transnational work at Dark Lab more easily than ever. With so many urgent global questions, information traveled at the speed of what was then called Twitter. In workshops, lab members and I grappled over a common vocabulary and how to define words such as "indigenous" in the twenty-first century. The UN identifies Indigenous "peoples and nations [as] those which, having historical continuity with pre-invasion and pre-colonial societies that developed on their territories, consider themselves distinct from other sectors of the societies now prevailing on those territories, or parts of them."[32] As much as this definition attempts to cover, it is imperfect in whom it excludes from protections. Lab members and I embarked on making an atlas together that connected the futurism and struggles for land sovereignty of Turtle Island and Palestine. What does it mean about indigeneity that until 2012 the U.N. did not recognize the State of Palestine? It was clear that pushing beyond the United Nations' official definition was the first necessary step.

✳

"What is the river you were born closest to?"

✳

"Which river do you drink your water from every day?"

✳

In 2023, my research brought me to the South Pacific. During a hike, a mountain tour guide told me that in Tahiti the answers to these two questions say more about a person than their name ever could.

He was correct. To define relation in this way is to place ourselves in climate history. How many other people would have the same answer as you? This hydrological orientation signals a responsibility to the water as well as a recognition of how fresh-flowing streams sustain generations of life. It is less individualistic and exceptionalist, and of more of a collective spirit of being. In the cold dark heart of colonialist London, the winding Thames tells a story of these flows. Born just north of the river in Westminster, I began my life near the Thames. Today I drink my water from the reservoirs of the Croton System, Catskill System, and Delaware System. I live closest to the Hudson River, which is also, at its southernmost point, where in 1626 the auction block greeted African people coming from countries like present-day Angola, São Tomé, and Congo. Perhaps if we thought about the answers to these two crucial questions more often, we would feel an obligation to protect the waters that nourish us.

Members of my lab come together to imagine how the violent architecture of sugarcane refineries, mills, and old plantations could be possibly repurposed for anti-capitalist scientific aims. Beyond tourist attractions and state parks, we wonder, What if a network of labs for researching climate solutions and policy could be created to connect island nations ravaged by ecological degradation? The human toll of sugar unites so many societies across the globe. Has there been a full reckoning on the aftermath? We have concluded that our problems have the same blueprints whether in Hawai'i or Jamaica because our colonizers are cousins. The aftermath of sugar is bittersweet for both islands. Blue Mountain coffee and Kona coffee are so exorbitantly priced they can rarely be enjoyed on island. In Jamaica, we listen to breezy and rebellious island anthems in the same key, and we are nourished by the same foods, Otaheite apples, named for Tahiti, and roasted breadfruit, known as uru in Tahiti.[33] The message of reggae made its way to Hawai'i to become Jawaiian music, popular across the archipelago and the broader Pacific.

In Tahiti, seeing my long-braided hair, a Polynesian man asked if I was a Rasta. He started singing a Bob Marley song and spoke of his love for marijuana. I thought, What if we were to fight each other's

imperial overlords? An older white Frenchman with sleeves of Polynesian tattoos told me I must be Tahitian, though I insisted I was not.
Wearing a pareo, the traditional long Pacific Island cloth folded into
a garment, I knew I looked like the girls from Gauguin's paintings
to him. To certain men, collectors, we are objectified. These white
men transformed the rich oasis of islands into cold laboratories of
their dominion for biological and agricultural experimentation. My
lab is situated in direct opposition to the premise of the Western
lab because we understand how the island laboratory is a place of
retreat. It is a space to honor regenerative practices and inventions
that embrace the beauty of plant, animal, and other forms of nonhuman life. With new modes of storytelling, we articulate the climate crisis by using virtual reality platforms for worldbuilding. As
test subjects, we were not meant to survive beyond the lab. Thus,
our members completely redefine the premise of *research* itself.

The infrastructure of the plantation, including sugar mills, turbines, and barracks, is the result of insatiable greed that is a militaristic assault on the natural environment of Polynesian and Caribbean
islands. Europeans built directly on top of Native infrastructure to
perpetuate the myth that it did not exist. The Spanish constructed
the Iglesia de Nuestra Señora de los Remedios in the sixteenth
century directly on top of the sacred temple architectures of Tlachihualtepetl, Mexico, evidence that they knew the cosmic power.
European and U.S. empires have produced a series of nuclear experiments on Indigenous peoples who live on islands from Guam to
Chagos Islands to Tahiti. The experimentation mutated our DNA
and disrupted our kin structures. "We were not meant to survive,"
to echo Audre Lorde's poem "Litany for Survival."[34] Yet the test subjects survive beyond the abandoned lab.

Before I arrived on the New York archipelago, the British Isles
were my home, and they are where my father was born too. My
mother arrived from Xaymaca (Jamaica) with roots in the Fragrant
Harbor of Hong Kong that her father had called home. How many
kinds of flowers grow in an English country garden that are actually
from China? So many of these flowers were stolen by plant hunters who transplanted life from colonies like Sri Lanka and Jamaica.

While primroses (*Primula*) are found in the Northern Hemisphere, they are also found in the tropical mountains of Ethiopia, and Indonesia. Half of the species are from the Himalayas. Rhododendrons, common garden plants across the Western world, are the national flower of Nepal, not Scotland, where they bloom year-round. Foxgloves have been cultivated in England since the 1400s, but with species found in the Canary Islands and the Mediterranean the blurred lines of origin make it difficult to pinpoint exactly. What I have inherited of these ecologies as a Black British woman is full of flagrant and fragrant colonial contradictions. When I migrated to New York City as a child, I came to understand environmental racism intimately, though I did not have the words for it. I grew up in Queens, near JFK airport, where urban marshlands met the edge of a city shaped by white flight. As a Black woman in the United States, I understand that though I am an immigrant, this is not my first Atlantic crossing. I had been here before, centuries ago, because my African ancestors were here not by choice. To embrace these shorelines is, then, to embrace an ecological inheritance of many plants, flowers, and provisions for the future.

3

Natural History Museums, Shrines to Explorers

I am doing my best not to become a museum
of my self. I am doing my best to breathe in and out.[1]
—NATALIE DIAZ, "American Arithmetic"

It is very easy to get lost at the British Museum. I was four when it first happened to me. Looking upward in the crowds of people, I could not find my aunt's hand. I learned the museum is also a mausoleum, like many institutions of natural history. My aunt, cousins, and grandmother were visiting from America, which was very exciting because there was so much American popular culture on TV. I had let go of my aunt's hand because her hands were full anyway, minding five kids under the age of ten in one of London's busiest tourist attractions. I remember vividly, if not perfectly, the pattern of my yellow floral dress. In my four-year-old imagination, I was certain I would have to wear that dress forever because now my new home was the British Museum, and I would never be found. Lost, would I become a museum of myself? I was scared because I believed I would have to sleep in a sarcophagus each night, wrapped up with the Egyptian mummies when they closed the museum.

Museum basements are full of uncataloged holdings, a fact I first learned in the Netherlands in 2019 on a research fellowship at Leiden University. Ethnology museums display fetuses, shrunken heads, the remains of scalped people, dissected reproductive organs—all manner of macabre human trophies from around the world, often suspended in formaldehyde. Ethical questions of burial and repatriation are in continuous legal negotiations between nations that demand that what was stolen be returned. In some cases, these objects are restituted and even interred, but natural history museums are mausoleums regardless. So perhaps my instincts at the age of four about the danger of being collected as an object of empire were not so far off the mark.

Luckily, it was not long before a police officer found me. He gave me a stick of chewing gum, which I knew I technically shouldn't have accepted, but it kept me occupied until my aunt showed up in tears. She was grateful, holding on to four other children even as the police officer told her off. We all went to a Pizza Hut nearby, and I asked my aunt why she was crying. I could not have comprehended then how grateful she was that I was no longer lost. Neither of us could comprehend the magnitude of how Jamaican specimens underwrite the wealth derived for centuries by the British Museum. She was born a British subject in 1950s colonial Jamaica, and this museum was founded based on the Englishman Sir Hans Sloane's collection of 1,589 dried plant specimens, donated when he died in 1753. From 1687 to 1689, he collected and preserved not only botanical fragments, but live animals, shells, and rocks before sending them to London, some in bottles of rum. Having been the doctor of the governor of Jamaica, his inquiry was at first one in pursuit of medicinal knowledge. A tradition was born when Sloane published his findings in *Voyage to the Islands Madera, Barbados, Nieves, S. Christophers and Jamaica* in London in 1707.

Now having read that volume, I found myself back at the museum in 2023 for the second time ever. Just as lost in London as a naturalized American, I was not used to doing touristy things. My research had brought me to confront this building where I felt pulled toward the African objects, the Benin Bronzes. Most of the

estimated 350,000 objects are stolen ceremonial objects grouped on
the lower level in the Department of Africa, Oceania and the Americas. I saw sacred objects from West Africa and Caribbean masquer-
ades piled on top of each other here with references to the Notting
Hill Carnival and contemporary Black artists like Hew Locke and
El Anatsui.

Now I was ready to process what had always been in plain sight:
Sloane is the patron saint of this museum. As former president of the
Royal Society, Sloane's death marked the founding of the museum
via an Act of Parliament. These buildings are not so secular as they
proclaim. The museum's collection comprised 80,000 natural and
artificial rarities, and enslaved African labor on sugarcane plantations financed these purchases. Sloane's wife, Elizabeth Langley
Rose, was heiress to sugarcane acreage in St. Catherine, Jamaica.[2]
There are so many Jamaican contradictions in London. Since the
Windrush scandal beginning in 2018, the British government has
been deporting Jamaicans who have spent their whole lives in the
U.K. Sadly, I was reminded of the jingoism I heard from Black British people at a University College London Institute of the Americas
conference in 2017, parroting Donald Trump: "Make Britain Great
Again." These people were not scholars, but as they were community leaders with a strong anti-immigration bias against non-English
speakers, I wondered how we arrived here, so backward and ungen-
erous toward non-English-speaking migrants.

Jamaica is everywhere you turn in England. It lives in the opu-
lence of Sloane Square. From this tube stop on the Circle Line, I
followed Hans Sloane's trail, walking to the Chelsea Physic Garden,
where his statue looms in the center of the grounds. It was founded
for the study of medicine in 1673 and is the oldest botanic garden in
London. Sloane's statue is a marble shrine to a man who built his
life on what he extracted from Jamaica's wildlife laboratory. Such is
the foundation of most British wealth, that it was originally derived
from exploitation in the colonies. What would it mean for Black
people to "Make Britain Great Again," especially post-Brexit?

The logo for the Oxford Museum of Natural History is a Mauritian dodo bird, and it was the neighboring botanical gardens at

Oxford on which Lewis Carroll based *Alice's Adventures in Wonderland*. The Oxonian enshrined and embalmed the Mauritian bird and the history of colonialism that led to the bird's extinction. The island of Mauritius in the Indian Ocean does not own a dodo specimen, though the birds were once plentiful and prized. The fact that an insignia of extinction would be the Oxford logo shows the perverse logic of endangerment for the sake of capture and entombment that pervades modern conservationism. The history of the island laboratory of Mauritius and the loss of biodiversity form yet another case study of the labor of enslaved Africans and indentured Asians.

I also have vivid memories of a primary school trip to both the Natural History Museum in South Kensington and the Science Museum next door. When I visited as an adult, the exhibition design of the natural history museum struck me as being centered around climate crisis. The order in which you are encouraged to walk through exhibition rooms perpetuates myths of evolution-

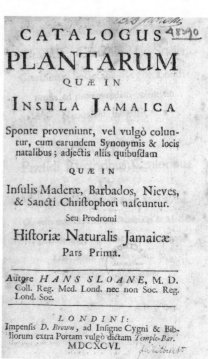

Hans Sloane,
frontispiece, 1696

CATALOGUS

PLANTARUM

QUÆ IN

INSULA JAMAICA

Sponte proveniunt, vel vulgò coluntur, cum earundem Synonymis & locis natalibus; adjectis aliis quibusdam

QUÆ IN

Insulis Maderæ, Barbados, Nieves, & Sancti Christophori nascuntur.

Seu Prodromi

Historiæ Naturalis Jamaicæ

Pars Prima.

Autore *HANS SLOANE*, M. D.
Coll. Reg. Med. Lond. nec non Soc. Reg. Lond. Soc.

LONDINI:
Impensis *D. Brown*, ad Insigne Cygni & Bibliorum extra Portam vulgò dictam *Temple-Bar.*
MDCXCVI.

ary progress. Africa is a continent of primitive origins. The geology hall features shimmering sapphires and rubies, gems stolen from India. There are melted artifacts from a treacherous volcano in Martinique that claimed the lives of 40,000 people in 1902 (the subject of chapter 9). Children are the primary audience, and it says much about the stories our society tells itself about climate. Since 2020, placards of acknowledgment—more than I have seen in the United States—read as asterisks admitting the role of the transatlantic trade of enslaved Africans and colonialism in South Asia. Theses addenda can easily be removed, and I suspect they will be one day. Their presence reminds me of the virtue signaling of those who are not Black and since 2020 have removed #BLM from their social media profile bios.

After leaving the Natural History Museum, I happened upon the Jamaica High Commission, and the former Royal School of Mines (the departments of Earth Science, Engineering, and Materials at Imperial College). It is not an accident that all these buildings are clustered in this part of London, laden with secret bureaucratic histories designed to dispossess Africa, Asia, and the Caribbean of their material wealth. European conquerors fully understood the value of what they were stealing and how advanced Native infrastructures were.

*

"You cannot get lost out here."

*

ENSHRINING NATURAL HISTORY

At any given time, at most 5 percent of a museum's collection might be on display. How do we repatriate what we cannot see? The only ethical answer is to return 100 percent of these remains. Displaying human remains is a common European practice, one upon which museums are founded as public institutions for the common good. "Facing Scrutiny, a Museum That Holds 12,000 Human Remains Changes Course," reads the *New York Times* headline about the

American Museum of Natural History.[3] Bones of enslaved African people, of Indigenous peoples, and of New Yorkers who died as recently as the 1940s are in the purgatory of the American Museum of Natural History's storage. Remains, bones, and skulls from countries across the world are documented as parts of the collection. This is a fraction of the estimated 100,000 remains of Native Americans that are currently held in U.S. museums. The piece was published on October 15, 2023—more than three years on from the revelations and public reckonings of 2020—but natural history, anthropology, and ethnology museums still refuse to expediently address the process for returning human bodies and body parts to their communities and countries of origin. Museum land acknowledgments came as quickly as black squares on Instagram. Diversity programming through one-off events recorded on Zoom remains in rotation at many museums across the world since 2020. However, structural change requires at least an upheaval of the acquisitions process, an audit of off-site inventory, and a refashioning of the metadata catalog.

Public museums and university museums alike find themselves in the legal quagmire of what happens next. In a 2021 press release, Harvard University admitted to holding the remains of 22,000 different individuals in its collections. Off-site storage facilities hold the remains of 33,000 people at the Smithsonian in D.C. The ethical breach of their foundational practices of embalming has become exposed and it is central to the colonial project of the West. The bodies in the bowels of Manhattan's Upper West Side reveal that the American Museum of Natural History has always been a crypt on unconsecrated grounds. The unholy plunder was the basis for eugenicist science, ethnology, and what they called human anthropology, including pseudoscience such as craniometry. Now anthropology department chairs from across the nation send me emails wanting my help to decolonize and to lure candidates of color to apply for jobs. And yet they refuse to do the intellectual work to address the colonial and racist origins of their fields of study. Museums are part of this complex of institutions at the heart of colonial knowledge production. They benefit from tax breaks and often advocate for conservation and green initiatives to combat climate crisis without acknowledging their role in propelling this crisis.

These institutions are haunted shrines. Just as Sloane was the patron saint explorer of the British Museum, at the American Museum of Natural History, Theodore Roosevelt is exalted. The building is literally an official monument to him as a New York statesman. Constructed in a Roman style, the rotunda and memorial hall to the American Museum of Natural History was authorized in 1924 by the New York State Legislature—specifically, Teddy Roosevelt's fifth cousin, Franklin D. Roosevelt, who was also New York governor and would become the nation's president. Life-size statues of the naturalist explorers John James Audubon, Daniel Boone, Meriwether Lewis, and William Clark stand on columns. Roosevelt is lionized as a naturalist and explorer—coded names for the bloodlust of conquistadors whose trophies are the stuff of mausoleums. The story a civilization tells itself about itself is enshrined in these rooms. Etched into the building's façade, the titles are a telling litany of imperialism under the guise of the Enlightenment: Ranchman, Scholar, Explorer, Scientist, Conservationist, Naturalist, Statesman, Author, Historian, Humanitarian, Soldier, and Patriot.

Tracing Roosevelt's biography alone reveals a sinister truth of American and European naturalists and their necrophilia. They were nature lovers who destroyed it at every opportunity. Roosevelt delighted in killing one of the last American bison: "As I rose above the hill, before he could go off, I put the bullet in his shoulder."[4] The curatorial logic of the Akeley Hall of African Mammals leading into the Hall of African Peoples illuminates nineteenth-century racial thinking and the U.S. colonial imagination. The museum proudly states that the naturalist namesake Carl Akeley "successfully petitioned the King of Belgium to create the first national park in Africa." For students of Belgian history and the occupation of the African continent, these words are chilling. The way a society organizes its scientific knowledge reveals its innermost hatred of the other. The perverse legacy is enshrined in museums of natural history, and there is nothing natural about it.

Taken together, Hans Sloane and Theodore Roosevelt illustrate the violent British and U.S. legacy of naturalism wherein the museum is a hunting trophy collection.

The bronze statue that has since been removed from the front steps of the American Natural History Museum depicted Theodore Roosevelt flanked by a Black man and a Native man. It is no surprise that museums of natural history are prime targets of anti-colonial protest and vandalism. Decolonize This Place movement protestors, who are based in New York City, splashed the statues with red paint in 2019 to signify blood spilled. Since they started in 2016, the group has organized anti–Columbus Day tours of the American Museum of Natural History. Their rallying cry has been "Rename. Remove. Respect." Beyond New York, bleeding stone is a genre of protest across empires, vandalizing monuments to genocide and colonizers.

In Martinique too, protestors splashed the statue of Joséphine Bonaparte in Fort-de-France with blood daily as an act of anti-colonial defiance, till it came tumbling down in the global Black Lives Matter momentum of 2020. Installed in 1859, the statue of the Martinican empress was beheaded, stone guillotined, in 1991. Wherever they are in the world, natural history museums are inevitably shrines, death cults of colonial worship. Collecting and cataloging are practices of control bolstered through knowledge production. Alexander von Humboldt is exalted as a god in Berlin, and while there are those in the United States who feel his scientific legacy has been neglected, a stone bust in his likeness stands across from the American Museum of Natural History near Central Park. In the U.K., Humboldt's legacy and his five-volume opus on scientific knowledge, *Cosmos: A Sketch of the Physical Description of the Universe* (1845–1862), is being celebrated and is reentering the university curriculum through a speculative lens, introducing the possibility of Humboldt as a queer icon because he never married.

*

"You cannot get lost out here."

*

We will always be lost in museums.

*

The American Museum of Natural History profits from a suggested admission fee of $25 per adult. How does the museum justify its tax status? Columbia University PhD candidates train in the museum and have offices that show the interconnected network of knowledge production, collecting, and colonial practices. The American Museum of Natural History hosted the Second International Eugenics Congress in 1921 and tied it into the museum's curatorial decisions. Ethical audits of cultural institutions are necessary to illuminate the racial crisis initiated by the invention of natural history and the natural sciences. People of color have been the test subjects to develop these academic disciplines. In the name of a science that is broken, many people have been treated as disposable for higher learning.

Look to the sordid history of medical schools and their experimentation on the living and the dead to understand how race has always been a tool of valuation and segregation. For the living descendants too, the pain is unresolved, whether or not they realize the fate of their ancestors. The consent of the dead in museum collections is entirely in crisis. Each specimen is positioned as a primitive ancestor to man, from the coral reef to the mongoose to the elephant to the shrunken heads of humans from the tropics. The corrupted scientific methods that justified kidnapping these corpses is part of the twisted logic at the heart of the West's environmentalism problem. Life is treated as disposable on a sliding scale of valuation, and yet in the climate crisis, we will all be saved or none of us will be. This is how the planet arrived at the brink of the total collapse of the natural world. The rising floods are unearthing the bodies of the disposed, metaphorically and materially, from unmarked graves and uncataloged museum collections. Those who have not been laid to rest are a tally of names we must continue to say.

While the Native American Graves Protection and Repatriation

Act of 1990, or NAGPRA, has set the precedent for American Indian burial grounds, archaeologic matters remain murky on how to legislate the handling of African American remains. There are marked and unmarked African burial grounds across New York. Wall Street as national park, Harlem African Burial Ground, Freeman Alley and the New Museum, Van Cortlandt Park in the Bronx. I was not imagining these frequencies of the dead: centuries of African presence across the archipelago of New York, from the auction block in the south of Manhattan to the farmlands of the Bronx.

On April 20, 2021, the Derek Chauvin verdict coincided with the global reckoning for Katricia Dotson and others murdered in the MOVE bombing of 1985 in West Philadelphia. I was devastated by the thought of the burned bones being bandied about in classrooms of institutions I had been proud to attend years ago. The state-sanctioned murder of Black people in the United States is an unmistakable pattern of violence. The police officer who murdered George Floyd was found guilty of unintentional second-degree murder, third-degree murder, and second-degree manslaughter on that day. The Penn Museum continues to be under siege and interrogation by the public for its human holdings. The museum's management of the remains from 1985 serves as a way to hold Katricia and other African American people hostage as targets of the state. They are in limbo, permanently in a state of improper burial. Say Her Name and Say Their Names, the rallying calls led by Black feminist law professor Kimberlé Crenshaw, tragically extend to many dead Black people.

In the Philadelphia and University of Pennsylvania community, people attest they have known about the devastating extent of the collections for some time. Still, the timing and the banality of the news was jarring for me. Professors at Princeton University and the University of Pennsylvania routinely used the scorched remains of African American children who were victims of government-sanctioned terrorism for teaching demonstrations. The question of justice is an unsettling one, and one for which university museums must answer. Their inventory and acquisitions undergo much less scrutiny than public museums. The lesson plan is clear when such sacrilegious acts are not only sanctioned by universities but have always been core to the Western syllabus of conquest: the canon. Whether the lab acts as one of living test subjects or of pathology analysis populated with the remains of the dead, the haunting is palpable. Diversity and inclusion initiatives never seem to begin here, by confronting the core intellectual question of burying the dead. Can they identify the remains in these collections? The metadata of these collections are corrupt and corrupted. Why would we expect practices with the living to be better? This is the immoral foundation of knowledge used to gatekeep and exclude people of color.

Reject the methodology of racial autopsy for the living and the dead. Refuse the dissection. In 1903, the *New-York Tribune* featured a photograph of a pyramid of bones stacked high by laborers who removed the bodies from an African cemetery in northern Manhattan. What are we to make of the pile? How many bodies does it comprise? Why did the workers form the bones into a pyramid? Why did the photojournalist take the picture? Why did the newspaper publish it? Evidence, perhaps. So many crime scenes like this one exist in plain sight. The improper burial and strange arrangement of the bones add insult to injury. Pathological disciplines such as anthropology and museum studies must face their role in the ongoing crisis. Their clinical methods are, by definition, extractive. The nation's oldest natural history collection, Philadelphia's Academy of Natural Sciences, was accessioned to Drexel University and includes a locker room full of ossified evidence of colonial crimes. These bones are the material basis for a science that was core to the American colonizing mission.

The naturalist's imagination was a militaristic fancy of conquest and capture. Climate imperialism goes hand in hand with progenitors like Theodore Roosevelt, Hans Sloane, John James Audubon, and Ian Fleming. These figures recur in my research across the Caribbean and histories of environmentalism. I cannot escape them. Serial killers with an unmistakable modus operandi, waging war on nature, which they claimed to love, created taxidermized shrines of conquest. A love language of embalmed possession, this grisly catalog contains human and nonhuman animals as trophies.

What could reparations for these crimes look like? Are diversity, equity, and inclusivity (DEI) initiatives willing to address meaningful reparative financial justice?

ENVIRONMENTALISM AND THE OMISSION OF RACE

Those who live in the Caribbean know intimately how the climate crisis is a racial crisis. Time can be measured in recurring cycles of Black rebellion but also in the increasing emergencies of hurricane season, which underscore how disposable Black life is to the state—even states run by Black people. One need only look to Hurricane Beryl in 2024 and all the records it broke so early in the season. Colorism and classism come into play, but it is also well known that white supremacy operates well even in the absence of white people. Nothing is intuitive about the system of race or racialization. The slippery way race functions in the Caribbean has much to teach the world about the perils of pigmentocracy. The racial regime is the related and violent logic of the plantation economy that U.S. eugenicists Davenport and Steggerda's *Race Crossing in Jamaica* shows us. Caribbean racism often functions by omission, by what is not said. Skin is a passport and yet multiculturalist mottoes such as Out of Many, One People commonly deny racial division, part of a rhetorical strategy in favor of nationality. Many radical ideological currents swirl and form across these islands because they have had to fight against homegrown conservatism.

The United States follows the same multiculturalist model of the "melting pot" of modernity. *E Pluribus Unum* translates as the same Jamaican motto, Out of Many, One People. These universalist politics extend to environmentalism and conservationism that colonizers fashioned as alibis for ecological extraction. Part of the power of race as a rubric is that the people in power deny its existence when convenient. Toni Morrison put it piercingly when she said "race" is an unspeakable thing, and so it has been in environmentalist circles. As Morrison puts it, "Suddenly . . . 'race' does not exist. For three hundred years black Americans insisted that 'race' was no usefully distinguishing factor in human relationships. During those same three centuries every academic discipline including theology, history, and natural science insisted 'race' was the determining factor in human development."[5] In predominantly white circles and societies, the mention of race shuts down conversations as the same circles celebrate universalism and lament the climate crisis. The omission is violent. As Morrison said in 1988, delivering the Tanner Lecture at the University of Michigan, simply the act of uttering the words "white" or "race" becomes taboo. Race functions as an invisible, violent force.

In 2017, when I wrote a book review of the nature field guide *Wildlife of the Caribbean,* I first confronted the racist rhetoric of naturalism. I published the review in *Caribbean Quarterly* based at the University of the West Indies, as I wanted to situate the book in the context of the region's history by reviewing it in the flagship journal. A review of this book in a scientific academic journal would have had a different mandate, but as a scholar of colonialism I could not help but hear ongoing colonial attitudes in the pages of a book that was supposed to be objective. Extinction events were blamed on Amerindians, while the book omitted the ecological impact of Columbus's arrival. I was astonished by this erasing of the Indigenous people of the region. Native peoples were depicted and described as overhunting mammals and overfishing the waters. What of the European invasion? How could it be neutral or impartial? Written for birders and hobbyists, the book is now a bestseller. It is not neutral or ethical, nor does it provide an accurate portrait of Caribbean biodiversity. Instead, it is a tool for colonial extraction and endangerment of wildlife. Some ecotourists arrive specifically to capture animals for the exotic-animal black market. Does the average bird-watcher care about the people who inhabit the islands they parachute between? Indeed, who is the average bird-watcher in the Caribbean?

Whose version of natural history do we turn to for answers? European natural history is a creation story that erases others. In July of 2021, the initiative Black in Natural History Museums was launched to address the contradictions of natural history museums. This project's aim is to share perspectives and stories that reimagine Black people's relationships with biodiversity through museum work and to find collective ways to articulate Black occupational struggles. Celebrating Black in Natural History Museums Week, the group provides mentorship with a focus on wellness and mental health. The lack of racial diversity in natural history museum staff is just one part of the problem they identify.

The more time I spend in scientific history archives, the more it becomes impossible to unsee Western science's role in perpetuating the climate crisis. Nothing is natural about the discourse of natural history, and its dehumanization is essential to nationalist histories and mythmaking. I ask myself: What is the purpose of the field guide or the nature pocketbook? Is it an identification manual for those who want to look past people and the impoverished conditions they live in? These "guidebooks" are unconcerned with the lives of people of color, the local culture, or the ecological impact of European colonialism on the flora and fauna. Their pages are goal-oriented, with no regard for the texture of the natural world that surrounds a prized plant or animal.

DIGITIZE THE WORLD:
ENDANGERED ARCHIVES, ENDANGERED SPECIES

The logic of the taxidermy impulse of cataloging has many costs. The price of this impulse is not only monetary but also existential when it comes to endangering the environment and hampering sustainability. It is not possible to digitize everything, despite what cloud services may promise. Cooling can represent up to 40 percent of a data storage center's energy consumption. Water is used to lower the temperature of servers working in overdrive. There is always an ecological cost to be factored in. When I visited the Museum für Naturkunde in Berlin in 2022 for an AI conference, the mission was to digitize the entire collection. From what I could see in the room of about a hundred people, I was the only Black person there. Museum officials celebrated a 3D printed bust of Humboldt, the patron saint of that natural history museum. I knew to be suspicious when I heard their mantra: "Digitize the world." The same logic of collecting the world is championed by institutions such as the British Library, which promote grants and fellowships to save endangered archives. The climate risks that threaten archives in the Global South are considered in a vacuum. Without questioning how the archives became threatened in the first instance, the cycle of classification, exploitation, and devaluation continues. The model is conservation and custodianship as an alibi for stealing materials for safekeeping. Why is it that developing countries are not able to afford to maintain these archives? Yes, climate crisis risks increase the precarity of the storage conditions for valuable artifacts, but where is the regard for the people living in these countries?

The Endangered Archives Programme is funded by the company Arcadia and is part of a charitable initiative ranging across ninety countries. The project's mandate also includes preserving endangered ecosystems and culture. Arcadia partners with Bloomberg Philanthropies and includes the digital preservation of the historical botanical collections at East African Herbarium Library at the National Museums of Kenya. The program requires a local archival partner and training for local staff as well as open access as its goal. Overlapping with the UNESCO World Heritage Centre, the twin aims of endangerment for natural and cultural history are noble on the surface. But without reckoning with the structural, geopolitical, and bureaucratic origins of the crisis, the problem will only be amplified. Such programs distract from the ongoing colonial condition by acting as saviors. NGOs form, maintain, and reinforce dependencies on institutions in the name of knowledge production.

The Netherlands, whose colonial tentacles spread across the world from New Zealand to Suriname to Southeast Asia, has taken the lead in repatriation, with an acknowledgment about colonial theft housed in museums and other crimes. In a 2020 decision by the Dutch Council for Culture chaired by renowned Surinamese Dutch human rights lawyer Lilian Gonçalves-Ho Kang You, it was recommended that the government unconditionally return objects seized by the Dutch under their regimes of colonialism across the globe. I had heard rumblings of these debates when I lived in the

Netherlands in 2019. I was told that the human remains being held in Dutch museums were left uncataloged on purpose. Would they ever receive justice or burial?

In July of 2023, the Dutch government announced that it would repatriate 478 looted colonial artifacts to Indonesia and Sri Lanka. The archives of the Vereenigde Oost-Indische Compagnie (VOC), or Dutch East India Company, show a detailed paper trail of documentation, perhaps a fraction of what has been stolen materially. The Rijksmuseum was one of the holders of objects thieved from Sri Lanka. These objects are slowly being returned, which sets an important legal precedent for millions more objects sited in Western European museums. European, American, and Canadian collections may never fully reveal their holdings, their trophies. Exclusive auction houses such as Sotheby's and Christie's deal every day in these morbid materials to the highest bidder. Smaller brokers of artifacts and fine art deal in such objects too, including carved tribal masks with ritual significance to Polynesians and Africans. It is an industry of colonial transaction from which many continue to profit in the circulation of objects including human hair, carved bone, and other organic materials. So much has been stolen intergenerationally, and the objects are the DNA evidence of colonial theft. The valuation of the auction block is never far from memory for colonized people and those whose ancestors were enslaved. The barter and trade in flesh is a not a metaphor for descendants; the trade is also in bones and other biomatter.

THE VIOLENCE OF KNOWLEDGE PRODUCTION
AND UNIVERSITY MUSEUMS

Museums profit from the bodies on display with the sale of each admission ticket as part of the tourist economy. The nonprofit tax write-offs have an exponential impact on the profit museums enjoy that these remains generate. Political power is derived from the power of the archive, its interpretation and its constitution, French philosopher Jacques Derrida reminds us.[6] What collections do university museums comprise? How does this relate to politi-

cal power for donors and other wealthy alumni? In a nether space
of tax exemption, university endowments have come under recent
scrutiny for being founded based on the sale of enslaved Africans
and the direct dispossession of Native territories. A vital interrogation of university museums and libraries is taking place. They carry
the tools to do the archival work that leads to the money trail. In
doing so, do they undermine or strengthen the intellectual mission
of light and truth? Tax-deductible donations made to universities
bolster blue-blood family legacies and are often used to launder rep-
utations. The acquisitions are celebrated with plaques. But in many
cases provenance is left blurry for certain alumni by curators who
are happy to grow the collection.

How many more unmarked graves will wait in legal limbo for
repatriation? In the interim, an interregnum on the status of nonprofit institutions—501(c)3s—would allow us to interrogate the
economy of philanthropy in the United States. Addressing the heart
of the common and public good, these museums of natural history are attached to universities and otherwise are in breach of the
tax statuses they have enjoyed. The interest they earn multiplies as
endowments swell, reinvested through hedge funds back into the
stock market to companies responsible for pollution due to fossil
fuels. If natural history museums were to pay back taxes on what
is owed, this would be the beginning of a global source for colonial
reparations. The tax breaks they receive are ill-gotten and can be
calculated as if they were for-profit businesses. If museums paid the
money toward mitigating the climate crisis, they could distribute
funds to the living descendants who will never be able to bury their
ancestors' remains. So-called developing countries that continue to
suffer pilfering from the Global North might be able to decide the
fate of their own cultural heritage and artifacts.

The scientific traditions of the world's majority have been discredited for centuries by the Western project of modern capitalism. So have the economic systems of the world's majority been
discredited in favor of the Protestant work ethic and what German
philosopher Max Weber called the "spirit of capitalism" in 1904.[7]
The West has repackaged much of this knowledge to profit from

it. Take, for instance, the pharmaceutical industry, the epicenter of which is headquartered along Route 1 in New Jersey—Lenape territory. Big Pharma is based on the exploits of ethnobotany, a colonial discipline designed to extract and catalog knowledge from Indigenous peoples while deeming them primitive. I am interested in the moment when botany becomes designated as ethnobotany— through the same colonial process of othering that makes musicology ethnomusicology, perhaps. These fields of study are pathology reports. Natural history museums showcase how the Western scientific method needs to be revised because it is killing us all slowly with its hypocrisy and double standards.

For this reason, I have found exciting spaces to experiment with art and design museums as part of the collaborative work of Dark Lab. These museums have been the fastest to critique colonial legacies, and because the mandate of an art museum is more abstract than that of a natural history museum, curatorial strategies can be creatively deployed. Still, art remains another murky market worthy of interrogation. I am interested in what can be staged in reclaiming these spaces for activations where we can listen to collections. I am interested in dissecting the metadata used to catalog and tag objects in the colonial archive. I want to know who the standard archival finding aids were written for. Were these inventories written by archivists and librarians who look like me? Did they have descendants of these histories of violence like me in mind? If institutions are willing, there is great potential to make museums engines for climate justice amid the crisis, but this will require reckoning with race, the fraught politics of disinheritance, and the truth of how provenance leads us to what was stolen and is still being profited from.

BOOK II

Geological Time and Black Witness

4

Breathing Underwater

Full fathom five thy father lies;
Of his bones are coral made;
Those are pearls that were his eyes;
Nothing of him that doth fade,
But doth suffer a sea-change
Into something rich and strange.[1]

—WILLIAM SHAKESPEARE, *The Tempest*

Many underwater origin stories lead Black people to the Atlantic as a powerful realm of mourning and becoming. What if the first garden was underwater? The Middle Passage becomes a Black Atlantis, shrouded in futurity instead of suffering. It is a zone of transformation that anchors the order of Afrofuturist galaxies where Black people live free. A subaquatic Eden of sorts, it is a lost utopia. For music writer Greg Tate, this paradise recalls Nigerian Ìgbò lore, which he describes as part of the origin story of a place called Drexciya. Invented by the '90s house music duo of the same name, this place was imagined as an underwater Afrofuturist rave made up of the Black people who survived the horrors of the Middle Passage. Bodies that went overboard, into the Atlantic, survived. Tate notes the Ìgbò were "the Africans least preferred for the slave trade because they were known to jump off ships and drown themselves. There are even myths about them not drowning but walking or flying back to Africa."[2] Drexciya is part of a rich tradition of Black people embracing the poetry of the sea and the oceanographic knowledge that comes with it.

*

The people could swim.

*

Crossing into an intermediary space of Atlantic existence, people forced to leave West Africa's Gold Coast became baptized as Black in those waters. A color that absorbs all light became a race and a set of defiant and creative political practices, refusing the rule of chattel. When I began this line of work as a graduate student, I used the word "slave" often to describe the historical circumstances I was writing about, but now I balk at how cold the term is. Grammar is political, and I began to embrace in my studies that slavery is a condition forced onto people and the semantics of this. No one was ever a slave. When they were taken, Africans were abducted and inducted into an ecological world of Atlantic marine transformation, transmutation, and power. Slave trades are multiple across space and time, yes. But chattel shifted the oceans unlike any other trade in 1533 and did not formally end in the Atlantic world until 1888 (Brazil), and its temporal ripples have not stopped.

Africans were stolen specifically to transform the climate of the Americas. Mass enslavement of an epic proportion was needed for the desired extractive monocrop agriculture that led to a moral sea change. The stolen people became Black through the transformations of crossing the ocean together: a collective act. It was not a condition forced upon them; rather it was a politics of strength that emerged out of difference. Blackness is best described, to paraphrase Toni Morrison in *Sula,* as inventing "choice out of choicelessness." Because stolen lives were transported to cultivate stolen land, Black people understand what it means to continue living after multiple and overlapping time lines of apocalypse. The currents of antiblackness are global. When the European captors tried to drown the disobedient aboard ships of the Middle Passage, they did not know the people could swim.

Premature African death in the Mediterranean in the twenty-first century echoes the despair of centuries past in the Atlantic. Thousands of years from now, the Mediterranean Sea will become the Mediterranean Mountain Range, thanks to an accretionary wedge due to the Nubian Plate subducting under the Eurasian and Anatolian tectonic plates. Maybe Europeans in the future will hike to the African Continent as refugees in crisis? Tectonic plates will shift as sea levels rise. Whose bones will fossilize in the limestone mountain peaks of the far future? The gulfs between Black and white, between Southern Europe and Africa, between citizen and refugee, show us the arbitrary meaning of race based on the division of the continents and the decisions of European cartographers. Millennia from now, sedimented skeletons will form the biomatter of an unknown toll of Black life transformed and made brittle. On islands off the coast of Italy and Greece, such as Lampedusa, debris continuously washes up on shore. Shipwrecked Ivorians, Egyptians, Sudanese, and Mauritanians continue to travel in small boats and on rafts. Can we imagine a Drexciya where Black life continues in otherworldly forms for African victims of the Mediterranean too?

NGOs make estimates of the annual death toll in the Mediterranean, but these numbers are insufficient. There are no adequate metrics to measure this migrant crisis, except to say that the desperation is the result of centuries of European colonial design. Who does the crisis belong to? African people reluctantly flee their homelands because of political, climate, and racial crisis. The European mismanagement of the crisis forms yet another crisis for the Mediterranean to hold. At home across different African countries, droughts, floods, water scarcity, and shifting harvests caused by climate change are among the factors that make life impossible. Lake Chad is one such example; it has shrunk to one twentieth of its size since the 1960s.[3] How many Black people die at sea who do not wash up onshore? Their bones will become sedimented in the

Mediterranean Ridge of the far future. Desertification—a climate risk causing drought and famine—across Africa drives farmers to choose the Mediterranean Sea out of sheer desperation. But we cannot forget what the Conference of Berlin in 1885 did to carve up the continent before this with arbitrary lines for the scramble for Africa by European nations.

*

Black prophesying about crisis extends from African American Christian traditions to the Jamaican Rastafari religion. We can read this kindred sense in the lyrics to 1977's "Downpressor Man" by reggae singer Peter Tosh, who warns of the signs of Judgment Day. Punishment for the wicked is nigh as he interprets it from Revelations. "You gonna run to the rocks / The rocks will be melting. When you run to the sea / The sea will be boiling."[4] Our seas are rising because the water is boiling. The ice caps melt, releasing life-forms once frozen—plankton and bacteria from glaciers inhabit new lakes. Many apocalyptic Black religious traditions focus on Revelations rather than Genesis, and the signs they predict are ones of environmental ruin. Tosh reinterprets "Sinnerman," which Nina Simone made popular in 1965, but which retains its older origins as an African American spiritual.[5] Amid the crisis of global warming, in what ways can we read these lyrics as a climate reckoning? Today there are many signs of rapture, and it is hard to decouple the environmental degradation from the unbridled corporate greed that causes it.

*

Sea levels and temperatures are rising drastically, and most at risk are corals, marine animals found across the oceans that are incredibly sensitive to minute sea changes. The coral song is the siren song of the climate crisis because corals have witnessed much over geologic time. In Drexciya, Black folk live symbiotically with various underwater ecologies. When corals are healthy, the oceans are at a balanced equilibrium. Breathing with corals and imagining their perspective may open a deeper geologic timescale of crisis. Because they are made up of billions of polyps, it is easy to forget they are animals. It was originally erroneously believed by Western scientists that corals were rocks or plants. Made up of cnidae (stinging cells), the aggregate of neural cells helps the animals adapt to their surroundings. Corals bear closer relation to us as human beings, and I imagine them in community with the people of Drexciya. They are animate, capturing prey and using their nematocysts as defenses. Living, breathing, budding, sensing, hearing. Having survived ecological shocks, many corals across the globe are facing a major extinction event right now. As coral bleaching takes place, ghostly exoskeletons remain as a specter of the loud, vibrant life underwater reefs sustain.

The oldest recorded living corals are 4,200 years old. Ancient Romans imagined coral as looking vascular and veiny in form. When Perseus killed the gorgon, who had the power to petrify men, the red coral of the Mediterranean was her hardened blood that spilled onto seaweed. Across the Atlantic, sea whip coral (*Leptogorgia*

Sea whip, soft coral, gorgonian

virgulata) has spread its carnivorous tendrils and thrives in the salinity of shallow water. It is able to adapt to hard bottom substratum
and is prominent on the coast of South Carolina. The habitat of the
sea whip coral is the port city of Charleston, a gateway of Atlantic
transformation for Black people in the United States and the Caribbean. In 1861, the coastal areas of the Carolinas were over 85 percent
African. *Leptogorgia virgulata* has been found from the Chesapeake
Bay to the Gulf of Mexico to Brazil, where the trade of enslaved
Africans extended via the Dutch, the Spanish, the Danish, the Swedish, the Portuguese, the French, and the British.

There are black corals too. Not exactly black; they are so named
because they produce darkly colored finger protrusions. As old
as 4,200 years, black corals are the oldest living skeletal-accreting
marine organism and possibly one of the oldest living animals on
the earth. Roughly half of the Africans enslaved in British North
America traveled through the Charleston, South Carolina, market.
Peering downward as they disembarked from the gangplank of the
docked cargo ships, did they notice the underwater red and black
branches? Black corals are witnesses to centuries of violent transactions of flesh, growing fleshy themselves along the western coast of
the Atlantic Ocean. Polyp growths communicate as a decentralized
network. Could the corals have coordinated sinking ships with their
stony outcroppings? Corals have eyes, ears, teeth, and intricate ways
of sensing danger. They have witnessed over half a millennium of
Atlantic tragedy. Embracing the wreckage of ships, corals often
build homes using the structures as scaffolding for new marine life.
Corals have digested this maritime history. As they grew on ships,
they formed a submerged transatlantic dungeon that protected possibilities of growth for other species.

Across what was British North America, corals are imperiled.
From 1996 to 2015, an estimated 53.8 percent of stony coral in the
Florida Keys has been lost due to climate stressors. In the Carib-
bean, corals have declined by 80 percent over the past fifty years.
What will be the fate of corals in 2050? Flesh lives by the rule of
regeneration. When a piece of coral is cut, another one grows

through a process called budding. When the sounds of healthy corals are played in dead and dying reefs, life returns. Listening to the corals tells a story of transformation and potential regeneration. Returning to the soundscape of Drexciya, what possible healing can occur through the sound waves of the Black Atlantic?

Coral larvae scout safe locations for other marine life, floating on currents until they find the right conditions for new life. The multicellular organisms search for a habitable future based on sensory details. They can determine color and sound to find healthy environments. The mysteries of flesh animate these marine invertebrate animals, creatures that hold the balance of ocean biodiversity in their undulating polyps. The cold Atlantic is an intermediary zone of rebirth through interspecies ecologies. Underwater valleys also require protection from the threats of extractive economies due to advances in deep-sea mining and geoengineering technologies. We are experiencing a new Gold Rush of prospecting and speculation that will ravage the seas. It is taking place in Guyana, a hotspot of new oil in deep-sea reserves. Have we yet recovered from the extractivist mining frenzy of 1849 and the Gold Rush? More than 295 feet of high-quality oil-bearing sandstone reservoirs were drilled off the coast of Guyana in 2015. Transnational corporations trawl the ocean floor for new sources of oil deep within the earth's core. Exxon celebrates their "ultra deepwater and carbonate exploration capabilities." Modern-day prospectors hope to find new ores as solutions to the energy crisis, which can only deepen the environmental degradation. How much more can the seabed bear, when the Atlantic seabed is already haunted by the weight of transatlantic atrocities?

Chattel slavery was engineered to leach the soil. Ship captains' logs and records from the fifteenth century note the sound of turtle shells hitting hulls in the Atlantic—such was the plenty of underwater species. The Cayman Islands were a popular stopover for turtle soup on the way to South America from Europe because the marine reptiles were so bountiful there. Turtles are now endangered, forced onto beaches that tourists have colonized and whose waste is routinely flushed into the sea. Bermuda, a coral island and still a British territory, was also a popular halfway stop to the Americas for naturalists.

Racial slavery was a world-ending genocidal event that unfolded over centuries, and yet West Africans never stopped coordinating rebellions across languages and dialects to subvert their sadistic Portuguese, Spanish, British, Dutch, American, French, and Scandinavian captors. Black mutinies never ended; they are still being fought on other temporal planes such as in Drexciya, in other galaxies of nonlinear possibilities and alternate futures. The people could fly; and in death they flew across the Atlantic back to their homelands in Africa, part of the afterlife. This Afro-Atlantic tradition lives in the pages of a childhood book that I adored but which also terrified me: *The People Could Fly* by Virginia Hamilton, published in the 1980s.[6] The book's tales and illustrations still live in my imagination. It anchored Black American children's belonging in a deep past and a far future. In death, flying home is *an* end, not *the* end, and so a supernatural transcendence is possible. In Gullah traditions of South Carolina and the Sea Islands, as with many Afro-diasporic religions, this type of transcendence is known as "traveling."[7] In this continuous journeying, ancestors and spirits live on under riverbeds and lakes in the U.S. South.

Born beyond the unknowable oblivion of the Middle Passage, jettisoned futures were full of regenerative possibility. Some had no choice but the fate of the water; others chose to jump. Symbiosis was multiple for the number of marine ecologies with which Africans found themselves communing. Atlantic waters are an alternative world, nexus of a fate different from the shores of the Americas—Charleston, Martinique, Rio de Janeiro. Across the ports of the Black diaspora, corals have borne witness to the transformations of interspecies ecologies.

The freedom movements of Black flight and submergence are palpable in diasporic stilt-walking traditions across the Black world. The playful acrobatic choreography is a carnival dance of survival across the ocean. The stilt-walking footwork of the *moko jumbie* of

Trinidad and other parts of the Eastern Caribbean renders ghosts tall enough to walk across the depths of the Middle Passage.

*

The people could swim.

*

Narrow-minded stereotypes scorn Black people and their ability to swim, yet a vast nautical and aquatic precolonial history of West African swimmers and surfers has surfaced. "Bones too dense to swim." Black children in the United States have endured spurious stereotypes. Many factors stand in the way of Black children learning to swim. Swimming is a deeply political matter, and certain at-risk communities are drowning faster. Swimming is a lifesaving skill, and for many Black communities, the buoyancy of the Black experience has meant finding new ways to reconnect with the oceans. Despite the stigma of swimming—including affordability and the history of Jim Crow segregation, in which public pools were a contested site of racial pollution—programs such as Black People Will Swim are forming important communities. The orientation of the future tense of the water here is important. These movements are the buoyancy of twenty-first-century blackness. Reclaiming the waters, a San Francisco collective called Black Like Water hosts an annual Black Surf Week. The event takes place on the shoreline of Kumeyaay Territory near the Scripps Institution of Oceanography, and the two places could not be more different.

On a Dark Lab site visit to San Diego in 2021, I was told about the underwater cities just beyond the horizon. Kumeyaay Territory is a holy land for many disparate Native peoples who make a pilgrimage here to the West. The contrast between the devotion and spirituality of the Kumeyaay Territory and the homogeneity of the wealthy conservative outpost La Jolla (whose border with Mexico is a mere ten-minute drive away) spoke volumes to me as a visitor from the East Coast.

While I still have not yet learned how to swim, travels to shorelines such as San Diego's have only increased my desire for oceanic knowledge. I was told that those who are disabled are welcomed to participate in Black Surf Week through a meditative practice of watching the waves while lying on the sand—a kind of mind surfing. Out in the blue, past San Diego, are kelp forests and an artificial reef called Wreck Alley, a diving destination. Following the waves within my line of sight, I wondered if Black mermaids of the Pacific resided here, too. The water spirits of the Black diaspora are the maestros of the coral symphony teaching lessons on symbiosis— teaching us how we will breathe beyond the apocalypse.

CORAL COLONIALITY AND COLLECTIVITY

Corals show us the power in collectivity. They survive because they live in aggregate formations, made up of individual polyps supporting one another. Creating a home for multiple forms of marine life, they are the very builders of islands. Coral hobbyists who understand this architecture intimately nonetheless collect the organisms, separating them from their natural habitats to become pets. The oldest dead corals date back 500 million years, and they were solitary beings. The oldest colonial reef-building corals go back 25 million years. When I first read about the coloniality of corals in zoological studies, I was struck by how differently the word "coloniality" is used in the humanities. The shared vocabulary of "coloniality" led me to Latin American scholar Aníbal Quijano, who developed the concept, which was later expanded by Walter Mignolo to describe how colonialism and modernity are bound together at the heart of the European global project.[8]

For biologists, *coloniality* describes something altogether different: how individual organisms that live and grow connect to each other. I choose to think of the two definitions in relation. Scientists and critical theorists with vastly separate remits are not in conversation enough even to debate the common keyword and etymology. How will we approach the climate crisis without reckoning that we face the same problem? At a 2023 climate change conference held in New York, I spoke about the value of interdisciplinary research and how scientists and humanists of color ought to be speaking directly to one another. I was disappointed to hear an oceanographer who had fled the natural sciences for anthropology describe how much easier the work was. At first, he seemed to champion the dire need for the humanities. Then he announced that scientists need humanists to translate scientific findings or else no one will know about them.

The humanities do not exist as public relations for the sciences. If we are not willing to respect each other's research as serious research, any future beyond climate crisis is futile. To me, the oceanographer-turned-anthropologist's plea was an extractive way to value humanist research, and one that I am now accustomed to hearing from social scientists. Their work is entirely dismissed by scientists as subjective. The myth of empiricism governs the hierarchy between the disciplines. The scientific method as we know it consistently denounces what is considered research. Scientifically speaking, the fact that knowledge produced by people worldwide is dismissed as unworthy of inclusion in "science" shows a racial bias.

Corporate bluewashing has coopted many of these technologies to create superficial fixes. Collective action can be modeled throughout nature, and biomimicry lessons can fuel potential climate solutions. There are nonhuman animal models of collectivity beneath the sea. How can we learn from these submarine, vibrant reef communities about how to build a more sustainable planet? With exoskeletons accumulating as calcium carbonate bedrocks, the reef models a method for living after the impending climate apocalypse. Adult corals are brittle and sessile—fixed in place—but they begin their lives free.

Corals are living contradictions, and they should push us to embrace our own paradoxes. Their grammar is both singular and plural, like the mycelia of fungi. Caribbean philosopher Édouard Glissant wrote of what it meant "to consent not to be a single being."[9] We are all multiple, defying categorization like the corals that teach us a new grammar of collectivity necessary for the future of survival beyond climate crisis. Corals are players in a symphony with a labyrinthine tangle of thousands of species, each complete with gonads, pharynxes, and gastric cavities. Their tentacles undulate as they sense the conditions of the water. They are attuned to the velocity of the current, the salinity, the temperature; they feel the acidification of the oceans. Corals propagate by budding and fragmenting; through both sexual and asexual reproduction, they multiply according to what is safe for the future. Corals show us how to be multiple and how to defy binary categorization. Each reef is a brain, a nervous system of neurons wired and fired together.

People who escaped transatlantic chattel slavery navigated oceanic routes in small boats and maintained knowledge of corals and survived thanks to the underwater network of corals—an underground railroad extended to the sea. Slavers targeted Black people's knowledge and expertise as swimmers and divers in West Africa.[10] Whatever the racist stereotypes about Black people and swimming aptitude are, it has been documented since the eighteenth century that some Africans who were abducted from the Gold Coast were stolen for their aquatic prowess, as historian Kevin Dawson explains in his book *Undercurrents of Power: Aquatic Culture in the African Diaspora*.[11] With some noted as diving sixty feet and remaining underwater for two to three minutes, they were experts at retrieving pearls from oysters. Sometimes Africans aboard ships were forced to dive for the amusement of European ship captains. Black people were no strangers to coral life. They knew the underwater symphony of crackling shrimp and blooming underwater life. They understood that the reef was alive and that these were not inanimate rocks.

THE FREEDOM OF BLACK UNDERWATER WORLDS

Underwater racial formations of the Middle Passage overlap with Black and Indigenous traditions of diving, swimming, fishing, and surfing. Those who were abducted and sold into the transatlantic trade were among record-breaking spear divers, swimmers, and fishers, adept at holding their breath for many minutes at a time. Climates changed people, their lungs expanding as they adapted to new ecosystems. A static and abstract imagining of plantation life flattens the range of environments in which Black people thrived, including coastal plantations. The harvest was different here: oysters and clams, not cotton and sugarcane. Native tribes lived with and sustained these mollusks with their own traditions of harvesting seaweed and kelp.

Regenerative aquaculture and seaside cultivation carry the potential to be climate strategies of repair. In Shinnecock Nation, tribe members are creating clam walls at Heady Creek and constructing oyster reefs, which will naturally help to break storm surges. Part of the once flourishing endemic ecosystem of New York City's shorelines, oysters also help to filter the water of impurities. This allows other marine ecosystems and microbial organisms to thrive. Shinnecock Nation's environmental department is directed by Shavonne Smith, who oversees the shellfish hatchery plan based on a philosophy of working *with* rather than against nature. As she puts it, "We have an inherent responsibility to protect the homeland."[12] The Indigenous environmentalist sentiment echoes how the Maroons of Jamaica describe the land as an inherent pact. Afro-Indigenous science and policy offer alternative priorities and approaches of climate-forward governance. Aquatic restoration along Long Island shorelines encourages estuarial biodiversity, in stark contrast to the increasing pollution due to Hamptons luxury tourism.[13] Shinnecock Nation's reservation is located along 1,000 acres of Long Island coastlines. Home to about half of the total of 1,589 members, the lands are the grounds of ongoing protest against incursions. The foundations of Native science in Southampton, where the Shinnecock Indian Territory is located, are being deployed and adapted as the tides grow higher for those coastal communities who find themselves the most endangered on the shoreline, at the edge of crisis. Many Black and Indigenous people across the diaspora are taking back the waters of the Americas by returning to aquatic traditions. Land Back is a radical movement for Indigenous sovereignty and climate futurity, and the Pacific-led Oceans Back similarly demands ecological repair.

The texture of sound is what drew me to the enigmatic 1990s house music duo James Stinson and Gerald Donald-Drexciya. A friend who is an artist, Andrea Chung, told me about how they inspired her Afrofuturist worlding. When I first learned to DJ, I was drawn to the bass and the excitement of experimenting with the effect of the low-pass filter (LPF). This feature cancels or filters out the high and mid frequencies to emphasize the bass. For me it evokes the sound and feeling of being underwater. Basslines reverberate across Black genres of music. I turned to DJ'ing ten years ago because I needed to get lost in something other than academic work while I pursued my PhD. The echoes of house music translate deep within the seabed and travel across different parts of the Black diaspora. When I first listened to Drexciya, what they conjured of the beats per minute of Black breath underwater resonated with me deeply. In 1989, Stinson recalls being struck by the idea: "It felt like a tidal wave washing across my brain."[14] The existence of the Atlantis of Greek mythology remains in question, as it is cited by only one source: Plato. Drexciya, however, lives as a Black underwater society beyond Atlantis in electronic music and extends a tradition of sonic worldbuilding.

Many Black artists have added their interpretation, being inspired by the marine worldbuilding. Visual artist Ellen Gallagher began contributing to the Afrofuturist aesthetic in 2001, depicting shimmering sea creatures in her ongoing series *Watery Ecstatic* inspired by Drexciya. Black faces in Gallagher's paintings can be discerned with white wisps of hair flowing against the current like translucent pieces of kelp. She calls these beings "wig embryos." Gallagher's Cape Verdean ancestry points to the multiple points of entry for African Atlantic islands and the painful layers of colonial experimentation performed by European powers. Her Drexciya also features what she calls "whale embryos." At play is a Black mathematics of nautical geometry, based on the fractal patterns of the nautilus. Other works from the series feature cutouts of eels that symbolize underwater vitality and blackness. Using cut paper and animation, the multimodal series uses color to tell an Afro-diasporic story. Gallagher says, "in water, all that is dead turns white."[15] Much as for the bleached coral, whiteness signifies the ghostly loss of life and vibrancy underwater. As well as Chung and Gallagher, other

Black diasporic visual artists who have taken influence from Drexciya include Ayana V. Jackson, Firelei Báez, and Alisha Wormsley. Some Black scholars have called for a moratorium on Black underwater Afrofuturist imaginaries because they think it has gotten out of control, but Drexciya has a life of its own.

The Black worldbuilding interpretation of sound creates a submarine race of dwellers that first evolved during the Middle Passage. Between the mythic and nature, Drexciya's album *Journey of the Deep Sea Dweller IV* guides us through what they call the unknown aquazone. The people of Drexciya were gifted with the ability to breathe underwater, having been born from the wombs of pregnant African women who leaped or were pushed from cargo ships that sailed as part of the transatlantic trade. The lungs of the offspring adapted and the offspring became amphibious, forming a subaquatic civilization. Stinson and Donald embarked on sonic ecological worldmaking using house music as an architecture of retreat for Black people breathing underwater. Their tracks include "Black Sea," "Wave Jumper," and "Bubble Metropolis."

The Black inflection of the Atlantis myth has given rise to an aesthetic movement, built on the theories advanced by British academic Paul Gilroy's transformational book *The Black Atlantic,* written in 1993. In it, he describes the Atlantic Ocean as an aqueous sonic gateway of echoes between Europe, Africa, and the Americas.[16] This theory by the Black British sociologist became a sonic portal for the duo Drexciya in Detroit. To Gilroy, the ocean was an antiphonal space of alternating recitations, in other words, of call-and-response.

Drexciya is a techno-ecological statement as much as a racial one. The electronic dance music rhythms of Britain, Berlin, and Detroit pulsated alongside house and jungle music, the time zone and time signature of Black Atlantic possibility. In this sonic underwater utopia, Black people rave through the bass-heavy rhythms of house music in the 1990s. Drexciya took Detroit electronic music underwater because it was born underwater. It was a return to a primordial uterine space that connected the genre to other Black electronic subgenres: techno, bashment, drum and bass, jungle, dub, and garage. These Black diaspora genres emphasized the lower frequencies, remaining attuned to a bass-heavy aesthetic of grounding.

Corals narrate what has happened and what is to come. With the acceleration of the environmental crisis, breathing underwater is a necessity and a certain future for us all. It is projected by the National Oceanic and Atmospheric Administration (NOAA) that by 2050, the earth's coastal regions will be submerged in an additional twelve inches of water.[17] Corals hug the equator, forming a belt across tropical regions. As such, many of the colonized and formerly colonized nations of the world who suffer the worst consequences of global warming are having to protect the reefs. Many communities of color across the tropical belt depend on reefs, and yet the rhetoric of biodiversity in climate activism has always conspicuously excluded racial diversity. Often environmentalists in developed countries are the first to blame local communities for overfishing instead of pointing to the threats that societal unrest pose to reef health.

CORALS AND THE JAGGED TAXONOMY OF RACE

Coral outcroppings appear in the writing of author Ian Fleming, who was a nature hobbyist. He describes a type of corals that were known as "niggerheads" in his 1958 James Bond novel *Dr. No*. Wintering in Jamaica inspired him to pen the spy thriller using the backdrop of a tropical Eden. His world was full of racist characterizations that were part and parcel of Western scientific nomenclature of the time. Fleming's island retreat, GoldenEye, was his writing laboratory where he penned the masculinist colonial fantasies of Bond that were often as touristic as they were pornographic. He writes of the corals of Jamaica:

> Bond's only duty was to keep paddling. Quarrel did the steering. At the opening through the reef there was a swirl and suck of conflicting currents and they were in amongst the jagged niggerheads and coral trees, bared like fangs by the swell. Bond could feel the strength of Quarrel's great sweeps with the paddle as the heavy

craft wallowed and plunged. Again and again Bond's own paddle thudded against rock, and once he had to hold on as the canoe hit a buried mass of brain coral and slid off again.[18]

The ugly term is now considered archaic, but it was a bona fide nautical word used to describe the sea's jagged geology. White supremacy has its nomenclatures, part of a fossil record imprinted in Western science. Racism and anti-black racial epithets are embedded in the taxonomy of European systems of categorization. The natural environment was seized by colonizers who subjugated its people as a labor force.

The Alaskan tundra is described as being "studded with niggerheads" in a 1941 article called "Photographic Foray" by Dow V. Baxter, a naturalist who traveled to Alaska.[19] The hatred imbued in the vocabulary of ecological extraction translates from Alaska to the Caribbean to Australia. European sailors feared that these sharp geological coral formations would sink their ships. Sailors feared that these corals would tear into their wooden hulls and described them as having fanged teeth. British explorers said the outcroppings resembled the heads of Black people, thus their hateful name. How can we turn away from the correlation of differing outcomes when it comes to race and climate, as if climate degradation is due only to

Niggerhead coral, North West Island, Queensland, Australia, ca. 1931

causation? The term "niggerhead" abruptly stopped me when I first read it silently on the page in *Dr. No.*

Noted scientist Rachel Carson also used the term in her 1951 book *The Sea Around Us*: "Landward is the almost unbroken wall of submerged reefs where the big niggerhead corals send their solid bulks up to within a fathom or two of the surface."[20] Ships in Miami must carefully navigate the coral islands of the Florida Keys, sailing between the reef and the Gulf Stream in the scene Carson describes. The coral jaws of the Atlantic, Pacific, and Indian Oceans had the power to devour the shipping profits, a factor in the greedy calculations of transnational corporations from the time of the Dutch East India Company (VOC) to the present.

Carson's anthropomorphizing of the corals, like Fleming's, invests them with an animal-like agency that is deeply racial. The sea creatures are described as slightly above the water, peeping up from the tide. Corals were agents that shaped destinies, mooring mariners in search of new worlds. Their teeth protected islands, and pirates created maps using this to their advantage. They knew the maps could be read only with nautical knowledge of the tides and freshwater channels flowing in currents down from mountains where the reef cannot grow. Most corals require very saline water. Why does the colonial imagination map and project bodies of those it seeks to own onto the land? Is the answer as clear as the animus of racial injury that continues even now? Corals not only sit just beneath the surface of the ocean but also respond to moonlight and sunlight. In high tide and low tide, the aqueous terrain shifts.

I found other parts of the natural world that carry a racial epithet. A "niggerhead" is described as "a tangled mass of the roots and decayed remains of plants projecting from a swamp."[21] Elsewhere it is defined as "any various spherical cacti of the genera *Ferocactus* and *Echinocactus*." The cactus is described as the size of a cabbage and covered with large clawlike thorns. The term "negrohead" is also used to describe a species of tobacco. Racist place-names and scientific names pepper the natural world that Europe attempted to colonize. It is perhaps unsurprising that "niggerhead" extends to the extractive realm used to describe a pink, yellow, and green variety of the mineral tourmaline, a black gemstone often used to make jewelry.[22]

Colonial appraisals of the natural environment constantly value and devalue blackness. Archaic nautical and geological terms belie the racist taxonomy embedded in the ongoing European colonial project. It is a territorial strategy. Taxonomy steals land and lives through racist renaming. The currents and currency of blackness have a long transatlantic history. From the Gold Coast to guineas as a denomination of coin, there has always been a price on the head of blackness. The bounty is what is owed in reparations.

In 2022, I went on a holiday with the aim of relaxation; all too often my travel becomes a research trip or impromptu family reunion. I chose Sardinia thinking there would be nothing about this island that would spark my research interest. Soon I realized I could not escape the role of race in European national fictions. They say race is an American preoccupation, yet it is Europeans who created and enshrined the harmful system. The Sardinian flag, I soon learned, is divided into quadrants with the depiction of four heads of a Black man blindfolded. The blindfolded men are said to be four executed North African pirates. In some later depictions of the flag, the men are wearing the bandannas instead of being blind-

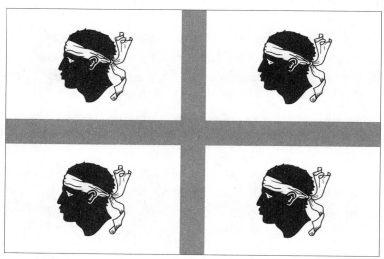

Flag of Sardinia

folded to represent the liberation of the island and Sardinian people. Many contradictory meanings have been historically projected
on the symbol of the Moor's head throughout European history.
Across the island, Italians proudly fly the flag and sell merchandise
with the insignia. The Moor's head has a long history across European heraldry, and why should it not be part of the natural world?
The ugly naming of the corals now made even more sense to me
in Sardinia. Europeans were obsessed with the value of Africans.
White supremacy depends on the continuous valuation of blackness at auction, enough to create a national identity around it.

Nevertheless, Dr. Joan Murrell Owens made scientific history by
countering the jagged taxonomy of anti-blackness in the nomenclature of geology in 1986; she had the chance to confront the racist
taxonomic tradition and she took it. In 1984, Owens became the first
Black American trained as a geologist and marine biologist. Like
me, she could not swim the depths of the Atlantic, but she too knew
the currents intimately. Her inherited sickle cell anemia prevented
her from conducting scuba diving research. Dr. Owens identified
three new species of coral from her research. She named one of
them after her husband, which may be the only coral species named
for a Black person. In her scientific naming, she revisits the violent
grammar of taxonomy and attempts to commemorate a Black family: Black love.

Being able to breathe underwater is a political matter for many
Black people. Because of her sickle cell anemia—which is especially
prevalent in Black diasporic communities—there is limited oxygen
in Owens's blood. The sickle shape of her blood cells refuses to
hold as much oxygen as a fully spherical cell. In her dissertation,
she named three new species in the button coral genus from archival research alone: a powerful act, and an important moment in
any scientist's career. The invertebrate animals she identified can be
found in shallower water and resemble delicate snowflakes. They
are a mobile species of coral. Did they encounter black corals in
their journeys? What may these corals have witnessed of the Black
Pacific currents my grandparents traveled?

As a coral specialist, Owens understood intimately how questions of the oceans and science were racially determined. How many dollars of research and pharmaceutical investment go toward sickle cell anemia each year? The scientific community has long thought of it as a Black disease because it often affects people of African ancestry. Cystic fibrosis (CF), a devastating and rare inherited illness that primarily affects people of European descent, receives 3.5 times more funding from the National Institutes of Health and 440 times as much money from national foundations than sickle cell anemia.[23] While most Black people in the United States are likely to also be of European descent, these health statistics do not include them. In many medical schools, CF is still being incorrectly taught to be designated as a "white genetic disease."[24] Race always determines the outcome of scientific studies because we live in societies where it is a vital factor, outside the lab and inside the lab.

While Dr. Owens started her education at the historic Black university Fisk, there was not enough infrastructure or faculty to support research or training in the marine sciences. She transferred to

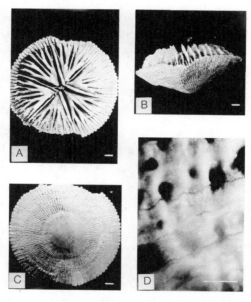

Photograph from Joan Murrell Owens's article. Pictures taken by her husband. Button polyps, zoanthus coral.

George Washington University in D.C. in 1973 at the age of thirty-seven, where, at the predominantly white institution, she was able to major in geology with a focus in zoology and paleontology. Corals were Owens's calling, and throughout her time at George Washington, she carried out pioneering research on their habits, diversity, and taxonomy and challenged fixed assumptions about their behavior. In 1976, she took a position as professor of geology at the historically Black institution Howard University. During her lab work at the Smithsonian Institution, Owens studied one of the rare varieties of coral that does not form colonies, discovering the new genus of deep-sea button corals.

When she received her PhD from George Washington University in 1984, her dissertation, "Microstructural Changes in the Scleractinian Families Micrabaciidae and Fungiidae and Their Taxonomic and Ecologic Implications," was based on the archival research she completed with the Smithsonian's 1880 Albatross archival collection. Embracing multispecies oceanic ecologies allowed for the possibility of cooperative biomimetic models of being. In a 1996 interview with Winni Warren, Owens said, "It's important because of its contributions to our understanding of the life of the sea and some of the ways in which ecology and evolution interact. Also, it is important to others working with deep-water organisms because of some of the relationships I think I indicated—if not proved—between water depth and availability of calcium carbonate that influenced the physical evolution of some organisms."[25] The social nature of coral life was important to Owens because she knew that the theory of evolution owes much to the coral as a model organism. She knew this was only the beginning of what could be discovered about how corals relate to one another. Corals listen and communicate with each other in ways unseen to humans and are hypersensitive to their surroundings. Thus, the marine invertebrates are a barometer of ocean health.

Corals are part of a vast underwater world that includes many other life-forms. In an 1846 letter, Frederick Douglass gestured to the natural world and currents of Black abolition. He wrote, "like sharks in the bloody wake of a slave ship."[26] It was a simple and speculative turn of phrase, yet it speaks volumes to the marine life of the Middle Passage. Sharks certainly would have smelled the blood and waited to feast from miles away. For enslaved people, that terror of jumping was the knowledge of what awaited them in the water even if they managed to swim in the shark-infested sea. Douglass was a witness to the moral economy of the slave trade, so he also bore witness to the interconnected, bloody crisis of climate ruin. The perils of the natural world of racial slavery and racial indenture were entwined within debates about land usage. Douglass's remark about Atlantic sharks opens a rarely considered world of ecological possibility in terms of how African people lived in relation to marine life. Navigating to avoid jagged coral reefs near shore, slavers and ship captains had to contend with countless other forms of marine life like barnacles that formed impediments to their ugly task.

Barnacles continuously latched on to these death ships, eroding the exterior of the floating caskets. The crustaceans held on to vessels with the strength of cement for the journeys back and forth from the Gold Coast. Rudders gradually became less precise in their steering, and interspecies ecologies were witnesses to each bloody transaction. Atlantic schooners grew thick with seaweed, grasses, and mold, and that inhibited steering. These marine ecologies were carefully factored into formulas for profit, slowing business down till the trade stopped legally in 1807. But we know the illegal traffic continued.

DARWIN'S PACIFIC CORAL PILGRIMAGE

While the Atlantic is home for me, the Pacific has also taught me many important lessons about coral reefs. I went to Tahiti in 2023 to listen to the sounds of corals underwater and soon realized that volcanoes and coral had brought Darwin there like sirens on the HMS *Beagle* in 1835. Darwin nurtured a fascination with corals after reading the Dutch and British East India Companies' survey maps. His first book, *The Structure and Distribution of Coral Reefs*, explored the marine invertebrate animals.[27]

I brought my hydrophone, an underwater microphone, to record the symphony of the reef in Moʻorea. I wanted to take the idea of the ecological witness a step further and see what I could learn from observing the sounds myself. This soundtrack has been a constant for centuries, familiar to those who dive, and we are losing it. Once this symphony is gone, we will never hear these underwater songs again.

What did Darwin fail to hear of the heralding of Black freedom in the nineteenth century? I wonder: How could he ignore the protectors of the reefs he studied? In the backdrop of the voyage of the *Beagle*, which, from 1834 to 1838, would change the shape of natural science and history, the sea change of Black emancipation was taking place. A new labor scheme of indenture was invented in the wake of abolition to subdue millions of Indians, Chinese, Southeast Asians, Africans, and Madeirans into debt bondage over the nineteenth and early twentieth centuries. This regime was not abolished fully until the 1920s. The labor scheme, which often proceeded under the guise of indentured servitude, brought many Europeans to the Americas to labor on plantations. Consent was in deep crisis, though landowners claimed that the laborers had signed contracts.

Why, Darwin wished to know, were coral and volcanoes always found so close to one another? Corals were the missing link that led him across the tropical belt of the world, where he would develop his theory of evolution. Was Darwin really the first to see that all life-forms, from elephants to humans to barnacles to plankton, are interrelated? What sorts of parallel theories might have developed from other societies and traditions of science alongside evolution? Before his theory of evolution, Darwin hoped to discover the geological relationship between volcanoes and corals. He did this in the Pacific and Indian oceans by observing atolls, which are low-elevation ring-shaped coral reef islands that surround lagoons. His hypothesis, which turned out to be incorrect, was that atolls formed when volcanic islands sank and corals formed around them.

How much knowledge did he derive from people native to the locations where he sailed? Did he cite them or deride them? Upon meeting the queen of Tahiti, he commented that she was "an awkward large woman, without any beauty, gracefulness or dignity of manners."[28] In 1836, Darwin described the island of Mo'orea as a "picture in a frame." The barrier reef surrounding the island was one of the underwater phenomena he hoped to understand that connected zoology to geology. In his travels, Darwin excavated fossils and hypothesized the delicate equilibrium of biodiversity that coral ecosystems support throughout the tropics. His major epiphanies would famously take place on the Galapagos, the cluster of island laboratories of Ecuador spanning the Indian Ocean to the South Pacific.

In the Indian Ocean, Darwin pinpointed the Chagos Islands and Mauritius as important to his study of atolls. Would Darwin have crossed paths with illegal cargo ships carrying enslaved and abducted West Africans? Would Darwin have encountered ships carrying indentured Indians to these atolls? In the South Pacific, these labor regimes were also taking hold as Europeans claimed dominion of plantation islands. What might Darwin have known of the brutal scheme of blackbirding as he sailed through Polynesia? Pacific Islanders who were conscripted into this system often labored at gunpoint. Darwin largely avoids the scrutiny of such political and historical events in his writing, but he neither lived nor theorized in a vacuum. His theory of evolution impacted the lives of millions whom new generations of researchers further subjugated.

Though it was not his direct intention, Darwin's claim, "the survival of the fittest" became interpolated as a racist justification for racial cleansing. Yet in *The Descent of Man*, he recalled "a full-blooded negro with whom I happened once to be intimate." He wrote, "The American aborigines, Negroes and Europeans are as different from each other in mind as any three races that can be named; yet I was incessantly struck, whilst living with the Fuegians on board the 'Beagle,' with the many little traits of character, shewing [*sic*] how similar their minds were to ours; and so it was with a full-blooded negro with whom I happened once to be intimate."[29] The scale of the theory of evolution was willfully skewed for political aims. In the mid-nineteenth century, nations and colonies of the torrid zone were bound up in freedom debates. Francis Galton, Darwin's cousin, extrapolated on the theory to create the field of eugenics. Scientific racism undergirded white supremacy in its many nineteenth-century forms, including policy and public health. When Darwin set sail for the islands of the South Pacific and Indian Oceans, his mistake was much like Columbus's mistake. It was one of misrecognition. Darwin saw coral islands as a "problem," one that he believed he had solved by the end of his life. He wanted to understand why barrier reefs and atolls form where they do.

The choreography of coral polyps instructs us on how it is possible to breathe underwater collectively. Corals are fleshy and sensile beings, but Darwin did not have the capacity to see coral this way. There are no neat lines of descent in coral genealogy, much to the disappointment of taxonomists. Corals are the limestone mountains of the future and so their ballads are future mountain ballads. The karst territory is composed of fossils that tell a story of mass reef extinction in the Devonian period, and are the peaks of today. Darwin used the term "living fossils" in his writings. This is relevant to racial thought and violence, as science was used to commit violence in the European tradition. He remanded his dear teacher John into the category of living fossil simply because he was Afro-Guyanese.

When I visited the islands of Tahiti, Tetiaroa, and Mo'orea in the South Pacific, I confronted Darwin's desires and attempts at theorization on the origins of atolls. I understood how they have since been disproven. Darwin hypothesized that coral reefs formed a ring around eroded volcanoes, which then sank into the sea, leaving a circular formation of the atoll behind. Scientists today are campaigning to have these erroneous theories removed from textbooks. As of 2021, geologists André Droxler and Stéphan Jorry have shown that atolls form over hundreds of thousands of years of cyclic sea level changes and are unrelated to volcanoes.[30] Rising sea levels due to global warming threaten atolls because the ring-shaped islands thrive in low water levels. People who live on atolls are protectors of the reef who are now facing down the drowning of their homes. Underwater cities are not mythic after all; they are an inevitable fate for the planet, and one that approaches much sooner than anyone could have anticipated. While numerous great floods have shaped the scale of geologic time, this current shift is sudden. It is a crisis of epic proportions that will hit Polynesian and Caribbean countries first.

For Hindus, crossing large bodies of water meant losing caste, according to their religious beliefs. Losing caste was worse than being of the lowest caste; one would no longer have recognition within society. Beneath the dark waters known as the *kala pani* in Hindi, a different fate awaited. The Indian poet Khal Torabully, who is of Trinidadian and Mauritian parentage, is part of both the Atlantic and Indian Ocean diasporas, displaced from India. He articulates the experience as one of "coolitude," borrowing from the formulation in the African diaspora of *négritude*. For Torabully, the rescued items from the ocean floor represent his concept of "coolitude."[31] Dwelling on the seafloor of the Indian Ocean, he is attuned to the atolls of the Maldives threatened by petrochemical companies such as Royal Dutch Shell because of ocean drilling. In his poetry, he describes the flesh of coral as symbolic of the stages of *becoming* for nineteenth-century Indian laborers who were part of this brutal plantation system of exploitation. Inspired by how the Afro-Martinican poet Aimé Césaire draws on the cragginess of the rock to theorize blackness and négritude, Torabully saw a likeness in the coral for history of indenture. Hoping to take distance from essentializing blackness, Torabully shows how interconnected the poetics of climate and geology are for Black people and South Asian people. He attempts to reclaim the jagged anti-Asian taxonomy of the tropical climate of racial indenture.

Usually this kind of poetic imagining is reserved for European explorers like Darwin and Alexander von Humboldt, whom I have heard popularly described on BBC podcasts as the last men who knew everything. They were permitted to have grand theories underwritten by the financing to travel the world to test them out, on willing and unwilling subjects. Not much is made of their methods and interactions with people of color in those locations, which must have been core to the knowledge they produced. Imagine if researchers of color today were given the range to explore the

world and study climate strategies for repair. The aim of my lab
is to make the conditions of this possibility possible, by honoring
Black scientific methods and Indigenous sciences. I support climate
activists, poets, and artists who are on the front lines protecting
coral reefs in the Caribbean Sea. One of our artists in residence at
Dark Lab, Nadia Huggins from St. Vincent, has focused her photographic studies on coral and ash, much as Darwin was captivated by
reefs and volcanoes. For her there is a key difference in the scope of
what it means to take these ecologies as objects of study.

Traveling with Captain Fitzroy aboard the *Beagle* in 1831 was the
first time Darwin had seen the submarine animals he had studied
for so many years in Edinburgh. Did Darwin live in such a vacuum,
separate from world events? No. At the age of fifteen or sixteen, he
was instructed in taxidermy by a formerly enslaved Afro-Guyanese
man named John Edmonstone.[32] Few in Britain received firsthand
mentorship from an emancipated person who had been enslaved in
the West Indies. Darwin was a witness, whether he chose to reckon
with this or not, and it influenced his budding scientific lens for the
natural world, as Darwin biographer R. B. Freeman noted in a 1978
article of the *Royal Society Journal of the History of Science* called "Darwin's Negro Bird-Stuffer."[33] He draws from Darwin's autobiography: "A negro lived in Edinburgh, who had travelled with Waterton,
and gained his livelihood by stuffing birds, which he did excellently:
he gave me lessons for payment, and I used often to sit with him,
for he was a very pleasant and intelligent man."[34] The barter and
transactions were of Afro-Indigenous knowledge. Darwin appears
to have had respect for his Guyanese instructor.

Edmonstone's surname is derived from the plantation owner
Charles Edmonstone, who arrived in the South American British
colony from his native Scotland in 1780. He founded a timber estate,
and soon married the daughter of an Arawak chief named Princess
Minda. It is said that he learned Arawak ways of being in tune with
the natural environment of Guyana from her family. Edmonstone
studied Native science to become a skilled hunter of Maroons.
Tracking runaways was a sport to him and other British colonizers.
Charles Edmonstone left Guyana in 1817 to live in Scotland with his

wife, Helen, and two children. He died in 1827. In his will he leaves 1,000 GBP to "a mulatto child Jeanie residing with me." Could this have been a blood relation of his? A daughter? A granddaughter? Upon his death, Charles Edmonstone manumitted three people he had enslaved in Demerara named Catharine, Betty, and Cecillia in a common practice of slave owners, thought to be benevolent at the time. What, then, of the free status of John Edmonstone, Darwin's teacher? He was freed as soon as he arrived in Scotland, and eventually opened his own bird-stuffing shop in Edinburgh. We will return to Edmonstone's story and the taxidermy of birds and bird-watching practices in chapter 6.

Dead ends in the archive are never dead dead. This phrase "dead dead" is used in the Caribbean because there are many stages and categories of death, this being the final one. As Jamaican writer Marlon James puts it, "Dead people never stop talking and sometimes the living hear."[35] The dead of the transatlantic trade will never stop talking in the hundreds of distinct African languages and dialects they spoke. Against the grain of European scientific and anthropological observation, we must learn to hear the bias and what was more likely true than what is presented as fact. People of color are always rendered as background characters, unnamed except for a descriptive racial epithet to give the scene color and authenticity.

CAPTURING CORAL

As if the climate risks are not enough, coral poaching is another major risk factor for the marine animals. Corals are kept as pets, propagated in saltwater tanks by those in the West. The Coral Room at the British Museum was a popular attraction for patrons to observe the living and dead curious creatures in giant tanks. Detached from their natural habitats, coral bodies were specimens on display in Victorian London. Today there are still remnants of corals and other sea life in the Great Court of the museum, though none are submerged. Instead, the bleached skeletal remains sit on display, functioning solely to illustrate Darwin's theory of evolution. Juxtaposed with Amerindian artifacts in the British Museum,

The British Museum: *The Coral Room, with Visitors.* Wood engraving, 1847.

the curatorial logic positions both Native peoples and coral as primitive ancestors to the civilized Victorian-era man and woman.

It was common for early European naturalists to throw sticks of dynamite to blow up reefs so that polyp fragments and fish floated to the top. Cyanide was also a scientific method of harvesting coral. These destructive techniques were commonly used by French naturalist Jacques Cousteau, who proudly depicted them in his famous film *The Silent World*.[36] Cousteau was personally responsible for the underwater endangerment of many coral species, and yet he is celebrated as an advocate of the natural world. More indirectly, he encouraged a culture of hobbyist diving that furthered the idea of the individual viewer as amateur naturalist, and thus a figure likely to disturb the marine environment.

There is potential for sonic regenerative aquaculture in technologies that mimic underwater soundscapes. Playing the sounds of a healthy underwater reef will lead marine life to return to a bleached or dying coral.[37] When I began listening to reefs with a hydrophone in Tahiti, I came to understand them as a philharmonic. Among the interspecies symphony players are pistol shrimp making a firing sound. *Pop. Pop. Fizzle. Click.* Vibrations echo, from whale songs to dolphin calls. Low-frequency sound waves can travel hundreds of miles before dissipating. Climate lessons offer technologies of adaptation and repair. Do we deserve these quick-fix solutions without addressing the origins of the crisis? The cycle is doomed to repeat itself. Just as the problem is interlocked with capitalist hyperdrive, the solution must be interlocked with anti-capitalism.

*

I was surprised to happen upon a series of saltwater reef tanks at Pratt Institute, a private university in the heart of gentrified Brooklyn, in the summer of 2023. At the college's writing center, of all places, I was able to observe the ocean's diligent engineers up close. Thanks to the custodian of the indoor reef, a hobbyist, the majestic and colorful corals have been thriving here for over twenty years. It makes for a pleasant attraction for university students in need of a calm place to study or write. I arrived at Pratt with a Dark Lab intern just in time for a Brazilian dusky snake eel's feeding. Frozen tiny eels were fed to the large eel. We noticed that the eel was eating its own tail in the small tank, and it had become infected. A pink, fleshy ouroboros, we thought. That day, the New York City sky turned orange due to wildfires from Canada. My intern and I wondered if the eel was symbolic of the climate crisis, swimming in circles in a confined tank and left to mistake its own tail for unsuspecting prey.

From here, I learned more about communities of coral hobbyists at conventions across the United States such as Reefapalooza, America's largest saltwater aquarium show. Sellers and exhibitors with a shared love of corals congregate at locations such as the Meadowlands in New Jersey. Does this passion extend to the oceans or to those who protect the reefs in the Global South? Fraggers hope to breed their coral to create genetic diversity in home tanks, beyond what is possible with asexual reproduction or budding. Some proudly call themselves coralholics, adherents to an addictive and time-intensive hobby. On a podcast, a coralholic asked, "Is the tank just a box full of the ocean with a rotting pool of water?" Like the farm plot, coral-keeping as a hobby is a practice of localized monocrop aquaculture. Intensive re-aquascaping follows the segmenting logic of the plantation and the laboratory. Coral collectors are not motivated by the same questions as zoological researchers or conservationists. Coral propagation of this sort is a DIY project not intended to save the species or to reintroduce them to the open ocean. Rare breeders hope to contain corals from "exotic" locations, forming their own liquid laboratories.

On Long Island in New York, a white American man named Joe Yaiullo built a 20,000-gallon saltwater tank before beginning his career at the Coney Island Reef in Brooklyn in the 1990s. He started his own personal Atlantis in Riverhead, New York. The facility has since been renamed the Long Island Aquarium and is now open to the public. The tank is large enough that Yaiullo can scuba dive inside of it. There is such a chasm between his world and that of Black and Indigenous life on Long Island. For 10,000 years prior, the Montauk, the Massapequa, the Seatucket, and the Canarsie resided across Long Island. In these territories, Shinnecock Nation—past, present, future—continues to be loud with Algonquian echoes that defy the incursions of Hamptons tourism. Water sovereignty is as much an Indigenous rights matter as that of land, and the tribe, which is comprised of many Afro-Indigenous members, continues to fight for land and water access and ownership. Tribe members are organizing to buy land and beach access back, parcel by parcel.[38]

The shorelines continue to provide a bounty of oysters, scallops, and clams for Native tribes. These shells have been harvested for millennia on Sewanhacky (Long Island) to craft wampum. The bivalves also serve to filter the water of impurities. The beaded craftwork has historically been the currency of these Indigenous peoples of the Northeast, who polish the shades of purple and pearly white as part of a ritual. The reverence for marine life is part of a sacred aesthetic, and the artisanal craftmanship gets passed on intergenerationally. The Native American Calusa tribe in Florida were prominent shell collectors and protectors of the reefs.[39] Many coastal nations design technologies of seawall building in the United States as climate strategies, using oysters and clams for repair. The mollusks also filter the water, increasing biodiversity for other marine species.

Climate amnesia suggests that the peril of reefs is a new cause for concern. Yet reading back issues of *National Geographic* from the 1950s, I have noticed journalists and photographers voice and depict concerns of coral endangerment off the coast of Florida that sound similar to the warnings we hear today. It is a myth that the peril is new. The cycle of forgetting allows for new violent acts of discovery. More pressing to the 1950s *National Geographic* reporters is the question of how to best capture the majesty of reefs accurately without being underwater. Many have attempted to record a true representation through various technologies. Our crises require oceanic timescales of moral reckoning with the toll of the atrocities of human traffic on the lives stolen and the impact on the ecological world. The divide between human and nonhuman life is blurred by the ocean and its temporality. Histories of transatlantic slavery may seem ossified and brittle, yet the past is plural. The lives of those Africans and Asians who were abducted are rich and multiple.

REGENERATION BEYOND THE FUTURE OF BLEACHED CORAL

Amid the ocean's detritus, new life-forms evolve. Corals can grow about a centimeter a year, but trash from beach resorts accumulates faster. Sunscreen from well-meaning tourists dissolves into the water, producing toxic conditions for corals. Coral bleaching leads to the proliferation of slime and bacteria. Death is pervasive, and the glowing white color sucks life away from the reef. Parrotfish are vital for the balance of the ecosystem because they eat algae blooms. However, because of local tastes, fishermen hunt and eat these creatures. As corals become endangered, marine scientists are acknowledging there is no panacea. Bioengineers are adapting coral DNA to withstand warmer temperatures, but these efforts are ultimately futile because climate is the larger issue. Attempts to trick or reverse engineer coral will not be fruitful because they signal to other marine life where it is safe to call home. Even if sonic technologies are used to regenerate reefs by playing the sound of healthy coral, this tactic is merely a shallow decoy.

Corporate bluewashing, superficial restoration campaigns that corporations have engineered to distract from the magnitude of the underwater crisis so vacationers can snorkel with a clear conscience, is yet another peril threatening coral life. Bluewashing functions through a logic of scientific empiricism used to obfuscate profit margins and accelerate a mission of ecological extraction. Marketing tactics tout shallow commitments to clean energy and green initiatives; meanwhile, corporations deflect the ecological extraction they continue to perpetuate. Cruise ship companies advertise that they are turning off their lights at night and say that they won't stop asking important questions about the environment, even if there are no answers. Deceitful corporate practices of bluewashing at all-inclusive resorts across the Caribbean and the Pacific have led to increased financialization and premature planetary death.

Whitewashed agendas shift blame to local communities of color as the easy scapegoat for pollution. Meanwhile, tourists from the Global North assuage their guilt by taking part in ecotourism. Corals hear by sensing the pressure and shifting of waves. Subtle signals help them navigate and choose somewhere to anchor as a permanent home. The larvae can detect color, and they grow in crevices, seeking safety. Photosynthetic corals form a mouth and tentacles and grow toward the sun. Like the regenerative budding and building of coral, transformative climate justice is possible. Atolls are just one of the many coral formations that show us how they are portals of possibility. Whether Atlantis exists, the processional of the bleached coral is spreading rapidly across the ocean.

Since the 1990s, the regeneration of the Belize Barrier Reef has been a beacon of hope. Placencia, located in the Central American country, is the second largest reef system in the world—an enormous success story for restoration after mass bleaching, going from 6 percent to 60 percent live coral. Since June 2018, UNESCO has delisted the Belizean reef as a site of world heritage in danger. The regeneration movement was community-led and is the result of decades of planting of thousands of coral fragments among reefs that hurricanes had wrecked. The nation's famed Great Blue Hole brings ecotourism. It is an atoll around a giant sinkhole, now protected as a UNESCO World Heritage Site and as part of the Belize Barrier Reef Reserve System. However, on the website Private Island Online, one can buy islands, as the likes of Leonardo DiCaprio, Jay-Z, Beyoncé, and Shakira have. Reef islands begin at prices upward of $68,000 USD. Belize welcomes foreign investment. Herein lies the fallacy of property ownership as it existed in 1492, made contemporary by billionaires who frequently claim to be environmentalists. What, truly, is the desert island? Uninhabited and uninhabitable to whom? From Marlon Brando to Richard Branson, even the most benevolent and benign billionaires have put nature in peril with multimillion-dollar island grabs. How can these transactions be valid?

In an effort to regenerate corals, artificial reefs have been constructed, from 2001 to 2010, out of 2,500 decommissioned subway cars dumped off the coast of Georgia. But this has led to further environmental degradation. The municipal and corporate aim has been tourism, which boosts recreational fishing and scuba diving. The Atlanta Metro rail network, MARTA, has dumped its defunct cars into the ocean, where they disintegrated just months after they were submerged. Sunken stainless steel led to corrosion and black reef risk.

*

Joining the chorus are the humpback whales who have returned just off the coast of Manhattan's harbor. Members of my lab have witnessed whales and their songs in the summertime. In the 1970s and 1980s, when New York was at its most polluted, we could never have encountered this variety of sea life. There are signs of hope, in larger part due to the inactivity of freight, which came to a brief standstill during the beginning of the Covid pandemic before going back to normal. We heard the echo of a whale call wash over us as the animal's tail splashed down onto the waves in July of 2023. One of the researchers in my lab said it poignantly on the ship:

*

"It was the sound we did not know we needed to hear."

*

The whale-watching experience was healing in its immensity: life returning to the Atlantic. We saw three whales that summer day in a majestic aquatic choreography. I thought across deep time: Who else had been immersed in this symphony of humpback whale song? The guide told us that the migratory pattern of these whales went from the Caribbean to New York and back. She said they wintered in the Dominican Republic. Knowing this island to be Hispaniola, I heard the omission of Haiti's ecology and waters as relevant to the whales. I heard the cycles of Black freedom struggles across the entire island, now silenced. There was no room for Haiti in this ecological fun fact. Where does the aqueous border lie for marine conversation? Where does the island split in two? The Santuario de Mamíferos Marinos de la República Dominicana, established in 1986, is a haven for marine life. Movements of manatees, whales, and sharks are recorded, but what of Haitian waters? A whale-watching port exists at Cap-Haïtien, where there is a consistent presence of sperm whales in the Gulf of Gonâve. Again, ecological sanctity depends on Afro-Indigenous sovereignty. Local economies for fisherpeople rely on a balance of ecosystems amid turmoil. In the marine environment lies the key to a sustainable economic future for Haiti and its local fishers.

Those who fought for the first Black republic knew of the music of those humpback whales. Witness to liberation as much as to the routes of agony and enslavement, the natural world cannot be disentangled from Black Atlantic history. The presence of the whales is political and so too will be their extinction—and our own. Their excreta provide oceanic fertilizer and rich food sources, supplying a delicate balance of marine microbial ecosystems in an intimate cycle. Guano, the excreta of seabirds, as it turns out, also holds regenerative potential that could save coral reefs, which we will explore in the next chapter. Ultimately, hope rests on the caesura of capitalist expansion, which the world witnessed briefly during the first part of the Covid pandemic. It was then that the whales returned and began to sing again. The world does not have to drown if we synchronize our breathing. Adaptive strategies of survival are possible, as coral sound waves show us. And listening to corals, the oldest living animals, shows us alternatives for living beyond and breathing underwater.

5

⁕

Guano Destinies

Why are we having all these people from shithole countries come here?[1]

—DONALD TRUMP, 2018

The German and Irish millions, like the Negro, have a great deal of guano in their destiny. They are ferried over the Atlantic and carted over America, to ditch and to drudge, to make corn cheap, and then to lie down prematurely to make a spot of green grass on the prairie.[2]

—RALPH WALDO EMERSON, 1860

The sedimentation of coral reefs in the process of transforming into limestone lie atop the compacted deposits of thousands of years of guano—the seabird fecal matter of cormorants, blue-footed boobies, and pelicans. Islands across the Caribbean Sea and Pacific Ocean rich in the powerful fertilizer have been a powder keg of contestation since the nineteenth century. Guano was mined as if it was a precious metal, carted off and shipped away. It was imported to restore the broken soil of Europe and the Eastern Seaboard of the United States, depleted by the monocrop agriculture Columbus's arrival set into motion. With the guano trade, once again, islands became colonial laboratories of biological experimentation. The bird dung was a life-giving substance and a life-taking one that impacted millions of laborers of color forced to mine the toxic substance. From the microbial to the geopolitical, the history of guano is a key chapter, and one that historians and scientists often neglect as part of the climate crisis.

Rebellious echoes are embedded in the few mountains of guano islands that still exist. Most were depleted in the nineteenth century, when millions of African American, Indigenous, Polynesian, and Chinese laborers were forced to mine it from the earth. The echoes of their collective refusal to work resound across the Pacific and Caribbean. Many guano miners took their own lives, trapped on those islands. Hundreds of thousands of laborers from Baltimore to Haiti to Peru to Hawai'i protested against the backbreaking work into which they were conscripted. Some were held at gunpoint to dig with pickaxes. Some turned those pickaxes against their sadistic overseers. Some plotted their escape against their oppressors, who exploded heaps of guano with dynamite that released poisonous clouds of ammonia dust. Rebellion against the toxic regime of white overseers was loud. Yet we fail to hear this labor history as a climate history. Why haven't the voices of the oppressed who refused to become fertilizer themselves been amplified?

~~Too much guano.~~

Ralph Waldo Emerson wrote in his notes on nineteenth century labor, "too much guano," and then crossed it out. As Nell Irvin Painter, a renowned scholar of U.S. history notable for her research on the nineteenth-century South, points out in her archival analysis of Emerson's notes in his journal drafts of the famous speech "Fate," guano captured the American thinker's imagination as early as 1851. Emerson's consciousness was colored by the Fugitive Slave Act of 1850. There is perhaps too much evidence when it comes to the consequences for the climate after the nineteenth-century guano trade. The same can be said of the magnitude of the climate crisis—there is far too much evidence. Would another shocking statistic about climate ruin shift global environmental policy? The cycle of forgetting these labor uprisings reveals how the climate crisis deepens as corporate interests racialize and then expend human life. Labor histories of organizing and dissent led by people of color do not serve the expansion of extractivist capitalism, and so a capitalist society attempts to disappear them. That the guano rebellions took place in minor and major forms of resistance is as certain as the fact that there are numerous unrecorded mutinies and uprisings across the globe. As the Trinidadian theorist C. L. R. James wrote, "The only place where Negroes did not revolt is in the pages of capitalist historians."[3] Through guano comes a philosophy of labor, geology, and race that surfaces with answers about why the climate is in crisis. Guano interrupts history as we have been taught it by colonial institutions because in it there is too much evidence of violence and refusal. Guano forms a gap in the geological record.

Gregory T. Cushman, author of *Guano and the Opening of the Pacific World: A Global Ecological History*, notes that during the nineteenth century all one had to do was commandeer a labor force to shovel guano, a statement typical of the glib approach of nature writing from a Western tradition on the matter of labor and race.[4] A former fellow at the Rachel Carson Center for Environment and Society in Munich, Cushman is now an associate professor at the University of Arizona who writes in an American and Western European tradition of writing Western environmentalist histories. Often, economic history or chemistry perspectives govern the global storytelling of guano. These points of view discount the individual lives and narratives of the miners from China, the South Pacific, and the Black diaspora who died mining guano. Though Cushman claims to teach and work on the human dimensions of climate change, he elides people of color by focusing on the Anthropocene to tell a top-down geopolitical history. Western climate innocence rests on colonial omissions and the lack of attention to the valuation of whiteness in the extractive economies of the past. In the guano trade, racism became sedimented in the Atlantic and

Pacific worlds as life became disposable according to race, all for the cause of imperial and economic expansion. The nineteenth-century world redefined race according to the shifting Atlantic schema based on abolition movements and the emergence of new shades of unfreedom.

It was common for guano miners to develop histoplasmosis, a respiratory fungal infection, after being surrounded by feces. The guano transformed their bodies, and their immune defenses raged against clouds of microbial pathogens. The fungal spores spread throughout the body, causing patients to suffer from enlarged spleens and livers, gastrointestinal bleeding, lip and mouth ulcerations, and decreased blood platelet formation, leading to bone marrow failure. The fatal disease resembles tuberculosis, and the Black, Chinese, and Indigenous laborers' eyes were often described in records as being crusted shut from the hot, arid weather. It was common for laborers to lose their sense of smell and to have their respiratory tracts corroded by the guano dust.

Guano connects these labor regimes as circuits of ecological degradation that fueled U.S. colonialism. The 1856 Guano Islands Act that President Millard Fillmore developed reified the American ecological identity as one invested as an individual colonizer with the backing of the U.S. military.[5] While the thirteenth president of the United States assumed the position by default after the death of his predecessor Zachary Taylor, Fillmore shifted the U.S. position as an explicitly colonial power when he endowed *every* American citizen with the right to annex guano islands as U.S. territories. Fillmore passed the Guano Islands Act in response to the British-Peruvian treaty that prohibited the direct importation of guano from Peru. Manifest Destiny now reached beyond the coasts to faraway islands, based on the suspicion that there might be guano to extract. Within a few years, the United States had claimed and incorporated into its empire more than 200 guano islands in the Pacific Ocean and the Caribbean Sea, all of which were supposedly uninhabited and unclaimed. To this day, the U.S. military defends land claims made by any U.S. citizen to said remote islands.

SHITHOLE COUNTRIES:
FROM EMERSON TO TRUMP

Guano is the missing link between the Agricultural Age and the Industrial Revolution. The eco-politics of every labor struggle is a climate struggle. In the nineteenth century, at the peak of U.S. colonialism, Ralph Waldo Emerson's guano rhetoric paralleled contemporary currents of white supremacy in the name of environmentalism. In his 1860 speech "Fate," Emerson mentioned how "[t]he German and Irish millions, like the Negro, have a great deal of guano in their destiny. They are ferried over the Atlantic and carted over America, to ditch and to drudge, to make corn cheap, and then to lie down prematurely to make a spot of green grass on the prairie." This passage exemplifies how nature writing is often about colonial ambition and supremacy. Ideals of United States white supremacy have shifted as the hierarchy of who is deemed expendable—determined by race and class—evolves and shifts with national labor demands.

While Emerson spoke of guano destinies, U.S. president Donald Trump concerned himself with the valuation of human life from "shithole countries." In 2018, he lumped Haiti in with African countries and spoke candidly about his desired white supremacist immigration policy. What more dehumanizing rhetoric than to compare people to feces? Disposability is the engine of capitalism, and this rhetoric shows the logic at the heart of American values. Trump lamented that there wasn't more migration from countries such as Norway, with good immigrants to populate the future gene pool of the United States. Emerson favored Anglo-Saxons. This type of proto-Aryan eugenicist rhetoric is not only anti-black; it also belies racial formation as an evolving dynamic from the nineteenth to the twenty-first centuries. Emerson believed that someone like Trump, of German stock, had guano in his destiny. WASPs believed that Germans, as one of the disposable white races, should meet their fate and lie down to die—to become a patch of grass on the prairie. Anglo-Saxon exceptionalism excluded Germans,

Irish people, Italians, and other whites. As a commodity and a figure of speech, guano gives us a lens for the excremental poetics of U.S. statecraft.

The tiny offshore island of Navassa (claimed today by the United States and Haiti) reveals the ongoing colonial logic of ecological degradation as entwined with white supremacy and against Black liberation.

✳

"Shit! We're sinking."[6]

✳

Geological extractive practices supply a violent hierarchy of racialization. Entangled in the saga of guano, or white gold, as it was called, lies the rhetoric of white supremacy. American ecological theories that exalt a fictional Anglo-Saxon history are a part of the bedrock of the nation and its logic for seizing territories from non-white people. The scatological logic of matter, mattering, decomposing, and existential meaning is always already eschatological.

Few know that there is an ongoing contested land dispute between Haiti and the United States that began in 1857. A National Wildlife Refuge was established in 1999, using paternalist conservationism to claim the Caribbean island of Navassa for U.S. colonial expansion. When the white American Captain Coleman first made claims for the island of Navassa, did he not know its other names, or that Columbus's men had landed here? Peter Duncan later made claims to the island. The Haitian government disputed the claims each time. Known as Île de Navasse in French or La Navaze in Kreyòl Ayisyen (the Haitian language), the guano island is a key made of millennia of accumulated bird droppings. The French had laid claim to La Navasse since the Peace of Ryswick in 1697, when they settled their dispute with Spain to share Saint Domingue (now Haiti and the Dominican Republic).[7] Haiti has held claim to Navassa since 1801, but the United States refused to acknowledge the existence of the world's first Black republic until 1862.

*

Guano islands have been treated as the outhouses of U.S. empire. Cruise ships dump billions of gallons of sludge, comprised of feces and food scraps, into the ocean each year. To the United States, the Caribbean islands have historically been a toxic waste dumping ground. In 1986, the City of Philadelphia loaded 14,000 tons of rubbish on a barge hoping to dump the incinerated ash into the waters of the Bahamas, the Dominican Republic, Panama, Bermuda, Guinea-Bissau, and the Dutch Antilles. Each nation refused. The barge dumped 4,000 tons in January of 1988 near Gonaïves in Haiti, saying it was "topsoil fertilizer."[8] Once the Haitian government learned of the toxicity of the substance, they ordered the incinerated ash to be reloaded, but the barge had already left.

*

Located between Hispaniola, Jamaica, and Cuba in the Windward Channel, Navassa is considered strategically important because of its geography. It is a shelter island for red-footed boobies whose deposits of phosphate-rich droppings could be used as fertilizer. In the 1860s, Peter Duncan, an American sea captain, sold the island to an American guano farmer in Jamaica, who later sold it to what became the Navassa Phosphate Company for $200,000 USD (the currency of the time).[9] Uninhabited but for the red-footed boobies and goats, the island is a symbol of the Haitian struggle for Black sovereignty. In 1504, Columbus had been stranded on Xaymaca for six months before the Arawak stopped feeding him and his men. It was his fourth trip back to what he believed was China. The Amerindians determined they were unsatisfied with what was being traded by the Europeans. Columbus's men made it to Navassa on the way to Spain, naming the small island for Navaz, meaning plain or field.

*

After I delivered my keynote address at the Second National Conference: Justice in Geoscience at the American Geological Union (AGU) in 2022, a Black geoscientist told me about the town of

Navassa in North Carolina. This fact set me further on my research trail in search of guano, environmental racism, and echoes of Black rebellion. Navassa Guano Factory was founded in 1869 near Wilmington, North Carolina. Named for the infamous Caribbean Island, the historically Black community was incorporated in 1977 and lies at the nexus of multiple factories and former plantations. Wastewater poisons the seagrasses. Even with a pledge by the United States Environmental Protection Agency (EPA) to clean the site, residents continue to wait without any answers or restitution regarding the carcinogenic creosote they have imbibed. The chemical is derived from coal tar that was used for a wood treatment plant in Navassa. As of August 2023, eighty-seven acres of the Superfund land in Navassa are being sold to the highest bidder.[10] Yet again, the guano fate of a Black community is left in the hands of capitalist profiteers. Precisely because there was no possible escape from guano islands beyond swimming, the consent of such labor schemes is thrown into crisis through the geography of isolation. Still, laborers creatively devised means of sustenance like fishing and eating bird eggs. Islands are common locations for penal colonies and other carceral structures that exploit forced labor. Navassa was no different.

*

Is a shithole a toilet or a strategic military outpost? To Emerson, not only are certain people born to labor and disintegrate; their decomposing bodies are fated to become more fertilizer for a new nation, a white nation. His was a deathly environmentalism that exalted white progress at the expense of the weaker races. The Emersonian environmentalist tradition aligns with Trump's political rhetoric in 2018. Embedded in the nineteenth-century guano industry were the latent geological conditions for the twin projects of race-making and nation-building.

*

Trump promotes an American anti-intellectualism that is meant not to be read but spoken. It is the epitome of what Richard Hofstadter

defined as the "resentment of the life of the mind, and those who
are considered to represent it; a disposition to constantly minimize
the value of that life." By contrast, Emerson's liberal intellectual-
ism developed through a lecture circuit where he advanced his ideas
about nature and American identity. An orator influenced by the
pedagogy of divinity in which he was trained, Emerson wrote and
edited his lectures to be published—quite the opposite of Trump
(although perhaps X, formerly known as Twitter, is simply a different publishing platform for Trump). Emerson's essay "Fate" from
The Conduct of Life was not published until ten years after he delivered a draft of it as a speech. Critics of Emerson's time denounced
him as a secondhand writer and as too abstract. In 1984, U.S. critic
Harold Bloom dubbed Emerson Mr. America, a canonical figure.[11]
Trump's Americanist argument for more immigrants from Nordic
countries makes explicit that he is not anti-immigrant at all, but
specifically anti-black. He later tried to retract the statements, but
enough people in the room noted what he had said.

*

Trump's wife, Melania Trump, is a white European immigrant
from Slovenia who reportedly modeled in the United States prior
to being granted a work visa but has not met the consequences or
scrutiny for breaking the law that millions of Black Haitians face.[12]
Although they are legal refugees under *Temporary Protected Status*
(TPS), they sit in limbo, a stigmatized purgatory underscored by
centuries of unresolved guano disputes. Today, multiple mythologies of whiteness cohere into a *permanent protected status* for white
immigrants in the United States that follows an epidermal logic at
the Department of Homeland Security. The ongoing racial, political, and social meanings of Haiti to both Trump and Emerson
reveal how American identity is linked to the underdevelopment of
Africa. Haiti's Black African sovereignty and progress still looms as
a threat to the United States. Anti-Haitianism runs deep in the U.S.
In the 1980s, Haitian bodies were portrayed as vectors of contagion
in association with Africa, as witnessed in the AIDS epidemic and
the naming of the 4H disease. Trump espouses racist notions of

degeneracy that stem from this era. He portrays Haitian life as disposable and Haitian people as carrying the threat of contagion for the nation. It was not until 1862 that the U.S. reluctantly acknowledged Haiti as a liberated nation-state, though it had declared its independence to the rest of the world since 1804.

*

Abstraction of climate rhetoric is the violence that Emerson and others of his time espoused as the American idyll, another justification for colonial expansion—an extension of the belief that some races were naturally more "expensive" than others. As Nell Irvin Painter puts it, Emerson "blithely consigned these 'guano races' to faceless hard work, death, and then service to the greater American good as mere fertilizer." Declaring himself in opposition to transatlantic slavery does not necessarily make him a true abolitionist in my eyes. My research studying the historical ledger of racial indenture shows how many white abolitionists advocated for the kidnapping and indenture of Africans instead of enslavement, yet again, on the same plantations. Forever fated to become guano for the expensive races.

*

Shifting back to the labor context of the United States and racial slavery, we ought to consider the incarcerated activist and writer George Jackson, who understood these layers well.[13] Taking him as a climate thinker begs a spatial understanding of how environmental thought shaped his philosophy. Jackson spoke of the carceral poetics of the African American experience as one of transfigurations. In *Soledad Brother,* he plots himself into the time line of racial capitalism to write about fertilizer. His prison letters describe the process of racialization of Black people through labor, and how human extraction fuels the engine of United States capitalism. Jackson further imagines the transformation of his own captive body, fated to eventually become more fertilizer for Manifest Destiny.

*

"My recall is nearly perfect, time had faded nothing. I recall the very first kidnap. I've lived through the passage, died on the passage, lain in the unmarked, shallow graves of the millions who fertilized the Amerikan soil with their corpses; cotton and corn growing out of my chest, 'unto the third and fourth generation,' the tenth, the hundredth,"[14] writes Jackson. The dead demand a future audience from the soil in which they are buried. George Jackson demanded a future reader when he wrote these words, complete with a Biblical allusion to Exodus ("unto the third and fourth generation"). He is part of the infinite number of generations living in the wake of the transatlantic slave trade. As an institution, racial indenture is a part of that wake of unfreedom. Jackson describes the kidnap as though he were there: he lives the repetition of the Middle Passage for the millions of enslaved Africans who died, a tragedy underscored by the carceral logic of San Quentin, where he was murdered in 1971. More broadly, Jackson's narrative—speaking from the cell that is a tomb—represents the long, captive history of incarceration and surveillance of Black people.

*

Trump's dehumanizing comments were spoken exactly eight years to the day after the Haitian earthquake of 2010, a natural disaster that escalated into a global humanitarian public health crisis. From the earthquake relief arrived United Nations peacekeepers, boots on the ground, vectors of pathogens and sexually transmitted diseases. Cholera further destabilized the impoverished nation, resulting in 820,000 estimated cases. The United Nations exacerbated the crisis by bringing cholera with their peacekeepers (from Nepal) as spreaders negligently releasing sewage. We return to the valuation of life and fecal matter, the refuse and refused. The outbreak has been traced to the bacterial contamination of the Meye Tributary System of the Artibonite River due to human activity.

The management of the crisis became the next crisis, for which Black people paid the price. The white peacekeepers raped Black Haitian girls as young as eleven and abandoned children born from the violence. So many children were born that, according to Sabine Lee of the journal *International Peacekeeping,* Haitians nick-named them petits MINUSTAHS, using the acronym for the U.N. mission.[15] Where is the restitution before the next inevitable earth-quake? Haiti's fault lines converge on the question of crushing debt and shifting tectonic plates. Haiti has been treated like a dumping ground for pollution and left defenseless in crises—climate, eco-nomic, political—for centuries.

The premature death of poor Germans, Irish, and African Americans—those who till the soil—is required as a sacrifice; their decomposing bodies are the substrate that made Westward Expan-sion possible. Another term for Manifest Destiny, the notion of mov-ing West as ordained by God from sea to shining sea required the genocide via displacement of Native tribes during the nineteenth century. It is the fate of the lesser races, said Emerson, to become guano for the advancement of favored Northern or Saxon races. Poor Germans, as he called them, were his prototypical example of people fated to be peasant farmers. U.S. diplomat Benjamin Franklin famously denounced German immigrants too, saying in 1753: "Those who come hither are generally of the most ignorant Stupid Sort of their own Nation."[16] American transcendentalism that Emerson made famous rests on these racialist myths and cribs much of its spirituality from Hindu tenets, again without crediting belief systems of people of color, wherein lies the American fiction of this enlightened school of environmentalist thought.

South Asian cosmology and sacred texts were major influences on Emerson's transcendentalism. An American Orientalist interpre-tation of the religion and its philosophy toward the natural world played a significant role in the development of U.S. ecological phi-losophies and poetics. Emerson writes of the Hindoo following the wheel. By contrast, he writes that the Anglo-Saxon has a hunger for the earth and "love of possessing land." A rhetorical reverence for nature undergirded political thought and statecraft, which served only to disguise violent extermination practices. Without interro-gating the language of environmentalism in the origins of strands of naturalist, racial, and evolutionary thinking, the geological layers that cement racial ideologies will only continue to accumulate and compound.

From geoscientists of color, I have learned about the beauty of ice sheets, volcanic flows, cyanobacteria retrieved from the ocean floor, and the information these geological sources provide about former climate crises. In 2022 I attended a conference for geologists of color; it was the second conference of its sort since 1972. My presentation centered the history of guano from the perspective of rebellion, using my notion of racial sedimentation: the dynamic process of racialization told through overlapping layers that make up the Western Hemisphere.[17]

During my keynote lecture, I played recordings of mountain ballads from around the globe—Appalachia to China to the Blue Mountains. I also showcased how sediment makes up the very human biomatter of history. The conference organizers highlighted three themes: Archive, Urgent, Imaginary. Scripting for the future, the geoscientists created a community report for the year 2072 containing strategies and checkpoints for broadening diversity in the geosciences.

A DIAMOND AS BIG AS HAITI'S SOVEREIGN DEBT

The legal distinction between sovereignty and property is at the heart of nineteenth-century debates and definitions of personhood determined by race, gender, and property ownership. Haiti's calculable sovereign debt is an index of climate injustice. The period of U.S. invasion and occupation of Haiti, described officially as the years from 1915 to 1934, stretches far beyond when we calculate the U.S. sphere of influence over Haiti. The reality is that the U.S. has always had a vexed and exploitative relationship with Haiti. U.S. president Woodrow Wilson led the initial military operation after the assassination of the Haitian president in July of 1915, which echoes contemporary political and climate unrest in the island nation. U.S. presidents have not veiled their ambitions to position naval bases on the island of Haiti and the Dominican Republic (Hispaniola). In 1868, at a pivotal time of Black liberation in the United States, then American president Andrew Johnson suggested the annexation of the island of Hispaniola.

In devaluing and indebting Haiti, the United States and other Western nations proved the power of Haiti's existence. It reminds me of F. Scott Fitzgerald's 1922 novella *A Diamond as Big as the Ritz,* which I teach as a parable of geological storytelling about race, markets, scarcity. Ultimately, Black power is the message of Black freedom, which Fitzgerald shows us was a secret that white plantation owners and industrialists wanted to keep for as long as possible. To subdue their labor forces, emancipation was hidden. A fantastical diamond mountain located in the wilds of Montana serves to destabilize the world economy. Money loses value, and markets collapse. Juneteenth, as it is celebrated as part of Texas history, is yet another example that echoes like the secret of Haiti's freedom.

In Grand Cayman, too, I learned how Black emancipation was kept a secret through the story of a woman named Long Celia. She suffered fifty public lashings for spreading the rumors of Black emancipation in 1820. Celia, an enslaved African, was put on trial, to be decided by a jury of twelve white slaveholding men, for stoking revolution and inciting rebellion on the island. "Slavery be dead, we be free," became the anthem for Black Caymanians celebrating the formal close of a brutal labor regime.

After the Navassa Island riots, Haiti reclaimed the territory from U.S. business interests, but the Haitian officials did not know how to operate the machinery, and the mining of guano ceased. Haiti sold Navassa in 1900. The *New York Times* headline reads, "ISLAND SOLD AT AUCTION; Navassa, in Caribbean Sea, Goes to Only Bidder for $25,000. Transaction to be Investigated for Stockholders of the Company that Owned the Property."[18] In 1913, Congress appropriated the island by decree of U.S. president Woodrow Wilson, who declared, "Island of Navassa in the West Indies be and the same is hereby reserved for lighthouse purposes, such reservation being deemed necessary in the public interests." Wilson mobilized the marines to occupy Haiti beginning in 1915, and the lighthouse on Navassa was completed in 1917.

Through guano, the ongoing contestation for Black freedom and sovereignty extends across the hemisphere. Among the United States Miscellaneous Caribbean Islands and the United States Minor Outlying Islands, Navassa exists as property. The New York Botanical Garden in the Bronx sends researchers to the remote island to observe isolated mosses, lichens, and lizards. The naturalist's mission is a nationalist mission, a cover for military usurpation of local and Indigenous claims to the land. Conservation is an alibi. In a 1998 scientific expedition led by the Center for Marine Conservation in Washington, D.C., the island was described as "a unique preserve of Caribbean biodiversity." Aside from a few extinctions covered below, the island's land and offshore ecosystems have mostly survived the twentieth century.

DX-listeners are a modern-day decentralized community of amateur explorers who seek to set up radio signals in obscure and forbidden locations; the more remote, unsanctioned, and storied, the better. A running list of high-value targets for DX-peditions to set up radio stations is regularly tallied and updated; it includes North Korea's demilitarized zone, Lulu Bay, and Redonda Island (another guano island). Before it was removed in 2015, Navassa— otherwise known as Devil's Island because it is so hard to survive here—had been number two on the list. The CIA World Factbook describes the island, which is primarily made of coral and limestone, as about nine times the size of the National Mall in Washington, D.C. The frame of reference for the scale speaks magnitudes about the U.S. colonial appetite. Do Guantanamo escapees, one hundred miles south of Cuba, wash up on the shores of Devil's Island? What of Haitian refugees? Transient Haitian fishermen and others are known to camp on the island from time to time. An estimated 300 to 400 Haitian artisanal fishers frequent Navassa Island when they are not fishing close to the mainland of Haiti. Western conservationists blame the subsistence Haitian fishermen for exploiting the

reefs that fringe the guano island, a typically misappropriated rationale when the risk factors are geopolitical and structural.

PERU: CHINESE AND INDIGENOUS LABOR, ANDEAN AGRICULTURAL SCIENCE

Shifting now from the guano islands of the Caribbean Sea to Pacific
territories, Peru was another focal point of the guano wars. The
puzzle of soil exhaustion and the regeneration of nutrients has long
confounded and fascinated European thinkers. In the 1860s, Karl
Marx described soil degradation as one of the many perverse processes of capitalism's metabolism, yielding starvation and immis-
eration.[19] Yet while the crisis of replenishing the nutrients of the
soil has historically been a limiting factor to the hyperdrive of Western colonial expansionist desires, Indigenous agricultural traditions
have long centered cycles of restoration and equilibrium. With such
methods of renewal as crop rotation and slash-and-burn agriculture, the soil had time to rest. To these communities, human life
was just one small part of nature, and an overarching goal of harmony dictated their agrarian philosophies of cultivation, sustainability, and variation.

Long before the nineteenth-century guano craze, the prized
substance had meaning to many South Americans, Caribbeans, and
Central Americans. Guano has been considered a sacred material by
many peoples—its uses were not limited to gunpowder and fertilizer, per the Western imagination. Its accumulation points to a deep
and mysterious record of geological time in Jamaica dating to 4,300
years ago. As discussed in chapter 1, in 2021, the undigested material in the bat feces in the Home Away from Home Cave shows the
changing diets of the flying mammals over millennia. The Antillean
ghost-faced bat lives in this ancient home, hitherto undisturbed. In
the fifteenth century, observers noted changes in the carbon composition of the sticky brown bat guano, speculated to be related to
mass sugarcane propagation on the island.

Archaeological evidence shows Andean societies may have been

utilizing guano as a fertilizer as early as 100 CE. The Wari and the
Tiwanaku used guano even before the dawn of the Inca Empire.
They lived in fertile mountain slopes where they developed technologies of the soil. The Wari of Huari of the south-central Andean
and coastal areas are understood as an economic network along the
Pacific Coast.[20] As mountain residents, they were experts in tilling
arid sloping land. Before Francisco Pizarro and the Spanish invasion
of Peru in 1526, Indigenous agrarian traditions of cultivation valued
guano's regenerative properties across terraced slopes. The pathogens that arrived as part of a genocide have erased the scientific
methods that thrived before.

In what may have been the first recorded piece of climate legislation during the fourteenth century, Inca rulers decreed that hunting
guano-producing birds was an offense punishable by death.[21] Not
only did the Inca law protect the birds, but it also took into account
long-term, sustainable agricultural cycles of cultivation.

Although European colonizers discredited them as a justification for usurping the land, Indigenous agrarian methods of the Inca
long preexisted Alexander von Humboldt and his "discovery." The
Prussian geographer and naturalist first studied guano deposits in
Callao on the Peruvian seaside in November of 1802. European and
Western thinkers were mesmerized by the secrets of the Andes,
especially those of agriculture. In contrast to Alexander von Humboldt, we can look to the Marxist writing of Polish Jewish dissident,
anti-war activist Rosa Luxemburg for an environmentalist view-
point that centers South American Amerindians.

Luxemburg turned to Peru for wisdom regarding how societies
are organized. In her writings and teaching on political economy,
her interpretation of Marx looked to South America and the Caribbean for case studies of proto-Communist societies as she saw them.
From her readings on ethnology, she determined that Inca ways of
life and collectivity held promise for future forms of governance in
Europe. While she held these models as examples, she also believed
the failure of these societies was fated. She would eventually teach
a class on what she called the curious science of political economy.

Luxemburg speculated on societal organizations before the arrival of Columbus and connected her hypotheses to the politics of wage labor and debt, using ethnological sources to draw her conclusions. Luxemburg was one of the first women in Europe to earn a doctor of law degree in political economy. In her recasting of the Incas as an age-old agrarian society, she argued that theirs was successful because of a communistic basis. In her critique, "Introduction to Political Economy," Luxemburg glossed South American history to draw communist parallels toward a global vision for the system of governance. She likely began writing the text in 1907 as part of her lecturing duties, pausing in 1912 before being assassinated in 1919. Her critique was published posthumously in 1925 by editor Paul Levi. She writes:

> Indeed, on the distant continent of South America, among the Amerindians, living traces were found of a communism so far-reaching as seemed quite unknown in Europe: there were immense common buildings, where whole clans lived in common quarters with a common burial place. After agrarian communism had been discovered as a peculiarity of the Germanic people, then as something Slavic, Indian, Arab-Kabyle, or ancient Mexican, as the marvel state of the Peruvian Inca and in many more "specific" races of people in all parts of the world, the conclusion was unavoidable that this village communism was not at all a "peculiarity" of a particular race of people or part of the world, but rather the general and typical form of human society at a certain level of cultural development.[22]

Luxemburg emphasized that there was "complete common property under the paternal theocratic government of generous despots."[23] Studying their organization of society, she argued, helped contemporary theorists to be less Eurocentric in their assumptions and to understand what she called the curious science of "political economy" through a more global lens.

Humboldt's name appears across Latin America and is embedded into the scientific nomenclature. The *pajaro niño*, which burrows nests in guano, was renamed the Humboldt penguin. Even as Humboldt becomes the namesake, the role of Spanish Jesuit missionary and naturalist José de Acosta displaces Inca knowledge. Acosta's disease, an altitude sickness, shows how ill-adjusted colonials were to the landscape. In 1570, while traversing the Andes, Acosta deduced the air was too thin to breathe. In 1591, in his *Historia natural y moral de las Indias,* Acosta wrote about these same Indians that they were a "good-natured people who are always ready to prove themselves of service to the Europeans; a people who, in their behavior, show such a touching harmlessness and sincerity, that those not completely stripped of all humanity could not treat them in any other way than with tenderness and love" in the same manner as in state slavery or bondage.

The common racial calculus was an early Americanist ecological one that continues to shape U.S. climate policy. The uneven distribution of the Federal Emergency Management Agency (FEMA) is but one example. In 2005, when Kanye West said, "George Bush doesn't care about Black people" during the aftermath of Hurricane Katrina, the former U.S. president described it as an "all-time low" and "one of the most disgusting moments in [his] presidency."[24] A shocking articulation for the nation and the world, the statement belied the fact that the devaluation of Black life is at the core of national myth-building, from the founding fathers of the United States to our recent presidents.

Crucially missing is the analysis of guano's trade in racial, colonial, and climate historical terms. Of the ships carrying indentured Chinese guano diggers from 1847 to 1874, a recorded 76 mutinied. Others never made it to shore: in 1855, Chinese laborers bound for Peru died of suffocation; and in 1871, 600 Chinese migrants aboard the infamous *Dolores Ugarte,* likewise bound for Peru, died in a catastrophic fire at sea (a tragedy of indenture that Frederick Douglass reported on in 1871[25]). The fire burned off the coast of China, sixty miles south of Hong Kong. One man named Leung Ashew who escaped the ship said that as he left the burning wreck, he saw blood ooze from the sides of the hold where the men were.

Douglass expounded his beliefs on valuation and the institution of racial indenture. The stratification of race in the Americas cannot be detached from the auction block, a scaffolding built on top of the Native burial ground, and it is also the fate of the gallows.

> Human nature is the same now as then. The Coolie Trade is giving us examples of this unchanged character. The rights of a Coolie in California, in Peru, in Jamaica, in Trinidad, and on board the vessels bearing them to these countries, are scarcely more guarded than were those of the Negro slaves brought to our shores a century ago. The sufferings of these people while in transit are almost as heart-rending as any that attended the African slave trade. For the manner of procuring Coolies, for the inhumanity to which they are subjected, and of all that appertains to one feature of this new effort to supply certain parts of the world with cheap labor, we cannot do better than to refer our readers to the quiet and evidently truthful statement in another column of one of the Coolies rescued from the ship *Dolores Ugarte,* on board which ship six hundred Coolies perished by fire, deserted and left to their fate by captain and crew.[26]

The perverse duality of guano restored life and nutrients to the soil, and yet it brought death to those who mined it.

Racial formation is a dynamic, shifting, ongoing process across the Americas. Yet through the exploitative labor of debt peonage and chattel slavery, race became *fixed* in the Western Hemisphere—and these laborers were reduced to the categories of Indigenous, African, and Chinese. It must be said this process of racialization was uneven and deeply asymmetrical in life expectancy. Indigenous and Chinese people sometimes had the right to purchase land after their contracts expired, and therefore had access to white capital and mobility in ways that enslaved Africans did not. Racial crisis meant the exhaustion of the body and the soil.

Though out of sight of the mainland, kindred island labor flows of hundreds of thousands of debt-contracted Chinese laborers connected histories across Cuba, Trinidad, Jamaica, British Guiana, Brazil, Peru, and many other countries throughout the nineteenth century.[27] Chinese migrants traveled across both the Atlantic and the Pacific to perform the dangerous, agonizing tasks others refused to tolerate. Laborers were often forced to work seventeen-hour days by overseers in arduous conditions, living in tents or bamboo shacks. They died prematurely, some by suicide and some by manual labor so severe it physically broke down their bodies. Still others died slow, protracted deaths of lung failure or other respiratory illness. The mountains of guano were stinking gold mines where cheap labor was enforced; free Chinese people were deemed undesirables by law. Chinese families were not.

The newly independent Peruvian government signed lucrative treaties with the British to secure mining rights. They hoped to either expel Chinese workers who had completed their contracts or to reenlist them. These so-called Chinese bachelors were seen as leftovers, a social problem. In an 1877 account, mining engineer and British colonial official A. J. Duffield observed, "The free Chinamen stirred up their enslaved brethren to revolt; explained to them— which was perfectly true—that according to Peruvian law they could not be held in bondage, and if they escaped, they could not be recaptured. Many attempts at escape were made and many murders were the result."[28] Duffield's slippage here between "enslaved" to describe the indentured Chinese population versus "free" in his initial designation speaks volumes about the shades of unfreedom in the Pacific and Atlantic worlds in the nineteenth century. Regardless, neither the Peruvian government nor the British authority wanted to be liable for Chinese laborers as citizens if they managed to survive and complete their contracts.

Indentureship was a vastly different labor institution than chattel slavery, but it is significant that the system often functioned through a repurposing of the same ships and the same barracks. The debt contract was written to protect passengers, but in the legal clauses, we hear the echoes of the Middle Passage and its architecture. Detailed diagrams depicting ship ventilation show how indentured Asian laborers were racialized differently from enslaved Africans. Race was spatially determined by the square footage each indentured person was allotted to breathe. (Even this slim legal protection was not granted or considered for enslaved Africans, as they were deemed chattel.) Long after the Guano Era, the brutal labor regime of racial indenture continued, even into the 1920s in the Caribbean; this transitional period of racialized labor was the last gasp of the sugar economy before it collapsed.

Guano fantasies are ultimately racial fantasies. While Emerson deploys guano as a rhetorical device, Fleming uses it to bolster his plot. Both writers show how race works as a hierarchy and system of division and valuation. If freedom is ultimately a matter of property and ownership, how should we live in relation to the land? Is it even possible to own land? What validity do the deeds to waters, airspace, subterranean strata, and the shoreline provide? The soil of stolen land, guano-strewn or not, was exhausted by stolen lives expended. Across the hemisphere, premature death of racialized peoples was the expense paid for so-called expansionist progress. Manifest Destiny is an ecological mandate for extraction as much as a philosophy that justified genocide in the West.

Guano provides a portal into the eroded history of these mutilated bodies of the enslaved, the indentured, the conscripted, and the indebted. Yet there is so little preserved of the laborers of the Guano Era that I have had to sift for archival fragments. A landscape photograph of guano diggers by U.S. Civil War photographer Alexander Gardner from the series *Rays of Sunlight from South America* in 1865 (*top of next page*) shows the outline of the figures of the indentured Chinese miners. Gardner captioned the work, "Chinamen working guano—Great Heap—Chincha Islands." The racial slur to describe Chinese men that was used casually in the nineteenth century is just one indication of the colonial perspective's dehumanizing vantage. The guano mountains echo with geology as a storyteller.

The heap had already been exploded with dynamite, which unsettled phosphate and formed toxic clouds of guano. This haze was the typical molecular atmosphere of the Chincha Islands during the period of extraction. Peruvian government guards manned the area to prevent laborers from jumping or falling from hazardous heights into the noxious pits of guano. These mines were uniformly located on remote islands, meaning there was little chance of escape for fugitive laborers. A. J. Duffield describes the Chinese

Alexander Gardner, *Rays of Sunlight from South America,* 1865

indentured in Peru: "Escape for the Chinaman is next to impossible; he can only free himself from the horrible condition which he finds himself in by using his braces [suspenders] or his silken scarf for a halter, or the more quiet way of an overdose of opium." Duffield's cruel depiction uses racial slurs to describe the dire labor conditions that drove men to suicide with their own garments. Whether this was true or not, the imagery of Chinese men committing suicide during the period of racial indenture was common. Other writers describe Chinese men resorting to hanging themselves by their queues, the traditional braided hairstyle, from both Chinese and European perspectives. In this simple description, Chinese death and free will are illustrated through the very materials that the West desired and extracted from the East. Given the context of the Opium Wars, this image is even more charged as the backdrop of the guano wars. Each commodity indexes a history of abuse of global proportion. The scale of premature death is difficult to calculate, let alone comprehend or mourn.

The photographs do not show us the stifling heat and brutality of guano-mining, and in fact the title *Rays of Sunlight* romanticizes the weather. It was precisely the lack of rainfall that made the perfect conditions for the best quality of guano. In his 1872 poem "Guano Song," German poet and novelist Joseph Victor von Scheffel describes the climate: "a mountain it rises, and whitened / By rays of a tropical sun."[29] With no mention of Chinese, Black, or Indigenous guano diggers, Scheffel describes the seabirds as the laborers. The birds become the protagonists, though more so as mascots. It was common in nineteenth-century merchant literature to reify the bird as a perfect factory, converting dung to gold. By the late 1870s, nearly all the guano—again, the natural habitat of the guanay cormorant birds—had been depleted, precluding any possibility for renewal off the coast of the Peruvian mainland.

Europeans said that Chinese miners almost ate the birds and their eggs into extinction from the islands. Should starving laborers *not* have eaten the billion-dollar birds? This is the common rhetoric against Chinese immigrants. The casting of blame on laborers for this ecological disruption is perverse, especially without reckoning with the larger environmental rupture of the European- and U.S.-led guano extraction. With the bird populations dwindling, Peru instated legislation to protect the guanay cormorant from hunters—echoing the Inca law from centuries prior—but it was too little, too late. How will twenty-first-century climate legislation be able to protect all of our planet's endangered animals?

"I EXPECT YOU TO DIE": JAMES BOND'S GUANO DESTINY

Ian Fleming, like Emerson, was fascinated by guano's organic, economic, and geopolitical powers. Fleming lamented Jamaica's independence in 1962 because, to him, the island was one of the "blessed corners of the Empire." In his 1958 spy thriller *Dr. No*, Fleming describes Dr. No, a man of German and Chinese heritage, as born out of wedlock, on "the wrong side of the blanket." A guano mogul, his destiny is guano. Fleming's editor, William Plomer, said as much in 1957: "I got so fond of Dr. No, I was quite sorry to see him vanish under a mound of excreta."[30]

As an avid naturalist, Fleming set Dr. No's lair of Crab Key in a layered Jamaican geology made of guano dreams and the accumulation of limestone. In 1943, Dr. No purchases the key from the British for £10,000 with gold he stole during the Tong gang wars of New York in Chinatown, when he was a treasurer for one of the crime syndicates. Dr. No escapes Manhattan's Chinatown to Jamaica via Harlem and Milwaukee. With his acumen for financial forecasting, Dr. No establishes Crab Key as the headquarters for his villainous guano-mining operations. He says to Bond:

> The sole problem is the cost of the labor. It was 1942. The simple Cuban and Jamaican laborer was earning ten shillings a week cutting cane. I tempted a hundred of them over to the island by paying them twelve shillings a week. With guano at fifty dollars a ton I was well placed. But on one condition that the wages remained constant. I ensured by isolating my community from world inflation. Harsh methods have had to be used from time to time, but the result is that my men are content with their wages because they are the highest wages they have ever known.[31]

Predicting inflation during World War II, Dr. No induces a workforce of Afro-Cubans and Afro-Jamaicans to labor in his guano mine. For Fleming, the transformational power or alchemy of guano makes it a compelling plot device. The supervillain Dr. No is a self-made man between two cultures, a chimeric figure who made something from what was apparently nothing.

Fleming constructed *Dr. No*'s triangulated labor structure based on the British plantation economy and even invented the Afro-Chinese "Chigroes" as a buffer race between free and unfree. No one can leave the guano island without Dr. No's permission, as was the case with Navassa, the Chincha Islands, and Hawai'i. So how free are they? (How voluntary was Asian indentured labor in the British West Indies?) The Americas became the destiny for so many who were cast away. Fleming positions the tough, reliable "Chigroes," or his "Chinese negro" henchmen, as overseers to discipline the "simple" Cubans and Jamaicans. They, like Dr. No, are of mixed racial heritage, rejects of the Black world and the Chinese world. They enforce a racialized, divided hierarchy of plantation-style labor on the guano mine, an arrangement echoing the real-life plight of indentured Chinese laborers in nineteenth-century Jamaica.

Fleming's guano dream of Chinese political and economic influence in the Caribbean was startlingly prophetic. Today, Chinese mining companies extract bauxite and other precious minerals from the earth's crust in the Caribbean. Caribbean newspapers are in a constant flurry of suspicion about Chinese currency flowing through the region, as are U.S. and Western European media, recalling the rhetorical tenor of the currents of "Yellow Peril" propagated by Britain and France in the nineteenth century. Yellow Peril describes the fearmongering done to suggest that there was an imminent military threat by hordes of East and Southeast Asian invaders to Western civilization and values. (The English writer Sax Rohmer's evil, criminally genius Dr. Fu Manchu of *The Mystery of Dr. Fu Manchu* [1913]—who epitomizes the subject that causes anti-Asian fear—is a predecessor to Fleming's *Dr. No*.) In 2022, novelist Percival Everett published a book by the name of *Dr. No*, slyly remixing some of the Afro-Asian themes of Fleming's spy thriller. In it, he asks important moral questions by staging the absurd about villainy.[32] Dr. No is a mathematics professor who is an expert in nothing. Where Fleming gave us the anti-Asian, anti-black allegory of Navassa mixed with the criminal element of the Tongs as a global syndicate, Everett asks what today's supervillains would look like.

In truth, suspicious foreign currencies from *across* the globe circulate through offshore Caribbean bank accounts. In *Dr. No*, Crab Key mirrors and foreshadows the contemporary autonomous zones of labor and manufacturing known as the Jamaican Free Zones. In 2001 there were reports that these financial spaces of exemption served to undercut and disenfranchise the local majority-Black Jamaican workforce. While this was true, the migrant Chinese laborers working for subsistence wages in U.S.-subsidized sweatshops did not fare much better in 2000. To this day, multiple forms of racialized labor exploitation echo labor institutions of the nineteenth century, pitting Black and Asian people against one another in the Caribbean.

In 2014, the contested island space just off the coast of Jamaica—the Goat Islands—became a flashpoint of national anxiety around Chinese investment and debts in Jamaica (both governmental and private). With fierce argument on both sides of the Jamaican parliament, the contracting firm China Harbour Engineering Company's proposal to build a shipment port or duty-free international trade depot on the islands was narrowly rejected. Caribbean climate activists have rallied fiercely against what they see as Chinese incursions under which the natural environment will suffer in the name of development. Subsequently the Goat Islands were declared a wildlife sanctuary. The fight against climate capitalism in Latin America and the Caribbean is not over and is as much about future sovereignty as it is about race and imperialism. Once again, race is spatially determined, and even more geopolitically delineated.

HAWAI'I: THE REFUSE AND REFUSAL
OF PACIFIC GUANO ECONOMIES

While most of the guano for global supply was sourced from the Chincha Islands, the United States continually sought other repositories. In a related chapter of Pacific labor history, the colonial practice of blackbirding during the 1860s formed a racialized regime under which thousands of Indigenous peoples from Pacific islands were abducted at gunpoint and conscripted as the Chinese were from their homelands to toil on guano islands. Blackbirding was one of many labor schemes that attempted to reinvent the violent regime of chattel slavery, but held laborers at gunpoint to ditch and drudge the precious manure. The laborers came from Papua New Guinea, Fiji, New Caledonia, and the Solomon Islands and were known as Kanakas or South Sea Islanders.[33] Blackbirded men were forced to labor across the globe, including on cotton and sugar plantations in Queensland, Australia.

Laborers were deemed expendable under the system of indenture, shackled by impossible debt. Degrees of unfree labor across the globe are enforced by systems of dependency and the ongoing exportation of labor. Which nations and which people are worthy of taking on debt? The IMF and World Bank decide. Which communities are deemed creditworthy of being financed for solar loans? These are climate questions. Ultimately, this accelerating financialization of climate solutions imperils the future. Premature death looms as extractivist capitalism continues to mine the remains of the dead. What, after all, are fossil fuels but the accumulated biomatter of the ancestors, organic material transformed? Our fate is sealed as the fossils of the future. Should the dead fuel the future, as Emerson believed was the natural order?

Fertilizer replenishes the earth as part of the cycle of life and death. For Emerson, Fleming, and Jackson, fertilizer served as a reckoning with burial, a way of reifying the predestined fate of certain people. But guano's transformational power was more than the colonizer could have understood. It led to rebellions of racialized people from far corners of the earth who planted themselves in new soil. These African, Chinese, and Indigenous cultivators brought with them new sciences, gods, customs, and traditions that would transform the hemisphere. Even during the U.S. Civil War, confederate soldiers used guano to make gunpowder. Its powers were many. The United States shows its contradictions and the political foundation necessary to solidify the myth. It is important not to collapse the differences between specific histories. Distinct histories of Indigenous, African, Chinese, South Asian, Southeast Asian, and Middle Eastern presences coexist and overlap through labor uprisings and dissent in the Americas. Disintegrated bones are a layer in the strata of the hemisphere, racialized layers of history, peoples, and labor. The logic of some races being more expensive than others continues to shape the uneven consequences and outcomes of the climate crisis on communities of color across the globe.

By the end of the century, the guano mountains that had been the habitats of the guanay cormorant were decimated. The alchemy of Black and Indigenous science has long known there is gold in the restorative power of manure. Embracing the history of shit is the means of regenerating the future. Possibility may just be born here in developing climate strategies for agricultural sovereignty, which requires land back. In guano, the echoes of refusal are loud. It is said, shit is the great equalizer of history. Guano teaches us these lessons, the most important being the category of labor and how it should be defined by having the inalienable right to say no.

6

Colonialism,
the Birder's Companion Guide

Hours before George Floyd was pronounced dead on May 25, 2020, Christian Cooper and Amy Cooper collided in what would become known as the Central Park bird-watching incident.[1] Who was watching whom? Climate and racial crisis converged as, in the video Mr. Cooper filmed of a histrionic and deceitful performance, Amy Cooper cried wolf, calling the police to arrest Mr. Cooper. The video went viral, with over forty million views, and viewers cannot decouple its circulation from the virality of the filmed police murder of George Floyd. Most news reports noted "(no relation)" in their coverage of the two names at stake, clumsily trying to articulate the American conundrum of shared last names between African American and European American families. No coincidence. Cooper versus Cooper lays bare the invisible, bloody branches of so many U.S. family trees. In many families, mine included, there are Black and white relatives who share a last name but live worlds apart because of what race determines socially. Historically, there are shadow families, like the Hemingses, with overlapping lineages that connect Africa and Europe.[2]

Contrasting the two viral filmed confrontations in the beginning of the Covid-19 pandemic: in one a life was stolen, in the other a life was saved by the camera phone as a witness. The technology of an execution, the technology of an exoneration. As the sadistic Minnesota police officer Derek Chauvin kneeled on Floyd's neck for nine minutes and twenty-nine seconds. The simple act of breathing took on new global meaning because of the Black Lives Matter movement. Respiration in the outdoors by Black people seeking sanctuary in nature, as Christian Cooper did on that fateful day, became a radical act. On a cellular level, the question of Black life and premature death was also underscored by the unequal outcomes and statistics of the Covid pandemic. Floyd's autopsy showed he was infected with the virus at the time of his death. While it was not the cause of his death, the uneven consequences of the pandemic are clearly still being determined by race and class. Access to health insurance is stratified along race and class lines in the United States, as are the differences in medical treatment and bedside manner. What U.S. president Donald Trump called a "Chinese virus" coincided with Black Lives Mattering. This is not a coincidence but rather a continuation of a cycle of racial violence, born in America.

Bird-watching as a hobby came into the U.S. national consciousness for people stir-crazy during the pandemic lockdowns and because of restrictions on congregating. In some senses, bird-watching is like bird hunting without capture or kill. More commonly known as birding within its community of enthusiasts, it involves stalking and sometimes luring birds to observe them in their natural environment. Birders are often only interested in the specific birds they are seeking to tick off a list, as if in conquest. Caribbean birding guides have told me that many will not care to know about the surrounding natural or built environment. Even what insects the birds eat or plants they pollinate is often not of much interest to these tourists, though it could lead to finding the prized specimens. They see it as a waste of time when they are paying expensive prices for private bird tours through the rainforest.

Not Christian Cooper. He serves on the national board of the Audubon Society of New York City (which in 2024 was renamed the NYC Bird Alliance to distance itself from John James Audubon, a slave owner). Seeing Amy Cooper's unleashed dog, Christian requested that Amy follow the regulations of the Ramble, which mandate that dogs be leashed by their owners. The fabricated woodland area of Central Park is in many ways sacred and stretches thirty-six acres from the Seventy-Ninth Street transverse to Seventy-Third Street in what is known as the heart of the park. Although Christian Cooper was not harmed by the police when they arrived in the Ramble on May 25, 2020, the violence Amy Cooper incited on him by the state became a specter. The phone call very easily could have resulted in his death.

Christian should have been free to enjoy his pastime of bird-watching in the Ramble undisturbed. Coopers of no blood relation to each other: he was bird-watching, and she was walking her dog. Racial tensions between Black and white were projected onto non-human animals. The escalation of events demonstrated how nature and wildlife are contested political grounds for Black rights and existence.

With his own *National Geographic* TV show, *Extraordinary Birder with Christian Cooper,* debuting in 2023, Christian is now a prominent figure in naturalist circles. He travels the world connecting birding communities with a conservationist ethos. Advocating for climate justice as a part of racial justice, he has become a spokesperson for both movements. However, his visage is shadowed by the potential violence he nearly experienced, and the past layers of violence against Black people and queer people in Central Park and specifically in the Ramble. Cooper directly links bird-watching to the Black experience, recommending—to communities of color especially—a hobby that provides an escape. He has described it as meditative amid the stressors of what it means to be preyed upon as a Black person in the United States. The power of the camera saved him, though often hypervisibility can be a trap for people of color. His TV program is a platform for Black naturalists.

The weight of the evidence of the pre-crime is an aberration. The chilling tone of Amy Cooper's voice conjured Carolyn Bryant, another white woman who made false accusations with damning consequences. The lynching of Emmett Till took place after she claimed to hear the young man whistle after her in 1955. Decades later, Amy Cooper shrieked false accusations, among them: "I'm going to tell them there is an African American man threatening my life."

Bird-watching both endangered and saved Christian's life. Bird-watchers are meticulous documenters, quietly taking photographs of rare birds; perhaps that's why he thought to turn the camera on Amy that day. It is a voyeuristic sport. For Black birders, the question of double consciousness—a concept theorized by W. E. B. Du Bois—is an urgent one.[3] Du Bois described part of the Black American experience as one of watching oneself being watched, so that the veil became a powerful symbol of racial divisions in the United States. Black people have a second sight or sense of being surveilled because it is so prevalent and often a matter of life and death. Did bird-watching put Christian Cooper's life at risk that day? Or did the way it sharpened his instincts perhaps save his life? Powers of observation are core to birding. Cooper had honed his senses from the age of ten, when he began looking skyward for birds. He feels the hobby prepared him in uncanny ways to be how he describes himself: a "gay Black nerd with binoculars" in America.[4]

People of color have long had to subvert colonial practices for their own exploration of nature as hobbyists and scientists. In major and minor ways, coalitions of Black people across the sciences have been interrogating the role of race in the foundation of these academic disciplines. Black squares were quickly posted across institutional and personal accounts on Instagram and other social media platforms. A digital public relations disaster and performance of mourning for George Floyd, these squares soon came to signify the shallow commitments of lip service to diversity, equity, and inclusion across the world.

＊

Watching is an American pastime. Being seen is not.

＊

From the eighteenth-century Lantern Laws of New York City, we know this is true. These regulations required people of African descent and Indigenous people to carry candle lanterns while walking the streets after dark if they were not in the company of a white person. The Fugitive Slave Act of 1850 invested individual white Americans with the power to surveil and apprehend people of color. New York has a reputation for a deadly regime of policing Black people, long before and long after the abolition of racial slavery in the state, which took place in 1827. Heightened by the Covid-19 pandemic and global racial protest, the world watched. Through the camera phone lens of the seventeen-year-old onlooker Darnella Frazier, the world witnessed Floyd being beaten and killed by Minneapolis police officers. On May 25, with his permission, Christian's sister posted his three-minute video to the bird app Twitter. These transmissions would change the course of the Black Lives Matter movement.

By May 29, 2020, on Twitter, Black Birders Week was announced with a simple hashtag, an initiative to call attention to Floyd's murder in addition to those of Ahmaud Arbery and Breonna Taylor in the context of the Central Park bird-watching incident. #BlackBotanistsWeek, #BlackinNeuro, and #BlackinChem also emerged as hashtags. For Arbery, an athlete, running recreationally was a liability. For Taylor, sleeping inside one's locked apartment was considered to be a threat to police. For Christian, birding was a menace. Black scientists and scientists of color felt compelled to be loud. On social media there was a blossoming of Black nerd-dom, speaking out against racism. Black Birders Week was founded as a weeklong series of online events to highlight Black naturalists. It happened to coincide with the spring migration of birds in Central Park as people emerged from lockdown and found new hobbies.

Amy Cooper was the assailant, and Christian's power came from watching her and reminding her that the world would be his witness. In one of the two calls she made to the authorities, Amy lied directly, saying, "There is an African American man. I am in Central Park. He is recording me and threatening myself and my dog. I am being threatened by a man in the Ramble. Please send the cops immediately." There is irony in using the politically correct binomial nomenclature of *African American* to commit a hate crime. She said, he said. Cooper sets the record straight from his perspective in his memoir, *Better Living Through Birding,* expressing that he has always known what it feels like to be watched. Refusing to capitulate to Amy Cooper's acts of racial intimidation, he filmed her instead. Christian Cooper insists that the event was not a traumatic one for him.

Central Park's landscape architecture means that the design of the Ramble is shaded, giving an air of privacy. It is a part of the park haunted by many layers of racial violence and displacement. What is now named the Gate of the Exonerated for the Central Park Five looms. One of the five teenagers wrongfully imprisoned, Yusef Salaam, was elected as a New York City councilman in 2023. With a constructed stream, the Ramble is a pristine resting stop for migrating birds flying southward to the Caribbean and Latin America in springtime. Landscape architect Frederick Law Olmsted's vision of the natural environment is fabricated as an idyll. There is violence here against the land and its communities. Dynamite was used to clear schist; only certain geological formations were left intact as part of the landscaping for the vision of the majestic American park. Neighborhoods such as the historically Black Seneca Village were razed. Its name marks the Indigenous presence in the city in a way that it is simultaneously erased and suspended in the past. The Lenape people were expropriated of their land and have since been pushed to Oklahoma.

The Ramble is not far from here, the former Seneca Village, a space of sanctuary. It has also become a historic space of queer intimacy, known since the early twentieth century as a popular cruising locale. To ramble is to wander for pleasure without a definite route, a freedom not granted to Black people in the park in the nineteenth century. Accordingly, it has been targeted and policed as a space of anti-LGBT violence. In their exchange, it was only Amy Cooper who believed she could afford not to abide by the rules. She was incorrect. Dragging her dog away as she left the scene, Amy Cooper would be fired by her company the next day and would eventually be charged by the Manhattan district attorney for filing a false police report. She did not serve the potential one year of jail time, but was required to enroll in a diversity training as punishment in order for the charges to be dropped. A more appropriate punishment might have been to have her take on the labor of training for and leading the DEI course for other white people who do not see race until moments like these.

Who would believe Christian Cooper, if it was his word against that of a white woman? Most bird-watchers are avid photographers and videographers, using technologies of watching and recording evidence. Visuality and hypervisibility. Those three minutes circulated as if on a loop. Growing up on Long Island with roots in South Carolina, Christian was probably no stranger to understanding the danger of false accusations made by white women. Amy Cooper lowered her surgical mask to phone the police, demanding they come to her rescue in the Ramble.

Christian Cooper's new spotlight as a naturalist has inspired Black naturalists receiving more visibility and representation in other countries such as the U.K. Sudanese British cameraman and TV presenter Hamza Yassin's television program *Hamza: Strictly Birds of Prey* is a first for the bird-watching nation. For a Black man who lives in Scotland to be the face of birdlife in the British Isles is a departure from the ubiquity of figures such as David Attenborough. The intro to the show recounts Yassin's immigration story, how at one point he was living out of his car, though his parents are doctors. "Twitching" is a subsect of birding in which a hobbyist spends large sums to travel far distances to see a bird solely to tick it off a list, and Yassin has become the movement's representative in the U.K. But for how long? The obsessive preoccupation with documenting is in many ways a colonial behavior born from the desire for knowledge production. Worldwide ecological sciences became an unexpected terrain of attrition and a touchpoint in the global social movements sparked during and after 2020. Representation is one answer to the murder of George Floyd. Still, the lingering question for me is about life and death. As touching as the memorials were for the death of the beloved Eurasian eagle-owl Flaco in 2024, the question looms for me, Are birds safer than Black people in Central Park?

INVISIBLE BRANCHES OF THE AMERICAN FAMILY TREE

"It is a melancholy truth that even great men have their poor relations," wrote Charles Dickens. It is the sad truth for many in the Black diaspora that we share last names that lead to plantations. Our poorer relations are white enslavers or the descendants of those human traffickers and profiteers. Through the coincidence of the surname Cooper, the Central Park bird-watching incident unveils the invisible Black branches of the American family tree hidden in plain sight. Shared lineage between African Americans and European Americans is no surprise to Black people like Christian Cooper. It is only a surprise to people like Amy Cooper, who will never care to learn this chapter of American history. Whether of blood or not, the relation of chattel is indelible on the conscience of the Americas.

Black strategies of unnaming such as Malcolm X's are powerful acts of disowning the ugly binomial nomenclature of racial slavery. Often, but not always, our last names lead to the scene of abduction. This is why it is dangerous to invite all ancestors into the room. Science fiction writer Octavia Butler demonstrated this perverse genealogy in her 1977 book *Kindred*. She showed how time travel for Black people to the antebellum period would be violent because of confronting white ancestors who were enslavers. On May 25, 2020, when Cooper confronted Cooper—it was a chilling reminder across the hemisphere of how perverse family reunions can be across the United States, the Caribbean, and Latin America.

In my bloodlines, the myth of regicide runs deep and is speculated upon at every Goffe family reunion. Hanged, drawn, and quartered. The statutory punishment for high treason was established in the Kingdom of England in 1352. On the run, one of my fabled progenitors, patriarch of the Goffe clan, feared this punishment as he left England behind for New England. My surname is the same as that of one of the outlaws, William Goffe. Are we the descendants of one of the fifty-two judges who killed Charles I? On a tour at the Houses of Parliament, I witnessed firsthand the death warrant from 1649 with the signature of my putative ancestor. There are fifty-nine seals on the death warrant. Even if we are of blood relation, I would not claim William. I am not his. It is highly likely that white Goffes were plantation owners, and thus slaveholders, across the Caribbean. The onus should not be on me as a Black woman in 2025 to have to prove my relation to him or his white descendants. I also refuse to carry the shame of what it might mean if it were so. I come from a dissident genealogy of people stripped of their consent.

I love my distant relations who are white and Chinese; but despite our blood connection, a few continue to ask me for proof of our relation. It's tiresome. Of course, they are of African descent, too, though it is not perceptible to most. Yet they ask me to pull out the family tree. "How are we related again?" Yes, the tree is complicated, with inside and outside children. Generations removed. Siblings with the same mother and different fathers. But they can do this research themselves, as I have. It is U.S. history. It is British history. It is Chinese history in the Caribbean. Multiple books have been written on the topic. I am a professor; I am not *their* professor. They often turn to me for answers as if I am their Black ancestor. Race is a social construct that is socially coded, and our family tree proves that point. My Dutch cousins who identify more so as Asian ask whether being Black in America means I am afraid I'll be shot in the street because of what they have seen on the television about the Black Lives Matter movement. These questions may be valid, but they should not be asked of me as though Europe is not entrenched in racism, too. I was born in Europe, and I understand the ways in which race functions differently there than in the United States. When Europeans feign this ignorance, they—wittingly or not—efface the racist roots of that very continent, which presaged the violent founding of our own United States.

The crossroads of Goffe Street
and Dixwell Avenue in New
Haven, Connecticut

White fugitivity took different routes than paths of Black escape
in the early Americas. At Yale University, I learned about these complicated paths by following the trail of Goffe Street in New Haven.
Along with Dixwell and Whalley, the thoroughfare traces where
three men fled during the restoration of the monarchy. Charles II
sought the men who had sentenced his father to death by execution
in 1649. New England, England, and the Caribbean formed an interlocked triangle of commerce—molasses, sugar, enslaved Africans,
rum. Many sought refuge in the steady traffic of ships that moved
often between these ports during the seventeenth and eighteenth
centuries.

Attending Yale, I learned quickly from other students that Goffe
Street is a no-go zone. Students are discouraged from walking there
during the day, and especially at night. Free door-to-door shuttles
are available to prevent Yale students from becoming gunshot
wound statistics. What of the New Haven residents who know they
are not welcome beyond those gates unless it is to labor for a life-

time in the dining hall? Goffe Street is a Black street named for a white man who might have Black descendants. I did not know the rich Black history of Goffe Street then. At number 60 sits the Goffe Street Armory, now known as the New Haven Armory, which was built in the late 1920s. Now in disrepair, it served as a staging ground for the National Guard when, in 1970, they attacked the Black Panther Party in New Haven. Later, it transformed into a hub for the African American community of New Haven and hosted the Black Expo and events for the Black Coalition of Greater New Haven. It was once also a concert venue where performers like Stevie Wonder took the stage.

It is now part of the Connecticut Freedom Trail that honors the Civil Rights history of the state. The abandoned Goffe Street Armory is the target of community-engaged urban renewal initiatives. Is there green potential in such renovations led by Yale University faculty and friends of the Goffe Street Armory? Only time will tell. Why are infrastructures in Black neighborhoods allowed to fall into such disrepair? Who knows the fugitive tale of William Goffe? Who is aware of the Black Goffes who live across New England and the Caribbean? The ruinate architecture of Goffe Street reminds me of the unfinished Caribbean aesthetic, houses left dilapidated, half built, or destroyed by hurricane damage.

When I was pursuing my PhD, I wasn't thinking much about birds. I also was not much of a genealogist. I was laser-focused on finishing my dissertation. Still, I knew there were no coincidences of surnames. Perhaps subconsciously I did want to know where Goffe Street would lead. Which plantation? I left that to the obsessive habits of family genealogists who were looking for the family coat of arms. I had seen a coat of arms while conducting research in the Netherlands that featured depictions of enslaved Africans. Pedigree charts belie a colonial logic of inheritance of property, uneasy for people who were once property. Black people were bequeathed as chattel in wills.

In 2023, when I returned to New Haven years later to give a lecture in the English Department on Emerson and guano, I searched online for the best bird-watching locales and found the Regicides Trail. It was here that William Goffe and his father-in-law, Edward Whalley, hid in what has become known as Judges Cave. They arrived in New Haven in March 1661 as outlaws, and the geologic formation gave them shelter during the Stuart Restoration. Whalley died in 1675 in Hadley, Massachusetts Bay Colony. In the nineteenth century, there was speculation that William Goffe was the Angel of Hadley, though historians dispute this. They assume that Goffe died in 1679, because that's when his last documented letter to his wife was received. However, from 1664 to 1679, there is a mysterious gap in knowledge of his whereabouts. It is during this period that I could imagine him smuggling away to a Caribbean retreat from cold New England.

The fate of the British West Indies hung in the balance during the Restoration. Jamaica and Barbados, like the Thirteen Colonies, were British territories through which there was continuous transit. Since it was Oliver Cromwell who had claimed the territory of Jamaica, would the territory return to Spain? Barbados was the

Judges Cave, New Haven, Connecticut

wealthiest English colony in 1660 due to the transatlantic trade of enslaved Africans and of sugar. Many absconded to Barbados, and Jamaica was founded by Cromwell and his soldiers.

It was a crisp autumn day after I delivered my lecture; the golden and brown leaves were beautiful, blanketing the state park. I took in the surroundings and imagined Goffe and Whalley's flight. Seeing a sign warning of bears, I did not stay long. On the Uber ride back to campus I learned more than I ever could have sitting around the Harkness tables in seminars at Yale. The driver, an older African American man, told me about how he hadn't been to the trail in decades. As a child he had grown up there and spent every day playing with his brother and the neighborhood kids. They fished for bass in the reservoir and swam. He pointed out the toxic dump, where they also used to play and salvage things to eat. He said it wasn't called hiking back then, and there was no parental supervision. It was just being. He had the nickname Nature Boy, perhaps after the Nat King Cole song. The adventures all ended suddenly one summer, the driver told me, when he and his brother heard gunshots in the forest. Their parents would not allow them to return because it was rumored that devil worshippers were camping in the mountains. The driver told me about the urban ecologies of New Haven that he noticed driving at night: coyotes, moose, and fields with hundreds of wild turkeys. I could not have conceived of any of this as a student, because of the de facto color line between New Haven and Yale—Goffe Street was that dividing line. It was also the line that prevented me from exploring New England's natural majesty.

From the car window as we drove back to Yale's campus, I watched as birds of prey circled over the majestic West Rock—turkey vultures and peregrine falcons. Listening to the man's childhood memories, I knew that, even from a distance, this was also bird-watching. Just as the birds were watchful ecological witnesses to the guano wars in the last chapter, the birds here have borne witness to New England history, too. One day I hope I will return to hike to the summit of West Rock and witness the astounding view overlooking New Haven. What do these caves mean to Native nations? Who else carved refuge in these backwoods of Connecticut that we do not read about in the state curriculum? Yale now has an Indigenous land acknowledgment, a post-2020 commemoration of the violence on Black bodies. On an LCD screen at the medical school library where I was doing archival research, I read mention of "Mohegan, Mashantucket Pequot, Schaghticoke, Golden Hill Paugussett, Niantic, and Quinnipiac and other Algonquian speaking peoples."[5] This was not a plaque but a mobile TV screen, easily changed or rolled away. Institutions such as Yale proudly tout that they predate the nation, going back to 1736. As such, they are among the chief perpetrators of expropriating Native lands and dispossessing Native sovereignty, all with a $25 billion USD endowment.

When I graduated from Princeton into the recession and people asked me what was next, I said "Yale." White and Asian classmates who did not have jobs lined up warned me cynically not to get shot. They wanted to remind me that New Haven was associated with gunshot wound victims because it is a Black city. They wanted to diminish my accomplishments. Dining hall worker Corey Menafee reminded the world of what courage looks like when he took a broomstick handle and smashed a stained-glass panel in Yale University's Calhoun College. Menafee noted the window each day when he clocked in to work at the dining hall named for U.S. vice president John C. Calhoun, a South Carolina plantation owner and vehement defender of U.S. slavery. The stained-glass windows depicted Black people carrying sheaves of cotton above their heads; as a graduate student, I was a fellow affiliated with the college but hadn't noticed the tableau. I remember Black administrators arguing that Calhoun would be turning in his grave if he knew Black professors were in charge in his college at Yale. Was this speculation enough? Is representation ever enough, or is it merely an empty act of diversity hiring into high-ranking positions, stripped of the power to make structural change? In 2016, Menafee didn't mince words; he described the glass tableau as "very racist" and degrading, noting how the figures appeared to be smiling. Corey had worked at Yale since 2007 when he broke the window. He transformed the university more than I have seen any of Yale's deans do. Menafee's words echo plainly: "No employee should be subject to coming to work and seeing slave portraits on a daily basis."[6]

THE AMAZON: BIRDING AS COLONIAL PRACTICE

There are many genres of watching. By nature birding is a voyeuristic pursuit where pleasure is derived by watching. Binoculars might be one of the major distinctions between European colonial forms of bird-watching and Black, Indigenous, and Asian practices of observing birds. Birds belong to no nation, and yet nations claim them, even when they migrate. They are emblems, from the bald eagle to the Valkyrie, because nationalist myths are so often projected onto birds. Majestic and free birds represent something elusive, something watchful. If we understand them as the successors of dinosaurs that evolved to rule the skies, what does it mean to desire to control them? Birds are a missing link to understanding the former climate change and mass extinction of dinosaurs.

On a trip to Suriname, a country located between the Orinoco and Amazon Rivers, the meaning of birding opened my senses in unanticipated ways. I traveled to Paramaribo in 2018 for a conference on the legacy of racial slavery and indenture, and while there I arranged to meet cousins of mine for the first time. I had no knowledge of them for most of my life, nor did most of my immediate family. We did not know much of the former Dutch Caribbean, English-speaking colony in South America. But a long-lost cousin we had discovered a few years prior in Canada told us that we had many cousins in Paramaribo. When we met for the first time, we were astounded at our shared resemblance. I guess we should not have been surprised, as these were my mother's first cousins. One of them said, "It's amazing how you look like you have lived in Suriname your whole life, but you have never been here." Still, the fact that I am Black puzzled some of them. The clinical nature of Dutch colonialism translated perfectly. At a dinner party, one of my Surinamese cousins who is deaf said out loud what the others were thinking: "Why are they Black?"

These cousins are my grandfather's sister's children, and they are of African and Chinese heritage too. My grandfather, who grew up in Hong Kong, had an older sister born to a different mother. Also born in Jamaica and taken to Hong Kong, she made the Black Pacific journey in the 1920s as a two-year-old. A picture bride who did not look like the other Chinese brides who arrived from the boat in Paramaribo's harbor, she was betrothed by what the Dutch call marrying by the glove, *trouwen met de handschoen,* to a Chinese man in Suriname. Her husband was a stranger to her, but they built a life together and had many children—my mother's cousins, whom we were now meeting for the first time in 2018. Some of my cousins in Paramaribo are also of Amerindian, Javanese, and European heritage. They had questions about Jamaica and about China. Both seemed so far away to them, distant cousins. The iPhone feature that automatically detects and categorizes faces into albums alerted me that one of my Surinamese cousins in his twenties who I had just met looked to be a match to photos I had stored on my phone of my grandfather when he was in his twenties. Perhaps Apple has a secret algorithm for determining race, which is after all a sorting tool. But all I can say for certain is that Surinamese definitions of race differ from American and British understandings in ways that I will never want to fully understand. In the Netherlands I heard Dutch people proclaim that their forms of enslaving Africans were not as bad as the British's. These lies show how all systems of race are contorted, ugly, and ultimately lead to the eugenicist logic of genocide.

A Surinamese researcher I met while on a research fellowship in Amsterdam the following year explained to me the Dutch taxonomy of race and how it manifests in South America. It was a science, he said, that depended on the combination of three factors: the width of one's nose, the darkness of one's skin, and the texture of one's hair. People were very open about discussing these physical markers of difference. Much like the descriptive logic of bird-watching as a possessive colonial practice, I recognized the colonizer's impulse to categorize and collect hybrids and chimeras. To classify different and distinct types by taking notes and pictures. This is the metadata for cataloging the dead and taxidermized specimens of empire.

What a gift it was to meet Sean Dilrosun, the husband of one of my cousins, because by luck it turned out he is an expert wildlife guide and avid bird-watcher. From him, I learned how to be immersed in my natural surroundings, how to tune in to the frequencies of the trees, even in the city. Sean explained the new efforts taking place to introduce wildlife education for Surinamese children into the primary school curriculum. The majority of Suriname's population are people of color. Twenty-seven percent are Hindustani, 21.7 percent are Maroon, 15.7 percent are Afro-Surinamese Creole, 14 percent are Javanese, 13.4 percent are of mixed racial heritage, 3.8 percent are Amerindian, 1.5 percent are Chinese, 0.3 percent are white, and 2.5 percent are designated as other.[7] As he explained in his TEDxParamaribo Talk, Sean came to birding later in life, at the age of forty. So even though these are the country's demographics, he had encountered only white bird-watchers. As a person of color, it had special significance for Sean to take up this profession for his livelihood, learning about his native country on his own terms. On the way to the forest, he introduced me to fellow birding guide Fred Pansa, who is a Maroon. Most bird-watchers who arrived in the South American country, he said, were white tourists paying exorbitant amounts of money to be led straight from the airport runway to the country's interior in the Amazon.

In online reviews and reports, Dutch birders describe and evaluate the birding guides alongside the birds, sometimes with photos of Sean and others. One described Fred as a "young enthusiastic Sarramacan Bushnegro." Another noted Fred's incredible eyes and ears. In the thick of the rainforest where no planes can land in Suriname, the ecotourists expect to be led through the jungle to the home of one of the prized and rare birds, the Guianan cock-of-the-rock. The male of the species is a stout neon orange bird with a half-moon crest. The females are of a browner hue, with less vibrant colors—as is typical—to camouflage with their young in the shrubbery, away from predators. In 1766 Carl Linnaeus described this bird as being the type set of the genus *Rupicola*.

The more Sean explained to me about the Latin names of birds, the more I understood how racial taxonomies derived from the very same type of science. The biggest goal seems to be to discover a new species and name it. What more colonial action? Much of Linnaeus's work has been disproven, yet he is still exalted and we still use his faulty system. The notion of a "species" itself has also been disproven. It is acknowledged that his system falls apart with what we now understand of genomics. Animal and plant taxonomy have never been an exact science either. Among taxonomists, there are endless debates about what constitutes a new species. Taxonomists themselves fall into two general categories: lumpers and splitters, derived from what Darwin described as "hair-splitters."[8] Sean explained to me that lumpers tend to embrace difference as being less important than signature similarities, while splitters tend to emphasize precise definitions and designations in an effort to assign new subspecies. This is the logic of racial thinking, too. When new categories are agreed and voted upon, the original species becomes denoted by a repetition, or what is known as a tautonym. The genus and the species are designated as the same. Eugenics is a revenant that seeps its way back into contemporary culture even as much as people claim to disavow its racist consequences.

The logic of eugenics valorizes blood quantum and racial fractions (half, quarter, eighth, sixteenth). The arithmetic never quite adds up when you figure in the omissions of what is not worth mentioning. Which half is assumed, and which does not need naming? This language is casually embedded in the notion of pedigree and breeding when it comes to human life. The logic used to describe animal breeding is slippery and often extends to the language used for humans by the colonial governing authorities. In bird-watching, this language is apparent. The repetition of the tautonym intersects with the use of language to designate authenticity, often of race, ethnicity, and nationality. It is common to hear colloquialisms such as *Chinese* Chinese or *Indian* Indian for emphasis in diasporic communities, whereas a modification of the genus designates the diasporic subject as Chinese Jamaican. The reproductive politics of "breeding" cut differently for differently racialized people. Unlike Chinese women, enslaved women who were African were referred to as "breeding wenches."[9] Chinese women were viewed as protected. I see the racial mathematics in people's eyes as they try to deduce my blood quantum. I will not give them an answer, and I do not wish to have one. The answer would tell me more than I want to know about the specific violence of the plantation. Genetic testing companies like 23andMe further engender eugenicist ways of thinking and acting. Some use the data reports for selective breeding of racialized traits. It is a system of self-surveillance and DNA watching of which Charles Davenport would have been proud. Island experiments of breeding followed these scientific logics of animal husbandry and plant cross-pollination.

Though I did not have the chance on the 2018 research trip to see the magnificent orange cock-of-the-rock, I met my spark bird: the bird that first sparks one's interest in birding. For me it will always be the crimson-hooded manakin. Sean showed me the brightly colored male bird in his scope. With ginger caps, the tiny birds are described as small and chunky. I quickly pulled out my camera phone to record a video through the lens as instructed by Sean. He beckoned the bird closer, playing the sounds of mating calls from the Bluetooth speaker around his neck. I noticed that he was using

A Guianan cock-of-the-rock

the Cornell Lab of Ornithology app, noting that the coloniality of bird-watching was based in Ithaca, New York, the cold place I had just left.

When birders in the Caribbean learn that I was a Cornell professor, it is as if I'm royalty. At the Suriname conference, when a white Dutch man who was a professor learned that I was a professor at Cornell, he got up from the lunch table in disbelief that I had a tenure-track job at an Ivy League university. The idea that I was on the career path to have a university job from which I could not be fired was appalling to him. For him, knowledge production, so closely contained and gatekept, wasn't meant for people who look like me. Tenure does not exist in the Dutch or European higher education systems, but this was not his problem with me. He skipped the next day of the conference to go on an expedition to the interior in search of the cock-of-the-rock. It was hard not to feel like an exotic specimen followed around by white men at that conference. My spark bird was not revealed to me in Ithaca at the Lab of O, as it's called at Cornell, but in the Caribbean, where my ancestors probably knew entirely different names for these birds based on their melodious chirps, chitters, and twitters.

A crimson-hooded manakin

One day I intend to learn the names of the birds in Sranan Tongo. While I don't speak the language, its intonations feel kindred with Jamaican rhythms of speech such that I feel I know what they are talking about. Birding, I learned, is best when it is not only about birds but other contextual information on the animals and plants you did not go in search of. Sean and I looked upward into the canopy and saw what he told me was a female sloth peering down at us. She moved as slow as a snail, her delicate choreography beautifully intentional, preserving energy.

The joy of birding in Suriname's Amazon rainforest led me to other islands. Since then, I have had the chance to go to the Asa Wright Nature Centre in Trinidad and bird-watch with local guides, and with Black and Indigenous experts in the bird sanctuaries of Montserrat, Aruba, and Dominica.

On my 2022 trip to Dominica, I learned about European conservationists who arrived to rescue endangered parrots in the aftermath of Hurricane Maria in Dominica. The two most famous endemic birds of that island: the imperial amazon (*Amazona imperialis*) or sisserou parrot, and the jaco, or the red-necked amazon. The sisserou is named for the Indigenous warrior. Jaco is named for the African-born Maroon chief who escaped the Beaubois Estate. A racial hierarchy between the birds is delineated in terms of which is rarer and more valued. Government officials in Dominica have facilitated and allowed the rare parrots to be caged and taken off the island to countries like Germany. They may also have harvested eggs. On the illegal exotic animals and bird market, endangered pets such as these are prized and highly coveted. We may never know which of these Caribbean birds have been bred with their progeny living in temporary captivity for safekeeping in Europe. With every storm these birds are in peril, not only from the powerful winds but also from European and American conservationists who see an opportunity to steal these rare parrots.

While I hiked to the slopes of the Morne Diablotin National Park to wait for the sisserou parrots to glide across the treetops, I did not see the famed birds. Two of them adorn the Dominica crest and they are most often seen at dawn amid the early clouds. But I was not disappointed—what I had learned along the way was how valuable it is to take in the entire vista and to listen to the echoes of rebellion in the deep past. I had seen a family of jaco parrots and learned that they live for seventy-five to eighty years and mate for life. The Jaco Flats are the site of a Maroon settlement from the 1700s where 135 steps were constructed, carved into the rock by escaped Africans who had been enslaved under British rule.[10] The encampment was "discovered" in an attack by the governor of the island, George Robert Ainslie, in 1814, leading to the murder of Chief Jaco and his people. What atrocities and rebellions have Caribbean birds witnessed?

Returning to Charles Darwin's education in birds, the taxonomy of the birds of South America was taught to him by John Edmonstone, a Black man. Under the tutelage of white British naturalist Charles Waterton, Edmonstone studied taxidermy in his native Guyana. One source explains, "Waterton had a distinct preservation strategy that he passed on to Edmonstone, to soak his specimens in a sublimate of mercury and hollow them out so they appeared more lifelike. During these expeditions, Edmonstone also gained an expansive working knowledge of the flora and fauna of South America. Including swallows, water ouzels, and chaffinches—as well as one 15-foot-long boa constrictor." Edmonstone also taught Darwin techniques of bird preservation learned from Amerindians in Guyana. Who was teaching whom? Darwin valued his teacher's lessons but his thoughts about race were not veiled. In his memoir, he writes that a Negro could not learn as well as a mulatto.[11] How distanced was Darwin's thinking from dangerous racial thinking? Social Darwinism, premised on the survival of the fittest applied to human life, developed as based on eugenics. Yet we must question the heart of these assumptions, scientific methods, and techniques. Black and Indigenous science is discredited and relied upon as we look at Darwin's intellectual genealogy and influences.

Waterton was the son-in-law of Edmonstone's slaver. In a letter to his sister, Darwin explained, "I am going to learn to stuff birds, from a blackamoor I believe an old servant of Dr Duncan: it has the recommendation of cheapness, if it has nothing else."[12] Using the common slur to describe people of African descent, Darwin justifies the fact of learning from a Black man by discussing the "cheapness" of his tutelage as a rationale. Returning to the Edmonstone estate in 1820, Waterton noted how nature had reclaimed the grounds. Just three years after John Edmonstone left, we can hear the decay of the Great House. In Waterton's description of British empire's glorious infrastructure, I read unintended hope of nature's return:

> The house had been abandoned for some years. On arriving at the hill, the remembrance of scenes long past and gone naturally broke in upon the mind. All was changed: the house was in ruins and gradually sinking under the influence of the sun and rain; the roof had nearly fallen in; and the room, where once governors and generals had caroused, was now dismantled and tenanted by the vampire [bat].

Waterton reminisced on the valiant slavecatchers and the brutal regime of ecological extraction. Here the vampire bat becomes a hybrid witness again, inhabiting a house that knew the horrors of the whip. It remains a mystery as to why John Edmonstone was brought to Scotland by his master. Is it possible that Charles Edmonstone was his father? It is documented that Charles and his wife, Princess Minda, argued over property and ownership of enslaved Africans in Guyana. Scottish bureaucrat and soldier Thomas Staunton St. Clair describes enslaved Black women as always being naked at the estate, and the atmosphere it created.

John spoke to his pupil Darwin fondly of his childhood in the rainforest, describing the vibrantly colored feathers of tropical birds he knew intimately. He may have inspired the budding naturalist's future travels. Native modes of bird-watching and preservation in the Amazon produced a tradition of science that in turn influenced Darwin's pedagogy. This sense of imagination and bird-watching took him to his famous breakthrough with differentiating finches on the Galapagos. Darwin left the cold island for different islands of

Mbiri Creek, British Guiana

discovery and devised a grand unifying theory that would shift the world forever.

St. Clair's accounts are no different. The Scottish perspective is both that of a colonized island and one embedded across empire to do the bidding of the Crown through the management of insurgent populations. From his *A Residence in the West Indies and America, with a Narrative of the Expedition to the Island of Walcheren,* published in 1834, we can get a sense of John Edmonstone's home. In doing so, we can garner what it meant to be a Black, formerly enslaved naturalist in the nineteenth-century British Caribbean. His book's chapters document this journey: "Return to Demerara"; "Remarks on Vegetable Fecundity"; "Intended Insurrection of the Negroes"; "Hook and Bait for Alligators"; "Reported Approach of the Insurgents"; "Apprehension, Trial, and Execution of the Ringleaders"; "Practice of having Naked Female Domestics accounted for"; "Shooting and skinning a Serpent"; "Mr. Edmonstone's Settlement"; and "Beautiful Birds." To ignore the regime of racial slavery in this moment is to ignore the volume of Black insurrections and mutinies across the colonies. Edmonstone was proud of crushing Black rebellion using Indigenous knowledge of the land.

There is an eco-commentary in the 1834 sketch of the moment of emancipation, shown in the artist's depiction of the estate. The tree stumps represent the economy of colonial extraction. Apart from palm trees, the area has been deforested. The birds' natural habitat was decimated. The Black people in the foreground are barely clothed and carry axes. These could be potential weapons of rebellion, but instead they are the tools of ecological extraction here. In the background, people row a boat on the Mbiri Creek, a tributary of the Demerara River. This is sugar territory, and deforestation preceded such cultivation. Today the Amazon remains in peril due to practices of razing land for monocrop agriculture. In this image, Black parents and children are depicted as the laborers. The absence of whites is also a comment, as if overseers are not necessary to keep the Black people working.

The tall, slender white birds in the lower right corner are snowy egrets (*Egretta thula*), which we can tell from their crested heads. The heron-like birds have yellow feet, which they use to retrieve food from the mud of the river. Did John Edmonstone learn from these birds? Did he tell young Darwin stories of taxidermizing snowy egrets? Perhaps he told stories of other birds endemic to the Mbiri region, like the great kiskadee, the wattled jacana, the native toucan, the macaw, or the neotropic cormorant. Tales of these Guyanese birds would have excited young Darwin's imagination. Fences are also depicted in the drawing, showing the division of the surveyed and expropriated land. The artificial hand is at work on both the image and the land itself.

FOUR AND TWENTY BLACKBIRDS: KERRY JAMES MARSHALL'S PECKING ORDER

Following the Central Park bird-watching incident of 2020, painter Kerry James Marshall presented his series *Black and Part Black Birds in America*. Inspired by his longtime penchant for birding, Marshall confronts the American taxonomy of racial classification through the hobby. Displayed online for David Zwirner gallery during the early part of the Covid pandemic, the series is described as a "revision of the role of the Black figure in the Western art-historical canon."

Nonhuman animals become figured as a statement for Marshall, who said of the ongoing series that it is about the "pecking order." A colloquialism for human hierarchies, the pecking order for birds is literally observed among hens to establish dominance. As a basic pattern of social organization, each bird pecks another lower in the rung without fear of retaliation. It determines the order in which chickens access food, water, and dust-bathing within a coop. Marshall was deliberate in his response to Black Lives Matter through this ecological lens of painting and figuration as an avid bird-watcher. With his brushstrokes, he critiqued the system of valuation by nature hobbyists and the brutal formula of classification and pedigree that extend to race as a system of group-differentiation.

Marshall is known for his delicate use of chromatic black pigment to depict Black life, and he began the bird series as an engagement with John James Audubon's classic *Birds of America*. He has studied the canon intimately to understand the rules of exclusion. Who is an insider in the art world and who is an outsider? This question extends from Michelangelo to the present for him. Born in Birmingham, Alabama, in 1955, Marshall has a sense of the pecking order that is attuned to how the meaning of race has evolved in American life. No strangers to the white supremacist violence of the American South, his family moved to Indiana just months before the Birmingham bombing took place. The question of blackness and exclusion extends to John James Audubon and his genre-defining artistic production of birds.

Marshall's signature use of black paint reflects the invisibility of what is a silhouette of the marginalized, but not peripheral to American identity. The near-total black monochromes he has painted since the 1980s evolve with birdlife as the subject, outside of heteronormative whiteness and formal and social politics of representation for Black people and artists. His work gestures to the very unstable foundations of the United States and highlights the role of naturalism in national racial anxiety. Audubon's 435 watercolors, painted in 1827, have circulated as a veritable genre of their own. Marshall highlights the so-called human stain of miscegenation by highlighting a spectrum of blackness. As a commentary on the Cooper versus Cooper bird-watching incident, his paintings are a striking departure from Marshall's typical depictions of human figures.

As a bird-watcher himself, Marshall felt compelled to participate in Black Birders Week. In an interview with the *New Yorker,* he shared an anecdote about capturing a juvenile crow with his bare hands, sneaking up behind it. He tied one of the bird's legs to a milk crate and filmed it before feeding it and releasing it the next day. His method differs staggeringly from Audubon's capture-and-kill approach. Marshall recalled, "I'd always had a fantasy about a crow that was my friend, and would come to my call."[13] The types of black paint he uses as his signature style may at first may appear to be one color, but they are made up of many different shades of black pigment as a commentary on the depth of the color black and the absorption of light. Through what appear to be silhouettes—but on closer examination are fully dimensional figures—the politics of opacity shine through. Birds portend American omens in Marshall's work, as he tackles the American ideas of partial blackness, racial ambiguity, and the one-drop rule. Was Audubon's mother a Black woman? An enslaved Haitian woman? His figuration challenges the Linnaean binomial way of collecting; instead, taxonomies are reified by what appears in paint.

In March of 2023, the National Audubon Society voted to keep Audubon's name for the global society of amateur naturalists. The Bird Union and the Chicago Bird Alliance have changed their names, not wanting to be associated with Audubon. As of June 6, 2024, the New York chapter of the National Audubon Society renamed itself to the NYC Bird Alliance. They stated they wanted to be more "welcoming" and "to better reflect its values and mission of promoting bird conservation and habitat protection to New Yorkers and others of all backgrounds."[14] Audubon had been just an amateur, not much interested in preserving nature, but in taxidermizing it for profit. He was something of a John Crow with a morbid practice that made his style of depicting birds iconic. They appeared so lifelike because he kept them pinned in poses before killing them. He fictionalized ecological details and took liberties, and with this

knowledge, it is difficult to call him a nature lover. Though his compendium means a lot to bird lovers, they must question his cruel
treatment not only of nonhuman animals but also of people. Like
the John Crow, Audubon was a scavenger. He raided the graves of
hundreds of Native people, participating in the trade of skulls and
other human remains. Audubon's violent legacy as a slaver, grave
robber, and profiteer of the dead has not been enough for the society named for him to make a crucial change.

The American Ornithological Society, on the other hand, has
committed to replacing all bird names derived from people. The
decision was made to address birds such as Scott's oriole, named for
the confederate general Winfield Scott, who led the Trail of Tears
against Native peoples in the Southeast. The society states that this
form of naming can be harmful and exclusive and can detract from
the focus on bird appreciation. The Bird Names for Birds petition
and movement precipitated permanent changes. One of the found-
ers was inspired by the Central Park incident. The Entomological
Society followed suit with their Better Common Names Project.
Christian Cooper weighed in to say the descriptive names would be
better in helping birders visually identify the birds. But what of the
birds named after Indigenous and Maroon chiefs as heroes in the
Caribbean?

The circulation of the elephant folios shows how colonial knowledge traveled. In the ʻIolani Palace in Honolulu, a prized copy of
Birds of the Americas can be found. The volume became a veritable
genre, and it is arguably more valuable as an art object than as a
scientific one. Ornithologists like James Bond—a white man from
Pennsylvania—continued the tradition, writing a classic among birds
and Ian Fleming aficionados alike: the 1936 *Birds of the West Indies*.
The 460-page book was the first of its kind to make an argument for
the Caribbean as a cohesive region defined by bird migration. What
if nations were defined this way, by avian flight patterns? Bond
noted the distinction of the Guianan shield of South America and
the birds there, which are more connected to the Amazon. Accordingly, he also included Trinidad, Suriname, Guyana, and French
Guiana in this zone. In 1934, this became known as the Bond Line of

designation. Many followed in Bond's tradition. Fleming chose his name for the English Cold War hero because he thought it sounded natural: a strong English name for an anachronistic Saxon champion of a dying empire. An ornithologist is a good cover for a spy.

In the film *Die Another Day,* Bond (played by Pierce Brosnan) holds *Birds of the West Indies* and tells Jinx (Halle Berry) that he is an "ornithologist—only here for the birds." Bond met Ian Fleming at the famous estate GoldenEye once and confessed to the author that he had never read the books. Fleming, who was an amateur birdwatcher, must have delighted in an ornithologist's company on the estate he built in 1946 on Jamaica's north coast.

GoldenEye sits on the edge of a cliff overlooking a private beach, on a fifteen-acre parcel of land in Oracabessa that has a special significance for my family in Jamaica. Goffe land was spread all along the north coast of the island, where bananas were cultivated. One of my great-great-grandfathers, Alfred Constantine Goffe, owned this land, a region of the island that would later transform into a popular luxury tourism destination with Ocho Rios as its central hub. It was rare at the time for a dark-skinned Black man like my ancestor to have so much access to landownership and social power in Jamaica. His father, John Beecham Goffe, also of African heritage, had been born enslaved. Some speculate that he may have migrated to Jamaica from Barbados. Again, the question of disinheritance is a fascinating one that is necessarily ecological in the Caribbean because it is about land and property for people who were once sold as if they were property. Are the beaches the inheritance of Black Jamaicans? Tragically, no.

On my first trip to Jamaica as an adult with family, we were turned away from GoldenEye after having driven all day. While it had been more open to the public in the 1980s, today it is exorbitantly expensive to rent the lagoon villas and cottages. The namesake Fleming Villa has five bedrooms and can be rented for $15,270 USD per night from December 20 to January 3 in the high season. It is part of the luxury tourism on the North Coast that has stolen the land from people who were, themselves, stolen. Since GoldenEye today is too rich for my blood, I will be left to imagine the biodiversity of birdlife on the estate, which is probably kept in pristine ecological condition to attract high-paying tourists.

It has since come to light that Audubon was a fraud, which has led to an interrogation of his legacy—to say nothing of his speculated African lineage. Hence Marshall's usage of "part black birds" in the titles of his paintings. What is a part black bird? Why are black birds maligned while white birds like doves and swans are exalted? Ebony G. Patterson is another Black artist similarly engaged with race, nature, and Audubon's legacy. She reinvented Audubon's black birds in what has been described as the first immersive installation by an artist in residence at the New York Botanical Garden. In the patrician north of New York City, the contrasts are stark: colonial-era Dutch place-names exist in the predominantly Black and Latinx working-class borough of the Bronx. Her 2023 show entitled ". . . things come to thrive . . . in the shedding . . . in the molting . . ." activated the grounds. Patterson knew this and drew the perverse lineage to highlight ongoing colonial histories, all through the majesty of the vulture.

Patterson shows the John Crow history of Audubon in her multifaceted exhibition, which extends from the conservatory to the library. Audubon died in 1851 and never witnessed the Civil War. He died just at the height of the antebellum period, as Emerson was to deliver his speech on guano destiny. In the conservatory of the New York Botanical Garden and across the grounds, Patterson placed 400 fabricated black life-size vulture statues. Clusters of the glittering bird figurines populated the botanical garden and parts of the exterior in a haunting, macabre assemblage. Their presence evokes something dying or dead, so that the black birds almost form a funeral procession. They act as an omen for mass extinctions caused by the climate crisis. Patterson fabricated the birds as witnesses to the ecological crimes of colonialism, ecocide, and genocide. Black birds are demonized for their color, and they bear witness both to Marshall and to Patterson by subverting Audubon's text.

To make this work as an Afro-Caribbean artist is to conjure the folklore and proverbs of the John Crow in Patwa parlance. As Patterson casts them, the birds might be saviors. Their presence is unclear and ambiguous, but they thrive in the remains of what has been lost. She also fabricated phantom-like plants from translucent white plastic branches, which extended out from the living shrubbery across the gardens. Meant to signify a ghostly glow for the extinct plant species that will be lost in the twenty-first century, Patterson's work laments botanical disappearance. She has also expressed interest in what it means to leave something to die beautifully. Part of the exhibition features a fountain filled with a dark red liquid, perhaps gesturing to blood pooling or dark polluted waters. Two trousered human legs, feet in sneakers made of glass, are exposed, signifying the extinct human species. The vulture does not seem to pay the human body any mind. The birds are benign rather than sinister, exploring bushes with their beaks. Patterson thus rescripts birding as a verb by which the birds will survive us. She wrenches the tradition of art depicting birdlife from Audubon's cold hands and tacitly challenges naturalist colonial knowledge production and the very logic of the botanic garden as a colonial tool of containment that leads to these extinctions.

Birds are an ongoing theme in Patterson's work; in 2012, she used feathered creatures to depict intimacy among men who are

Ebony Patterson, New York Botanical Garden

lovers and fatherhood beyond biological relation. In the 2012 three-channel film installation, *The Observation: The Bush Cockerel Project, A Fictitious Historical Narrative,* two male figures wear feathered headdresses, one in black and one in white, while tending to a baby. The nineteen-minute video is a statement against the homophobic and transphobic violence that is normalized in Patterson's home country of Jamaica. Deep within the forests of the Caribbean, Patterson shows that there are refuges in the wilderness for queer intimacies, nurturing, and intimacy between men rearing children. The installation tackles masculinity and witness through the two birdlike fathers by subverting the heteronormativity of the Eden narrative.

WHITE FLIGHT, BLACK MOBILITY, WHITE FUGITIVITY

"They say when trouble comes close ranks, and so the white people did. But we were not in their ranks." Jean Rhys opens *Wide Sargasso Sea* with these words to encapsulate white flight in the Caribbean after Black emancipation in the British West Indies.[15] Refugees from the island we now know as Haiti arrived on the shores of Norfolk, Virginia, in droves in 1793. An influx of escaped white people, delivered by the French navy and comprising plantation owners and overseers, flooded the infant nation of the United States. These war criminals settled in Baltimore, New York, and Philadelphia and were able to adopt the mobility that white skin is given at the border. Taken as a global phenomenon, white flight is a useful frame for grappling with the fugitivity of plantation societies. For Audubon, as a man from what is now Haiti, the proximity to blackness threatens to contaminate his legacy. The identity of his mother is unknown, and historians believe she may have been an enslaved Black woman. The order of *partus sequitur ventrem* determines that if this were the case and if he was born of the womb of an enslaved person, he would have been enslaved.

The national animal of Haiti is a bird: the Hispaniolan trogon. It must be attuned to the centuries of deforestation that Haiti has suffered. Birds flock to safe places where food and cover are plentiful. Political boundaries follow the distributions of resources and racist immigration policies. As of 2022, a Dominican wall bisects the island that has been cleaved since the colonial period. President of the Dominican Republic Luis Abinader has suspended visas for Haitians. The trogon is an endangered bird, and it knows the island's original name: Ayiti. This is no more a border dispute than Russia and Ukraine's battle—or the Rhineland's, or Gaza's—is one about territory.

According to a Gallup poll in 1958, 80 percent of whites said they would leave their neighborhood if Black people moved in in large numbers. What correlations and contours of environmental racism accompany these data? The information shows the anxiety of whites and whiteness on the run because race is a shifting category in different nations. Because race is a social construct, whiteness has always been unstable. Migrating to Queens from London as a child, I experienced the aftermath of white flight. The generation before me did not admit just how severe the racial tension and climate had been in Rosedale. I did not remember this feeling of being at the scene of a former crime until in 2021, when I joined a virtual Cornell faculty writing group organized by a sociologist friend. He invited me to collaborate with his colleagues in the social sciences, all of whom were assistant professors glad to be given $600 each by the university. The payment seemed sufficient to motivate us to write and share drafts each week over Zoom, cameras on, microphones off.

As I am not a social scientist, I was the outlier in the faculty writing group. I was reminded that my island story is also a New York story in the data points they recited. Each week I learned that the data sets and ethnographies they were studying were somehow or other about Black people like me. Was I supposed to ignore this, and the racist assumptions being made about who is the researcher and who is the subject of research? As I gave them advice on revisions in Track Changes, should I have been happy about representation as an abstract data point? One white colleague wrote about Black immigrants who arrive on the shores of Italy and Greece, managed as a crisis of the undocumented. As an Afro-pean, and as a Black immigrant without the proper papers—though I migrated from the United Kingdom to the United States—this had partly been my American story. I was born in Europe and am of European descent, but I will never be European. I am a British citizen. The next week, a white colleague shared an article for peer-review submission on his analysis of a suburban white flight data set. He described inner-city youth as "risks" and "threats" to white suburban children. I grew up in what I now understand are the wetlands near JFK airport: for many, the first point of entry to the island metropolis. Queens remains an afterthought, but it will always be my first port of call.[16]

It dawned on me that I was the risk factor only when another island case study came to light in the context of BLM. The 1975 documentary *Rosedale: The Way It Is* by Bill Moyers depicted the grounds of social experimentation on Long Island. The social unrest was very much a consequence of Long Island shoring itself up as a white sanctuary of flight away from the inner city. As a student at the local Catholic school, did I know I was being protected from New York City public schools? Most of our teachers had Italian or Irish names. Keep the jobs, hate the children. One of the only Black teachers, my seventh-grade science teacher, Ms. Birchwood, whispered to me one day that my talents were being wasted in this school and it was time to get out.

I didn't know then about the book written by a professor at the University of Louisville, *Left Behind in Rosedale: Race Relations and the Collapse of Community Institutions.* Sociologist Scott Cummings documented the case study, which no doubt influenced urban policy and my own future. Urban policy is environmental policy. It has been found that impoverished and low-income neighborhoods are lacking in the biodiversity of birds. The ghetto of South Jamaica, once residential and almost all white, saw its crime rate skyrocket in the '70s, but the jury is still out on who was truly to blame—Black people or white. Return Our American Rights (ROAR), the home referral service of white buyers, fought to ensure a white future. They harassed white people, whom they called blockbusting families, for selling property to Black families. This was the Queens of the '70s. Growing up there in the early 2000s, it was still palpable to me that something had vacated Laurelton and Rosedale.

Migrating to the United States, I remained in the racial category of the problem minority, where Black people are often placed in the United Kingdom, too. Black children in the U.K. are stigmatized as the products of irresponsible reproducers, children responsible for knife crimes. In the U.S. the era and the myth of the crack baby predominated. "These children who are the most expensive babies ever born in America are going to overwhelm every social service delivery system they come in contact with throughout the rest of their lives," said Rep. George Miller of the Select Committee on Children. The government pathologized these Black children as a problem to be solved by truant prevention.

It was all a myth; none of it was true. Living in the flight path of JFK airport, I understood what the air and noise pollution of environmental racism meant before I could articulate the nuisance. I was part of the remains of an island experiment of white flight in New York City.

Had my mother felt the shrapnel of that phrase "Niggers Go to Hell" and its sting up close or from a distance, growing up in New York in the 1970s? The move from Crown Heights to Rosedale was a step up for her, but it was also a step directly into the line of fire in which Black middle-class families became target practice for white supremacists who were resentful because they were poor. Bill Moyers's 1975 documentary begins with a group of white men saying, "We're going to keep this neighborhood. We're going to keep Rosedale predominantly white the way it is. We're gonna fight it. This is our right too. You know they talk about minorities. Where the hell are rights for whites?" The far southeast pocket of Queens had been a refuge for those moving out of the Bronx. In the 1970s, working-class whites called it the last frontier in New York City. In a seven-bedroom home owned by an Afro-Caribbean family from London, a pipe bomb was placed with a note: "Nigger be warned. Viva Boston. KKK. We will get you. We have time. We will get your firstborn." I have met four Black professors who, like me, all happen to have grown up in Rosedale, Laurelton, or Springfield Gardens.

When I tell them the story, they are astonished. They did not know
how bad the vitriol had been, but they remember the smell of the
swamp. One of these professors happens to be a software engineer
at Dark Lab, and somehow we did not discover this commonality
in our biographies until long after we first began working together.
She is originally from Trinidad, and we both understood intimately
how all these island stories—of the Caribbean, New York, Britain—
are deeply geologically interconnected through colonialism.

I did not recognize the atmosphere of white rage because the
neighborhood is now predominantly Black. But I did acknowledge
the perverse bucolic logic of urban naming in New York City because
I had seen it growing up in the gray cement towers of South London. Springfield Gardens and Rosedale sounded to me like Cotton
Gardens, the flats where I grew up. It was built as part of a utopic
postwar vision between 1964 and 1968 by architect George Finch
through government subsidies to replace slum housing. Perhaps
they were aspirational projects, or an ode to the former ecologies
before the concrete jungle. Black people are confined to living in
these ghetto pastoral plots, council estates of brutalist architecture.
How does it feel to be a problem to be housed, solved? The entire
globe is riddled with the aesthetic of the unfinished infrastructure
of postwar city planning and racial segregation endemic of many
Black neighborhoods. Reliably, Black people find ways to turn these
architectures into buildings that dance and rooms that nourish.

JAMAICENESIS

Bird-watching may not be one of the most popular pastimes in
Jamaica, yet birds are spoken about in common parlance all the
time. To be called a John Crow or Jancro is one of the greatest
insults in the island nation. Someone who is the lowest form of
human life. The name is associated with ugliness, blackness, and
disgrace. Without redeeming qualities. The bird is a character that
looms large in the national imagination. The origin of this phrase
brings us to a peculiarity of colonial meaning of birds and naming.
The John Crow Mountains have carried the name for the peaks since

the 1820s, where the vultures glide the thermal currents, powerful columns of hot air. These birds of prey are native to the Americas. As an ode, Jamaican singer Derrick Morgan wrote the ska dance song the "John Crow Skank." Though some think of buzzards as synonymous with vultures, they are smaller. They will eat the dead but will often hunt and attack as well.

In Trinidad, the bird is known as the cobo or the corbeau, derived from the French surname meaning raven and "dark-haired." The three birds of prey are *Coragyps atratus, Sarcoramphus papa,* and *Cathartes aura,* and are known as the kings of the skies. Jamaicans often see them as ominous because they are scavengers. Phrases like "Ay, boy, like corbeau pee on yuh," or "You have corbeau luck or wha," denote bad luck and misfortune. The symbol of the bird is part of the Trinidad national gambling culture and a Chinese lottery game called Play Whe. Naturalists are pushing the public to rethink the reputation and role of the birds as important to the cycle of life and the removal of pathogens from the food chain. By breaking down carcasses, vultures remove these harmful microbes from the environment. The acidic pH of their digestive systems breaks down bacteria. Their bald heads, it is speculated, help to keep pathogens clear from their eyes because they do not have feathers here.

Traveling in Trinidad in 2018, I noted the pastime of keeping songbirds. Down de Islands (or DDI), I encountered local men walking with birdcages in hand full of brightly colored parakeets, finches, and canaries in Trinidad's Bocas del Dragon, or Dragons' Mouth. Birds compete against other birds in the practice known as "speed-singing" in Trinidad, Suriname, and Brazil. The cages in contests are placed on poles a foot apart; the first finch to reach fifty songs wins. It probably bears relation to the birding tradition in the Netherlands and Belgium, in a practice known as *vinkensport* or *vinkenzetting* where some finches can sing hundreds of birdcalls in a contest. Birds are selectively bred for the best singing capacity and aptitude for memory. To add to European approaches to birds, Victorians had a penchant for stealing the eggs of exotic birds, eating them into extinction. Ospreys all but disappeared from the United Kingdom due to these habits.

I first saw the *Buteo jamaicensis*, or red-tailed hawk, on Governors
Island during an Audubon Society bird-watching tour. The hawk—
called Chester—has white plumes and a puffy coating of feathers
on his chest. In his Linnaean name, *jamaicensis,* I felt him as distant
kin, especially when I saw a map of the Caribbean in his flight pattern. The hawk's locality type is Jamaica. The bird was first named
Falco jamaicensis by German naturalist Johann Friedrich Gmelin in
1788. What was its Arawak name before then? I will always belong
to Xaymaca, though the island will never belong to me. I will never
be Jamaican because I am not from there. I accept that. I have seen
these sentiments echoed by the Indigenous person and Ngarrind-
jeri writer Clyde Rigney, who says, "We don't think that the land
belongs to us. We think we belong to it."[17] Yet in my peregrinations,
I return whenever I can, drawn to the confounding island.

On the tour in New York, the Audubon guide joked that certain birds and dragonflies are migrants who like to summer in the
Caribbean. We learned about how the migratory patterns have
drastically shifted, due to diminishing factors of air, water, light, and
noise pollution.

These birds of prey have grown quite accustomed to the sky-
scrapers of Manhattan. Like tall cliffs, the concrete towers make a
perch for peregrine falcons. The Upper West Side, where I live now,
provides some of the best locations in Manhattan for birding. I saw
the *Buteo jamaicensis* in the high branches of a tree at Riverside Park.
Sitting near Edgar Allan Poe's rock on West Eighty-Third Street, I
took the sighting as an omen—a good one, I hoped. Ralph Waldo
Emerson wrote of "birds with auguries on their wings." What are
the signs on these birds' wings? Are they the harbingers of the climate crisis?

BIRDING FUTURES

In a newsreel made by *British Pathé* in 1960, brown-footed boobies
teem across the Morant Cays—offshore islands, off the southeast
coast of Jamaica. The breeze blows, and the birds hover as Afro-
Jamaican men delicately gather booby eggs to sell in the Kingston

Market. Then men look wistfully at the camera holding the tiny eggs. In the Morant Cays, I hear the reef islets and the vibrations of the history of the Morant Bay Rebellion, loud to any Jamaican. On October 11, 1865, preacher Paul Bogle led hundreds of Black people in protest against the injustices of the colonial society. They reached a courthouse that they burned to the ground as dissent spread across the parish of St. Thomas. Bogle, a national hero, was executed, although this was the uprising that could have brought Black Jamaicans their sovereignty. I hear the lesser-known rebellion of nineteenth-century Chinese debt laborers in these echoes as well in that parish and on the estate at Duckenfield Hall. It was the first recorded strike by paid Chinese laborers, who were successful in negotiating a six-day workweek.

From the rebellious echoes of the guano islands of Haiti, Hawai'i, and Peru in chapter 5, brown-footed boobies bear witness to the consequences of global commerce. Depositing guano to islands where they go to lay their eggs, the birds watched the strange, coordinated behavior of the guano diggers. While Germans sang guano songs from afar, worshipping the birds, Native peoples had different rituals of respect in relation to the feathered animals. Birds were an important part of the economy and nourish-

British Pathé—Morant Cay, 1960

ment for islands. Instead of naming the birds after themselves, they listened to what the birds called one another. These are their names. From these travels I have learned how to call for birds in their language, and by listening to them in the canopy I can sense the atmospheric changes when a storm is coming in the tropics.

In contrast to the white colonial tradition of bird-watching, there are numerous worlds of ways to coexist with avian life. There are Black, Indigenous, Pacific, Middle Eastern, and Asian approaches to observing that are more holistic about tuning in. Can we imagine studying birds without the invasive and militaristic gaze of binoculars? Without clipping birds' wings? By listening to the chatter of birds and the flutter of their flight patterns, we might understand something about adapting to the changing climate.

Birds watch humans as much as we watch them. Something of a Du Boisian double consciousness exists in how the laborers watch themselves being watched by the British cameramen, an instinct of the hunted. How can we describe the immensity of gathering of this many birds? Despite the islands being protected by the Morant and Pedro Cays Act of 1907, the sooty terns and brown noddies no longer inhabit these islands to this degree. Booby eggs, which used to be a plentiful delicacy in Jamaica, are no longer sold on the streets of Kingston. It was likely an offshore island like Morant Cay that inspired Ian Fleming's birding imagination when he wrote *Dr. No*, casting a bird-sanctuary island as a supervillain's secret lair.

By some grammatical quirk in English, a group of ravens is called an unkindness, a group of crows is a murder, a group of geese is a gaggle, and a group of hawks makes a kettle. In aggregate, birds demonstrate their collective power, which has long mystified humans jealous of their power of flight. What to call these rebellious formations of avian life? In the majesty of a murmuration, there is perhaps the model of a swarm choreography of collective climate action we need for the future.[18]

BOOK III

Life in the Garden after Eden

7

✳

The Curious Case of the Calcutta Mongoose in Jamaica

> If you read the old books of natural history, you will find they say that when the mongoose fights the snake and happens to get bitten, he runs off and eats some herb that cures him. That is not true.[1]
>
> —RUDYARD KIPLING, "Rikki-Tikki-Tavi," *The Jungle Book*

> I at once wrote to the Government of Jamaica, asking permission to obtain some Mungooses from India by the Coolie ships; but difficulties were placed in the way at first, and it was only after repeated and urgent solicitation on my part that Sir J. Grant gave the necessary orders to the emigration agent at Calcutta. In 1872, on the 13th February, by the East-Indian ship *Merchantman*, I received 9 of these animals, 4 males and 5 females, one large with young. I paid for them £9, "in reimbursement of cost attending the procuring and transmitting."[2]
>
> —W. B. ESPEUT, *Proceedings of the Zoological Society of London*, 1882

The Calcutta mongoose (*Herpestes auropunctatus*) was heavy with young when she was taken from India to Jamaica. She was not a stowaway; she was among nine intentionally imported by

W. B. Espeut, a white Jamaican plantation owner. She was part of
the transatlantic biological and economic experiments that formed
a desperate attempt to save what had been the multimillion-pound
sugar industry. In Jamaica, Espeut believed, the mongoose (endemic
to India) would eat the rats (endemic to Europe) that were eating
the sugarcane (endemic to India). The experiment, like many other
colonial machinations in the Caribbean, failed, but more on that
later. The traffic between Calcutta and Kingston was frequent, as
both were port cities within the British Empire because of other
British experiments. (The Trinidad Experiment came in 1804, followed by the Great Experiment in Mauritius in 1838.) The year was
1872 when Espeut arranged for the passage of four male and five
female mongooses aboard a "Coolie ship" called the *Merchantman*.[3]
Following this case, animal breeding became a central part of the
economy of experimentation during racial slavery and racial indenture, and it led to biodiversity crises across the Caribbean islands.
The mongooses, a nonnative predator, were imported to eat, not
to be eaten. The mongoose embodied a centuries-old character in
the popular imagination in India, which was translated to the Caribbean and United States.

Not only was the curious case of the Calcutta mongoose documented as one of the most devastating ecological acts to upset the
balance of a global island ecosystem, but the mongoose's introduction also traces the brutal human history of Asian indenture in the
Caribbean. One of the prototypical case studies of "invasive species," her tale illuminates the unscientific logics of the plantation
economy, and how we are still paying the price for these fallacies
today. In the same boats, human and animal labor experiments were
transported across the swift Atlantic currents from British India to
Jamaica. "Indeed, biologists regard the deliberate introduction of
the mongoose to the Caribbean as one of the greatest environmental disasters of its kind," notes scholar Louis Kirk McAuley,
"as it is responsible for the endangerment or extermination of
more species of mammals, birds and reptiles within a limited area
than any other animal deliberately introduced by man anywhere
in the world."[4] Who is to blame for this particular ecological crisis?

Espeut's grand mistake, like Columbus's, cascaded with many negative knock-on effects. It stands as a well-recorded case study—one example among many such unsanctioned acts—of plantation owners introducing "invasive species" into island ecosystems during the nineteenth century and prior. With each plantation owner fancying himself a scientist, a gentleman farmer, and a god over his estate, each plantation across the Americas was a laboratory of terror. The ecological consequences of these actions can scarcely be calculated. Understanding more about Espeut's indeterminate ethnic origins and backstory lends insight into how he viewed his role in shaping colonial policy and reordering the ecologies of the Caribbean plantation.

Social engineering for a desired outcome was second nature to white creoles, or white people born in the colonies. Bioengineering and gender played a significant role in plantation life. Espeut's idea of enlisting the mongoose was actually his wife's. Before arriving in Jamaica, she had grown up on an island too: Ceylon (now Sri Lanka), where she kept a pet mongoose. Espeut and his wife were part of the circuitry of white creoles across the British empire. Together they speculated that the mongoose would prey on the black rats that threatened the Jamaican sugarcane. The Espeuts were incorrect, and the Caribbean paid a mighty price. Even when animals are considered as significant actors in history, they are imagined as male. Due in part to Charles Darwin's theory of sexual selection (since disproven) and Victorian presumptions about women being passive, female animals are often missing from Western natural histories and imaginaries, tending to be marginalized, ignored, and misunderstood scientifically.[5] The original imported mongoose's progeny continue to propagate across the Caribbean islands today, a scourge to many.

In ways (though never explicitly stated in legislature), the importation of the Calcutta mongoose extended the logic and bidding of the Trinidad Experiment of 1804, which the Parliamentary Papers discussed in a secret memorandum. The experiment of racial labor substitution was recommended as one in which Chinese men would replace enslaved African laboring men and women on

the vast sugarcane plantations of Trinidad. Later, white plantation owners wanted to contract Chinese women as another part
of this colonial plan. It was believed that their reproductive potential would propagate a minority buffer race of middlemen Asians,
people between Black and white. White statesmen and plantation
owners postulated that the races would not mix or intermarry. Chinese women never arrived in numbers proportionate to men, much
to the disappointment of white creole planters like Espeut. They
wanted to see this reproductive experiment take place, but others
were invested in fantasies of racial mixing. Some planters believed
the races would mix and create a new race of African Asians better
suited to plantation labor than the African or the Asian. In his 1862
post-emancipation treatise on "coloured men" in the British West
Indies, British novelist and civil servant Anthony Trollope prophe-
sied the mixing of the blood of Asia and Africa in the formation of a
new subject. He wrote, "They will amalgamate if brought together,
all history teaches us."[6] The British colonial authority transported
human cargo and substituted labor, one race for another, bound by
debt. Animal and plant cargo traveled to be propagated, too. Colonials used divide-and-conquer tactics of biopower to discipline and
racialize through the machinery of nineteenth-century plantations
and mines.

The mongoose, itself a transplant from Asia, was racialized in
the British colonial imagination too. With origins in South Asia,
Indonesia, and Africa, the mongoose has often been mistaken for
a rodent, when it is actually a cousin of the lemur and the meer
kat. The word "mongoose" derives from the southern Indian languages *mungisa* in Telegu and *mungisi* in Kannada. The earliest
European name for the animal is recorded in Portuguese usage in
1685: *mangús*.

She carried a litter that would find homes on hundreds of Jamaican plantations, forever transforming the balance of the island
ecosystems of other parts of the British West Indies (and eventually even as far away as Hawai'i). Scientific literature describes
the mongoose as an opportunistic feeder, an omnivore—it eats
snakes, rodents, birds, turtle and bird eggs, and even crabs if they

are available—which has made it mythic. Hers is a legendary island story, from the tropical darkness of the ship to the sterile laboratory. Mongooses are diurnal, and thus they thrived because they never met the nocturnal rats. The rats thus continued to feast away at the raw cane in fields haunted by generations of torture under enslavement. The curse of monocrop agriculture had already poisoned the soil, depleting it of nutrients.

As one of the few white families who remained in Jamaica post-emancipation, the Espeuts grappled with the economic shift that the freeing of Black people across the British West Indies precipitated from 1834 until 1838. Much of the British Caribbean had absentee plantation owners, in any case, with white estate managers arriving from other colonial territories like Scotland, Ireland, and Wales. Plantation owners had been compensated for losses but would never regain the wealth extracted with a labor cost of zero. The order of the human auction block meant plantation owners paid other plantation owners so that the wealth circulated among the plantocracy through strategic marriages.

Espeut focused on the rats gnawing at the sweet and sharp roots of the sugarcane. And since he was already sending ships to Asia to transport indentured Chinese and Indian laborers to his estate, Buff Bay, in Jamaica, the mongoose was an easy investment. He was especially keen on the introduction of Chinese labor, which he demonstrated in a speech delivered to the colonial authority: "If Jamaica could get 20,000 to 30,000 Chinese labourers at work, I am satisfied, their labour in five years, would add materially to our production. Their presence would be an incalculable advantage, if for nothing else than the example of untiring industry, perseverance, and thrift, they would afford to our careless, indolent, and thriftless people."[7] He was an active proponent, not an absentee planter, as was the case with many white plantation owners across the West Indies. Espeut despised being ruled from afar, and knew he needed to participate in London politics to give voice to the remaining small white ruling class of Jamaica post-emancipation.

The furtive mongoose led to mass extinction events across flora and fauna, including the ground-nesting birds of Jamaica, which

Portrait of the mongoose
propagator W. B. Espeut

were decimated. Sometimes called "alien species" or "introduced predators," invasive species do not belong. Headlines about Asian carp in Mississippi, for instance, are often colored with xenophobic vernacular gesturing to long histories of anti-Asian sentiment about the influx of immigrants. Debates on migration and refugee crises run parallel to the politics and the fantasies projected onto introduced species. This jingoist rhetoric opens debates and philosophies of belonging and immigration but says more about those who invent the terminologies. The colonies of mongooses ate the eggs that were within easy reach as they made homes in underground burrows across the island, disrupted ecosystems, and created new natural orders of balance.

Yet how can we vilify the Calcutta mongoose simply for surviving? The travels of the mongoose only emphasize the myopic managerial class of white creoles dutifully serving in their role as administrators of empire, exiled from the seat of power in the metropolis. Espeut was a leader among the white businessmen and the scientific community of the British colony's plantocracy. He jockeyed in parliament for the destiny of Jamaica's economic progress as a modernizing colony in a globalizing world and was most interested in the sugar production of his plantations. Espeut had the hubris to believe that the fate of the colony was in his hands as an amateur scientist and entrepreneur. In looking for traces of the pregnant mongoose, I had to follow him. I found records in the Linnean Society, where he was a fellow, commanding an audience with the scientific community headquartered at the Royal Academy in London.

I became a member of the Linnean Society in 2020. Espeut's presence was noted in colonial debates on new industries and infrastructures. He expressed how he was obliged to any fellow of the brotherhood of the Linnean Society. To become a member then, and today, one needs recommendations from three existing members, at which point a vote takes place. He published scientific articles for the Royal Society, though he was just an amateur and nature hobbyist. Interest and friends in high places were the requirements to contribute to science as an "expert" in the nineteenth century. In Espeut's application he is described as "a gentleman attached to the study of Natural History, especially of tropical Botany." A shoo-in at the old boys' club, one of his recommenders was Kew director William Turner Thiselton-Dyer, who oversaw the Royal Botanic Gardens. Dyer performed his own Caribbean–South Asian biopolitical botanical experiments, introducing cacao from Trinidad to plantations in Sri Lanka. Kew Gardens holds Espeut's correspondence with the chief botanist in its archives.

Again, scientific racism is embedded in the naming practices of eradication. There are important lessons to be garnered in Espeut's folly regarding the ethics of animal control or human population control. The circuits of colonial scientific knowledge and agricultural science led to a network of agricultural and life science colleges. The University of the West Indies, St. Augustine, was founded in 1921 as the Imperial College of Tropical Agriculture to encourage the perfecting of these plantation experiments and bolster profits. Many white Caribbean environmentalists carry on in this tradition, including Espeut's own descendants in Jamaica. W. B. Espeut's great-great-grandson, Catholic deacon Peter Espeut, works in the wildlife conservancy sector today. The daily *Gleaner* often publishes Peter Espeut's outspoken and controversial writing; his letters to the editor frequently espouse conservative, racist, and homophobic views.

Over a century later, the Espeut family still represents the ruling class on the island and busies itself with managing the reproductive economy of Jamaica, clinging to regimes of the past. Despite his obvious and harmful biases, Espeut has been a columnist for twenty-five years, opining mainly on sociological and environmental issues. Of late, though, the deacon has made headlines for declaring abortion as murder. Espeut challenged the idea that people who can give birth have a right to choose, stating, "The Universal Declaration of Human Rights does not mention such rights, and the Jamaican Constitution does not confirm that right. I would say no, they do not have a right to abortion, safe or unsafe. Abortion is against the law. It is murder!"[8] He has also been vocal about what he describes as the "gay agenda," and asserts that Jamaica should not put aside its traditional values with "gay abandon." Peter Espeut the younger founded the Caribbean Coastal Area Management Foundation in 1997 to strengthen the island's fisheries, providing videos on fishery management and promoting environmental tourism.

Climate advocacy has become a patrician hobby in which environmentalists operate under the guise of scientific inquiry. Climate saviorism takes place across the Caribbean with paternalistic powers that maintain circuits of social governance through the performance of benevolence, philanthropy, and moral authority. These latter generations exist in a convenient, protected elite, able to critique society, its government, and the Black masses for environmental mismanagement. Much as with white Southerners in the United States, the wounds of the plantation past in the Caribbean are gaping. Populations in these regions seek reparations, as they are sought in the United States. Vigorous debates often point to the precedent that white plantation owners were paid reparations or compensation for forfeited property, enslaved Africans who were freed. Often white Caribbean descendants of slaveholders live on the same estates they inherited centuries before. In other instances, they migrated to other islands to assume a surrogate power that white skin commands across plantocracies. The power, political, economic, or otherwise, has not been redistributed, and the compensation they received from the British government for their slaves is still accumulating interest. University College London hosts an online database of the compensation granted to British slave owners, making a limited amount of this information available and decipherable to the public. The paper trail has been useful evidence that has become a finding tool for Caribbean genealogy, both of white people and Black people.

THE VEIL OF IGNORANCE: A COLLECTIVE CHOREOGRAPHY OF MOTHERING

What of the small Indian mongoose who transformed the ecosystems and microbial worlds of all those who encountered her? New studies show that when living in colonies, which can consist of as many as fifty mongooses, pregnant females all give birth on the same night in dens underground.[9] As a species, they form breeding societies. How did the lone pregnant mongoose adapt aboard the ship from India on the three-month transatlantic voyage to Jamaica? Work is usually distributed evenly in mongoose colonies, with shifts that alternate between watching for predators and hunting for food. Collective births are likely determined by the pheromone signaling of the dominant female of the colony. Mongooses can potentially give birth to up to three litters a year from the age of ten months, and they can live between six and ten years when not in captivity. Coordinated birthing protects the colony and blurs the biological ties of specific mothers to certain pups, known by the scientific term "veil of ignorance."

There is only one night when the colony is vulnerable, when the females collectively focus on birthing. The young are collectively breastfed for a month underground before growing strong enough to venture outside. By the time they are ready to hunt for themselves, their microbiome has developed significantly through a collection of shared milk antigens. Since so much of these birthing rituals take place underground, it has been difficult for scientists to observe them. Studies reported in the journals *Science* and *Nature*

Communications suggest that this choreography of birthing and subsequent mothering leads to even distribution of food resources.[10]
The mongoose pups most in need are breastfed regardless of parentage, and instead of favoring their own young, the group priori-
tizes feeding all the newcomers and decreases the risk of predation.
The role of the males is to defend the mongoose packs. W. B. Espeut
was, of course, ignorant of all of this when he ordered the transport of the animals to Jamaica.

But what did the pregnant mongoose witness before she was
captured? How did she and her pups manage to thrive, hopping
from island to island across the Caribbean and beyond? Infanticide
is common among mongooses that deviate from these patterns.
The perpetrators tend to be jealous mothers-to-be who have not
yet given birth (even if they are pregnant themselves). If mongoose
mothers in the colony happen to give birth at separate times, the
alpha females kill the pups of the other mothers. The moon is a factor here, as the rituals take place at night. Dominant females dictate
the nursery by eliminating competition for their offspring. Over-
all, more baby mongooses survive this way, though some are culled
when their mothers deviate from the mating and birth patterns.
It's ironic that governmental animal control solutions are ultimately
ones of culling. The mongooses survive *because* they have perfected
culling themselves to survive more effectively as a collective, rather
than as individuals. From this original experiment, the mongooses
spread from Jamaica to Anguilla, Hispaniola, Puerto Rico, the Virgin Islands, Trinidad and Tobago, and Cuba.

The "veil of ignorance": biological kinship ultimately means less
than the strength of the commune. After nesting and feeding, the
pups emerge from the den and become attached to adults, again
usually unrelated, who are called escorts. Also known as mongoose
mentors, the male adults show the young how to guard and hunt
over a two-month span. Juveniles follow the adult mongooses' every
move, tagging along for protection and pedagogy. Studies of the
mongoose mating and birthing patterns have yielded new knowledge about the synchronicity of rearing young for other mammal

species. There are lessons here in collective care, nurturing, and communal nourishment when we contend with climate precarity.

The true "veil of ignorance" was the colonial condition of bodily disavowal of Indian women, the dual intimacy and disdain, the desire, and the biopolitics of future nourishment. The intimacy and microbiome of the teat is a central part of the domestic inner workings of colonial biopower. British physicians advised white mothers not to breastfeed because the tropical climate was considered debilitating. The choreography of mothering was perverse and entangled on plantations, too. Mrs. Espeut was no stranger to the dynamics of surrogate mothering, having been raised in Sri Lanka before her marriage brought her to Jamaica. A perverse surrogacy between white mothers and South Asian mothers took place for centuries. She was probably nourished by the breast milk of Sri Lankan women. Tragically, Sri Lankan newborns were denied their mothers' colostrum, or the nutrient-rich first form of milk released by the mammary gland after giving birth, because their mothers were wet nurses for European infants instead. These were typically lactating low-caste Hindu and Muslim women, whom the colonizers called *ammahs*—a Hindi term that is more often used to refer to elderly respected mother figures, not mothers. White infants born in British India latched onto the nipple, while Indian infants could not latch at all. The surrogacy of suckling is a painful one for the *ayahs*, or nurses, and bearers who were also part of the intimate process of child-rearing by Indian people in the British colonial domestic sphere. Indian women were treated—and milked—almost as cattle, afforded less respect the more they nurtured white children in this surrogate arrangement of nourishment.

Indentured Indian women on ships were so malnourished that they could not feed their newborns. I have read records and rationales about why milk was not provided in the rations from Calcutta to Trinidad. The babies died of malnutrition aboard ships with only water to drink, which they could not digest because they were too young. The history of breastfeeding, pumping, and human milk banks reveals the imperial dynamics of power. Consent has always been fraught for mothers in the colonies amid the colliding micro-bial worlds and microbiomes. It is vastly different from the chore-ography of the mother mongooses, and I mention it here because Indian nourishment and stories of natural history sustained the Brit-ish colonial class for generations. Indian women's wombs have long been a political locus of exploitation by the West. The mothering was not only of milk and caretaking; women also carried out moth-ering through the act of storytelling. Ammahs acclimated Euro-peans to the tropics. Violent surrogacies divided by caste already shaped notions of bodily pollution and substitute mothering. Even today, white couples pay often-impoverished women subsistence fees to carry fetuses.[11] Again, consent is impossibly fraught amid the racial dynamics underwritten by the ongoing reproductive econ-omy of the colonial plantation, whether in Jamaica or Sri Lanka.

Colonial knowledge ignored and discredited local scientific knowledge and the contours of Indian natural history. The Hindu parables Mrs. Espeut heard from her Sri Lankan ayahs about mon-gooses had world-changing impacts. Her father, Colonel Armit (a name of Old French origin), was stationed in Ceylon as a Royal Engineer, and owned a pet mongoose that she had seen prey on rats. This anecdotal information was enough to shift ecosystems across space and time. Men born in Bombay and raised on the breast milk of Indian women, like Rudyard Kipling in 1865, have dictated the colonial imaginary for future generations. Certain strands of con-servationism act as if nature is the "white man's burden." Kipling's British colonial imagination was deeply informed by the extractive nature of suckling. Just as in chapter 5, where another nineteenth-century thinker (Emerson) derived his transcendentalist writings from South Asian and Hindu traditions, we see here the power of Indian philosophy appropriated by the West.

Kipling's notion of whiteness was inflected by his birth and early childhood in the colonies. His family held England dear as the motherland, but Kipling nursed from surrogate mothers with brown skin. He made a career of the stigma of the tropics, and then reaped the rewards of the biological and climate determinism he promoted. In Kipling's mongoose of "Rikki-Tikki-Tavi," British colonial knowledge of Indian ecologies spread across the Anglo-phone world disguised as a children's story. Indian nursemaids not only provided nourishment but also parented young children, telling bedtime stories from the Ramayana through the lens of India, Pakistan, Bangladesh, and Sri Lanka. India was another Eden to Europeans; perhaps it was the one Columbus had gone in search of, after all. What would the tally of combined reparations for India consist of from Portugal, France, and the Netherlands? In which currency could this repair ever be commensurate to the damage? The jewels in the British Crown—extracted from the Indian bedrock—should be one such ransom for climate reparations.

INDIAN NATURAL HISTORY AND THE PANCHATANTRA

Prior to the mongoose's circulation across the West Indies, Rudyard Kipling denounces the books of Indian natural history as untrue, as they contrast with his Victorian world in "Rikki-Tikki-Tavi" in *The Jungle Book* and *The Second Jungle Book*. Kipling's fictional mongoose becomes domesticated and transforms into an ally to man, despite his mother's warnings against white men. Rikki-Tikki-Tavi is personified as curious and loyal; he defeats the cobras Nagaina and Nag, who threaten the British family he lives with. The mongoose's name is derived from the onomatopoeia of his chattering war cry. For Kipling, as for Espeut's wife, the mongoose was a loyal companion to the expatriate British child from the danger Indian wildlife represented.

Turning to the Panchatantra, Sanskrit sacred texts offer but one of multiple views. They do not cover the range of the vernacular natural histories not recorded in writing. Alternative literacies exist across and have evolved over centuries. Turning to Sanskrit interpretations is not the corrective for Kipling's skewed point of view. However, the writings offer another vantage, though not definitive conclusions about Indian natural history and environmental knowledge. The earliest recorded copy of the Panchatantra dates to 300 BCE, but the Hindu text is much older in vernacular form. These texts offer an environmental perspective through animal fables and illustrations. The tale of the "Mongoose and the Brahmin" is one of the first appearances of the mongoose in Indian lore.

A Brahmin, or member of the caste of priests and teachers in India, domesticates a baby mongoose, raising it alongside his infant son. The Brahmin's wife is skeptical of the animal and its sharp teeth. One day, the mongoose saves the unattended baby from the poisonous venom of a cobra. The Brahmin's wife returns home to see the bloody mongoose, and, assuming it has killed her child, she kills it. Then she discovers that the mongoose saved her baby from the snake. The moral of the fable is to not make hasty decisions against those who are loyal to you. Kipling denounces such tales. *The Jungle Book* itself a confection of his colonial white creole childhood, it is a wonder what Kipling knew of Indian natural histories. The logo he adopted later in life shows how much Indian philosophies and ecologies meant to him; it depicts an Indian elephant with its trunk holding a lotus blossom and a swastika in the upper left corner. In Sanskrit, *svastik* (स्वस्तिक)—which is over 7,000 years old—signifies good fortune. Nazis appropriated the Hindu symbol for their white supremacist fictions of Aryan lineage.

"MASSA ESPEUT'S RATTA": BIOPOLITICS AND THE CHIMERAS OF EMPIRE

Black Jamaican laborers referred to the mongoose as Massa Espeut's Ratta, ironic, as it was imported to eliminate the rats. Espeut was more of a Pied Piper of Hamelin than a savior to anyone. Whenever he could, he mentioned what he believed to be his grandest achievement in the Calcutta mongoose. In a speech to the British colonial authority on the lack of timber in Jamaica, he says, ". . . and you must never forget that I live in the 'Home of the Mongoose.'"[12] The world never will. He was passionate about advancing Jamaican infrastructure into the modern age and delivered speeches on introducing timber production to the island and extending railways.

Upon reading Espeut's article in the Zoological Society of London journal published in 1882, members of the planter class of Hawai'i imported seventy-two mongooses directly from Jamaica to the Hāmākua Coast in 1883.[13] Population control assigned hierarchies of differential value and disposability, enforced sterility and enforced breeding, on the plantations of the early Americas. Imperial taxonomies are endless, anxious repetitions, performed to secure a fixed racial grammar. They are unstable taxonomies of empire and race that belie the colonial approach toward nonhuman animal life. The designs of colonial island plantation management are illuminated via the journey of the Calcutta mongoose to Jamaica through Cuba, Puerto Rico, Grenada, Trinidad, Barbados, and even as far afield as Hawai'i.

In Hawai'i, where the mongoose is known as 'lole manakuke, W. B. Espeut is shrouded in infamy, blamed for disrupting the natural environment of the archipelago and beyond. Mongooses are often referred to as a threat to humans and a vector of disease, and reports in Hawai'i note, "We have been told by local residents that mongoose meat is eaten for supposed medicinal qualities by some Chinese in Hawai'i. The danger of such a practice is evident from the occurrence of Trichinella and Leptospira in the mongoose in Hawai'i."[14] Chinese in Hawai'i and the Caribbean, mainly Hakka, were brought to the plantations to labor for sugarcane production, as was the case in Jamaica on Espeut's plantation in Buff Bay. The islands of Lāna'i and Kaua'i have never seen the mongoose or its parasites and hope to keep it that way. It is common to hear that the mongoose is "not welcome in Hawai'i," and residents are encouraged to control the problem.

An experiment of empire, the fantasies of the mongoose are undergirded by a deeper series of colonial fictions of Anglo racial "purity" that were unstable for Espeut. The tropics were a laboratory of genetic bioengineering as much as a space of racial degeneration, making and unmaking. Still, we rarely consider the pregnant mongoose and the choreography of mongoose mothering. The power to populate and to cross-pollinate was in the hands of creole plantation owners like Espeut. According to critic Ann Stoler, the use of French philosopher Michel Foucault's biopolitics illuminates "the notion that power relations are played out in how bodies are aggregated and individuated, healed, buried, made indistinguishable, and marked. His work provides not an abstract model but one analytic tool for asking grounded questions about whose bodies and selves were made vulnerable, when, why, and how—and whose were not."[15] How does this apply to the nonhuman animal world?

The 1804 Trinidad Experiment was designed as a means of statecraft to determine the intimacy of human populations by containing them. The plan both limited conscripted Asian laborers and enabled them to fantasize about their capacity to be bourgeois domestic subjects. The undisciplined intimacies of the plantation were blurred, though, leading to rebellious intimacies, which in turn engendered its collapse. At least in the case of Chinese indenture and the Trinidad Experiment, it failed. Anthony Trollope wrote of the Caribbean, "The Chinese and the Coolies—immigrants from India are always called Coolies—greatly excel the negro in intelligence, and partake, though in a limited degree, of the negro's physical abilities in a hot climate." This language shows the eugenicist logic of "hybrid degeneracy" or, alternately, "hybrid vigor." Whether the stereotyping is positive or negative, what is implied about mixture improving or degenerating offspring is problematic, as if they were animals.[16]

In the recorded proceedings of the Zoological Society of London, in "On the Acclimatization of the Indian Mungoos in Jamaica," Espeut describes expenditures of £200–300 a year on rat poison and traps to protect sugar cultivation. He describes the various types of rats that cause the ravaging of the crop. He finds terriers to be most effective in preying on the rats, although they often succumb to eye injuries due to the sharp blades of the sugarcane stalks. Espeut outlines efforts by planters in the past to eradicate the rats by introducing ferrets in the 1780s in Jamaica, and toads in Martinique and Barbados in 1844. He blames planter Anthony Davis for the "disagreeable" noise of the agua toad in Jamaica, which he portrays as what would today be called an "invasive species" that is "very destructive to poultry, chickens, and eggs."[17] Indeed, Davis's introduction irrevocably impacted the soundscape of the Caribbean and Hawai'i. Espeut acknowledges that the introduction of the "raffle ant" or *Formica omnivora*, from Cuba, by Sir Stamford Raffles, the president of the Zoological Society, helped to a small degree, but says the collateral damage inflicted by the vicious insects included young animals, birds, chickens, puppies, and even calves.

He calculates the "beneficial results of the introduction" of the mongoose as exceeding £150,000 a year. Citing the government botanist, Espeut quotes that the financial benefits of the mongoose are not less than £100,000 per year. Toward the end of his account he mentions, "Unfortunately, ground-nesting birds, the Quail and others, have been diminished; but the loss of poultry is not as great from the Mungoos as it was from rats, snakes &c. before the introduction of the former."[18] Espeut lists several other Jamaican planters who have followed suit, importing mongooses from India after him. But he emphasizes that most of these died without progeny. Who knows if this is true? Espeut asserts that the mongooses seen flourishing across the island, and that have been exported to other islands, are descendants of his nine original imported mongooses. He wanted to take sole credit for the rise of the mongoose in the Caribbean.

The speeches and reports evince the ways in which colonial knowledge was circulated and produced. It was significant that Espeut published his work with the authority of the Linnean Society, denoted by the postnominal F.L.S. Books within the archive of the society, such as *A catalogue of the plants of Jamaica & other English sugar-colonies ranged & digested according to the Linnaean System,* show how colonial biopower included both the flora and the fauna of the region. Correspondence between Espeut and other fellows regarding visitors to Jamaica from the British Museum shows the circuitry of the private club. Espeut writes, "I do not know him nor does he know of me," and speculates about a bat, saying, "It may be only an albino, or it may be a new species." He writes to obtain a specimen for dissection, asking for anyone interested in bats to be put in communication with him. Did he visit the Windsor Caves? What did Espeut know of the layered histories of guano in Jamaica, Navassa, Hawai'i, and Peru? The naturalist societies of Europe were a powerhouse of credentialing and producing colonial knowledge by amateurs with commercial interests.

Espeut concludes, "I question much if such enormous benefit has ever resulted from the introduction and acclimatization of any one animal, as that which has attended the Mungoos in Jamaica and the West Indies; and I marvel that Australia and New Zealand do not obtain this useful animal in order to destroy the plague of Rabbits in those countries."[19] Espeut recommends that the British ship the mongoose to other British colonies like Australia, and notes that they have been withheld in favor of places like Jamacia—an act he sees as stoking a kind of competition between colonies. In his speech on extending the railway for the intra-island transport of agriculture in Jamaica, he expresses resentment for the rationale that Australia is granted more money for a higher ratio of feet of railway per person because they have a majority white and thus presumed "civilized" population that needs the railroad. He felt that because it was a majority Black colony, Jamaica was being left behind in terms of infrastructural development by England.

Espeut was adamant in his advocacy for Chinese indenture. He arranged for the passage of several hundred laborers to his estate, Spring Garden, in 1884 via the German steamship the SS *Prinz Alexander*. In "The Timbers of Jamaica" Espeut wrote:

> My own belief is that the Chinaman is the labourer we require. He will certainly be the best and cheapest, in any such work as lumbering, combining, as he does, the excellent qualifications of strength, industry, energy, intelligence, and indomitable perseverance. . . . I am satisfied the presence of Chinese, or any real workers, will do more to educate, practically, our peasantry and lower orders, than ten times the cost of importing the Chinese will accomplish, if spent in the education—as it is called—of our youth under the mere Book System now in force.[20]

Indentured Chinese were used to discipline the newly freed Black laborers into submission. These were among Espeut's island schemes and labor experiments. The colonial design of creating a wedge between the Black majority and white minority by pitting them against one another was clear.

PEDIGREE AND WHITE FLIGHT

Trollope pitied the class of people W. B. Espeut belonged to—creole whites—in his observations of Caribbean societies post-emancipation. His 1859 book *The West Indies and the Spanish Main* is filled with notes on not only "Jamaica-Coloured men" and "Jamaica-Black men," but also "Jamaica-White men," creoles born in the islands. Trollope says it's time to challenge the unquestioned "ascendency" of white men over Black men, proposing that merit should be the determining factor. These sentiments echo Emerson's rhetoric about Black Haitians. Espeut's so-called benevolence toward Black people echoed this sentiment, too. He wrote, "Think of our people. We call them lazy, idle, thieves, ignorant, stupid, &c.; Whose is the fault? Is it wholly theirs? Fifty years ago, they were slaves, nearly as blind as the blind-born, nearly as dumb as the born mute. To think how little we have done for them often causes me sorrow; but I trust to live to see enough done for them to save me from feeling shame."[21] Using animal metaphors, Trollope describes the brutalizing effect of the institution of racial slavery on both the enslaved and the enslaver. Where the Black man is transformed into a "beast of burden," Trollope wrote, the white man becomes as "brutal and ferocious as a beast of prey." Enslavement, he believed, unleashed the basest instincts, with the white man transformed into something sadistic and monstrous.

Trollope describes the plight of the white creole, who endured the loss of profit in compensation for emancipated Africans. He writes that "as regards him himself, the old-fashioned Jamaica planter, the pure blooded white owner of the soil, I think his day in Jamaica is done."[22] He describes Jamaican whites as pleasant though "not given perhaps to much deep erudition." Noting the difference between European-born Englishmen and Frenchmen in the Caribbean, Trollope describes them as not "generally addicted to low pleasures."[23] The creole-born, those whites reared in the Caribbean, were subject to these base habits and primal instincts.

Considering W. B. Espeut's preoccupation with the taxonomic order and the social labor experimentation he carried out, there are several gaps in his family tree, as there are in Audubon's. Could Espeut possibly be descended from French human traffickers who fled Haiti to Jamaica during the revolution? There are a small number of French surnames like this in Jamaica, indicating the flight of the French in the 1780s. Although they may purport to be of the Norman aristocratic classes of England whose French names are derived from the eleventh-century conquest, another more likely possibility is that the families are descended from the Cedula of Population; in the 1783 edict, King Carlos III of Spain granted free land in Trinidad to white French Caribbean people who swore allegiance to the Crown.

Black freedom shaped the migrations and fortunes of the Espeut family throughout the nineteenth century, as it did for all Caribbean whites. *Burke's History of the Colonial Gentry* is proof of the colonial fictions of lineage that were penned when it was published between 1891 and 1895. The book states that Peter Espeut, the family patriarch, was the son of William Espeut and Caroline Marlton of Standing Hall in Hertfordshire, England. In the book, Peter Espeut was described as a British soldier who was stationed in Barbados (1779–1782), Martinique (1783–1785), and St. Domingue during the British occupation.[24] Yet no trace can be found of William Espeut and Caroline Marlton, and Standing Hall did not exist in Hertfordshire or elsewhere. The dates for Peter Espeut's service in Martinique as stated in the book do not line up with Britain's occupation of the island between 1762 and 1763. Peter Espeut was dead by the time the British Expeditionary Force set out for Saint-Domingue. Perhaps the deacon Peter Espeut is named for this patriarch and sees his role in contemporary Jamaica as continuing to hold court on environmental knowledge as part of the governing elite.

SLY MONGOOSE, THE PREDATOR

The rhetoric of "invasive species" during the nineteenth century exemplifies how climate debates need to shift. Ultimately, what colonizers said about the mongoose reflects more on who is saying it, and the sociopolitical moment. The animal has been personified and sexualized as the "sly mongoose" in Calypso tunes, jazz melodies, and Jamaican patwa proverbs. Many animals have had colonial ideas projected onto them; the mongoose is no different. The character of the sly mongoose is portrayed predominantly as a lascivious lothario, as if all mongooses were male. A savior? A trickster? A pest? A predator? The female mongoose was not the passive female prototype, as Darwin and the scientists who followed him so often assumed of female animals.

The lore of the mongoose extended in the twentieth century, as it led to the naming of the CIA Cuban Project: Operation Mongoose.[25] On November 4, 1961, U.S. officials assigned the top-secret name because there were thirty-two tasks and thirty-three living species of mongoose. The mongoose appears again in the militaristic history of the U.S. invasion of Grenada in 1982 with the terror squad dubbed the Mongoose Gang for the way they relentlessly wrought havoc on the island.

Saint Lucian poet Derek Walcott returned often to the trope of the mongoose. The mongoose provided a lesson that altered my pedagogy and citational practices. A syllabus of abuse came undone when a student told me in 2017 that reading Walcott's poetry had inspired a passion within her to write about the ecological beauty of the Caribbean and the haunted decay of plantation houses. Later in the semester, she told me she was devastated to discover the names of the people Derek Walcott exploited through a quick search online. I felt responsible as a literature professor. It was such a well-known fact that I had taken it for granted because there are so many predators in the academy. These were the open secrets of a Nobel Laureate granted impunity. Countless newspaper articles that I won't let take up space in these pages attest to these allegations of sexual harassment and misconduct. While Walcott's writings still constitute the classics of Caribbean letters, the gaps and contradictions have much to say about the flawed process of literary canonization. Teaching students this canon of authors without proper contextualization causes more potential harm than good, I have since decided. What does it cost to make room for other Caribbean writers on the syllabus who do not have Nobel and Pulitzer Prizes or a long list of allegations of indiscretions?

The tragedy of Caribbean letters is that sometimes the abused write of abuse so artfully because they have become the abusers. The tale of the mongoose taught me this, and it taught me how to navigate predators in academia, of which there are unfortunately many taxonomies. Junot Díaz took inspiration from the mongoose like Walcott did. He wrote himself into this Caribbean macho lineage alongside men of letters Walcott and V. S. Naipaul. But at whose expense? Now I understand that the mongoose was prey to these writers, too. She was fodder, Espeut's prey, and her legacy became prey, serving figures like Díaz and Walcott. These writers demonized the mongoose in the literary canon, saying more about themselves and their sexist fantasies of domination than about the reality of this special creature. The mongoose emerges at moments in the narrative where abuse against women takes place. I think about the communal spirit of the female mongooses, the commitment to raising as many litters as possible, the absolute determination these mothers employ to encourage maximum survival. There are other syllabi to write and teach, I've decided.

The mongoose has become a symbol of masculinity and a ferocious sexual appetite, as demonstrated in literature, popular culture, and music across the Greater Caribbean from the 1920s onward. Three Caribbean-born men—Díaz, Walcott, and Naipaul—who have each faced the ignominy of highly publicized accusations of sexual abuse and intimidation, become entangled in the literary imagination of what the mongoose represents. The mongoose appears famously as a supernatural hero in Díaz's 2008 Pulitzer Prize–winning *The Brief Wondrous life of Oscar Wao*. He calls the mongoose a time traveler, and we meet the animal in the blood-soaked sugarcane fields of 1980s Dominican Republic. The mongoose becomes a savior to a Black woman who has been beaten by the militia of dictator Rafael Trujillo's regime and left for dead. In Díaz's imagination, the animal becomes an Afro-diasporic sci-fi trickster figure who arrives in the Caribbean after a stopover in

India. Until 2018, it had given me great pleasure to teach this part of the book in Caribbean literature lectures to go on a tangent about island ecologies.

On April 9, 2018, in the *New Yorker,* Díaz published "The Silence: The Legacy of Childhood Trauma," and I felt compelled to assign it the next week. We had finished reading Díaz's short stories; this became the real-time syllabus and forum for students to discuss the controversy. Weeks later, accusations against Díaz flooded the press, as did letters in defense of his actions. Did this essay exonerate him from the swirling speculations? These questions highlighted many of the themes we were grappling with in that class on Caribbean writing, reggae, and routes. My students were not divided on the matter. Afterward, a flurry of attestations from scholars and writers came to Díaz's rescue, citing how his status as a prominent man of color made the tenor of these accusations different and predictable. Time will tell the verdict. The university MIT, his employer, determined that there was nothing actionable that would warrant firing the tenured creative writing professor. How long does it take a generation to forget? These are the very questions Díaz had so poignantly posed in his writings on the brutal impact of dictatorship on the national and racial psychology of societies across the Caribbean.

The mongoose reappeared to me as I looked for it in the writing of Díaz's hero, Walcott, who demonizes the mongoose to insult his former friend, V. S. Naipaul. He paints the mongoose as an "invasive species," and uses it to portray Indo-Caribbean people this way; he depicts them as disloyal and opportunistic feeders. Walcott accuses Naipaul of turning his back on the West Indies, the place of his birth, for the Mother Country of Britain after winning the Nobel Prize in 2001. Naipaul's award came a decade after Walcott became the first Caribbean-born author to win the Nobel. While Walcott's mongoose poem was never published, there is an audio recording of it from a famous Caribbean literary festival where he performed the vituperative attack on the Indo-Trinidadian author. Walcott warns the audience of his takedown: "It's going to be nasty." In Walcott's words, Naipaul is like the mongoose, "imported from India and trained to ferret snakes and elude Africans," who "takes its orders from the Raj." Walcott deploys the mongoose to challenge Naipaul's loyalty and to accuse him of anti-blackness.

He derides V. S. Naipaul as a clown and minstrel guilty of abandoning Trinidad.[26] Walcott claimed that Naipaul snivels because he is proud of being Asian, though he was not born in Asia. Bringing Naipaul back to the cane fields of Naipaul's hometown of Chaguanas, Trinidad, which Walcott referenced famously in his Nobel speech titled "The Antilles: Fragments of Epic Memory." Continuing on, Walcott said of Naipaul, "He screwed Negresses." He portrays Naipaul as a scoundrel, and literary critics as helpless chickens swept up by the mongoose.[27] Walcott denounces Naipaul, saying, "He doesn't like black men, but he loves black cunt."[28] Why recruit the mongoose for such an attack? Walcott knew the mongoose was such a familiar part of everyday Caribbean life and lore that it would conjure deep feelings. Some love the mongoose, but the animal is widely considered a nuisance, whether a good omen or a bad one.

Further evidence of the prominence of the mongoose in everyday Caribbean life are the several Jamaican folk sayings that feature Espeut's ratta, the ironic name plantation laborers gave the mongoose. The diurnal Indian rat that was expected to eradicate the nocturnal black rats. The proverbs fall into two categories. The first plays upon the fabled dynamic between the mongoose and chickens or other fowl. It references the extinction of endemic ground-nesting birds in Jamaica. The chicken is not endemic to the Caribbean, being an arrivant from the jungles of Southeast Asia. What is more Caribbean, though, than the soundtrack of roosters at all times of the day? "Fowl seh, 'I'm no business ina mongoose politics, for 'im no member a mongoose parliament.'" The chickens, which are the helpless prey of the mongoose, do not involve themselves in the affairs of the mongoose, which stands for those in power, because they do not concern him—nor does he have a vote. The taxonomy of the animals is important here in hierarchy and rank. The domesticated chickens in the henhouse become a signifier for the endemic Jamaican ground-nest birds who were devoured by the mongoose (their eggs taken as a delicious treat, too). Again, the anthropomorphizing sayings speak more of human projections onto the nonhuman animal world than of actual animal taxonomies. Another Jamaican saying goes, "Mongoose seh, 'If me had a penny me would lef' dis place.' Fowl seh, 'If me had a penny me would gi' i' to you.'" Again, we get a sense of the antagonism between the predator and the prey. The indigenous chicken would happily pay for the mongoose's return passage back to India, so to speak, as Walcott certainly would have for Naipaul.

The second set of Jamaican mongoose proverbs references the mongoose-human relationship. "Mongoose seh: man who cawn tek chance a no man at all," which G. Llewellyn Watson translates as, "The mongoose lives by the dictum that a person who does not take chances is not worthwhile."[29] The mongoose is not risk averse; like many migrants, he takes a chance in each new country. Another saying goes, "Mongoose seh, risk no man, even if 'im cawn run." This saying cautions against taking a chance with one's foes, even if they seem weakened. While it is said that the mongoose has no predators in the Caribbean, hence its effective survival, humankind is always the most formidable predator of the earth.

From folk saying to folk music, the mongoose has hitched rides all over the globe, traveling across the islands of the Caribbean in the twentieth century. The song "Sly Mongoose" is said to have originated in Trinidadian Carnival in the 1920s and was popularized around the world by Trinidadian singer Sam Manning in 1925. "Sly Mongoose" was so popular that it made its way into the jazz repertoires of Louis Armstrong, Ella Fitzgerald, and Charlie Parker. The song has also been performed by the Gaylads, Aston "Family Man" Barrett of the Wailers, Harry Belafonte, Ernest Ranglin, and Monty Alexander. Thus, the lore of the mongoose circulates as it continues its island-hopping.

The lyrics tell of the mongoose's flight from Jamaica, and the animal being caught in bed with the preacher's daughter. The anthropomorphized creature has relations with its "high brown mama."[30] The refrain tells of how the mongoose's name and his reputation travel abroad. Here it seems he is no longer being racialized as an Indo-Caribbean man, but as a Black man. The mongoose is gendered and sly in ways that are telling about human interference with the natural environment. The slick mongoose slides from Jamaica to Cuba to America, following the routes of migrant agricultural laborers of the time.[31] One of my great-grandfathers traveled these sugar routes; work songs traveled these routes.

From Jamaican music the mongoose traveled to the Bahamas with the Nassau String Band version called "Slide Mongoose Slide" in 1935. Returning to Trinidad, the Calypsonian singer named Lord Invader popularized "Sly Mongoose" again, just after World War II. He added the characters of Harry Houdini and Adolf Hitler to the song.[32] In pop culture of the era, the mongoose continues to be depicted as male and lecherous. "Sly Mongoose, everybody, only dog know your name / Yes, mongoose went in the White man's kitchen, / Took up one of his fattest chickens"[33] Again, the mongoose is a trickster who subverts the white man's order. Of other versions of the song popular in St. Croix, Dominica, St. Vincent, and Barbados, critic William C. White comments, "they know 'Sly Mongoose, Dogs Know Your Name,'" which he characterizes as the "the sad story of a raid on a white man's kitchen."[34] This version also references the 1880s preacher of Jamaican Revivalism Alexander Bedward, whose many followers united to resist colonial oppression and became Garveyites or Rastafarians. It is uncertain what business the mongoose had with Bedward's daughter. The mongoose is personified as a "trickster with a sexual proclivity" who steals the "fattest chickens" from the white man's kitchen, which portrays him through sexual innuendo like a fox in the henhouse.

INDIGENOUS KNOWLEDGE:
CHALLENGING "INVASIVE" RHETORIC

As we have seen, the mongoose has been blamed for a lot, but not everyone holds it as a scapegoat. Culling, which has been the primary method of management to control "invasive species," is not the only answer to the disasters that invasive species present. This animal control tactic of total annihilation has not worked to curtail the sly mongoose, who reproduces at a rate faster than it is killed. What other approaches are there to "invasive species" than eradication? The terms "native" and "endemic" are evoked in Western ecological rhetoric but have little to do with how many Native communities approach matters of biodiversity. "Native" attitudes to biology—of which there are a heterogenous range across the world—tend not to align with the scientific frame of "invasion" regarding natural life. The orientation of militarism and war against the natural environment seem to be more of a Western approach. The term "introduced species" may be a more relevant alternative. But if the treatment and the sentiment against the maligned plants and animals is the same, it makes no difference which name is used. Colonial scientific attitudes are "survival of the fittest": pit one against another, and villainize innocent plants and animals. Nor does it seem practical or effective in most contexts. Population control and animal control follow the logic of annihilation and segregated life. It stands to reason that a colonial solution would be proposed by those who introduced a colonial problem by playing God.

While marking the difference between native and non-native organisms dates to the eighteenth century in Western science, the militaristic nomenclature is a post–World War II interpretation. British scientist Charles Elton's 1958 book *The Ecology of Invasions by Animals and Plants* came to be definitive for twentieth-century ecology and centered on archipelagos and continents in its language of invasion.[35] The use of "invasion" dates only as far back as 1958, when the book introduced it as a call to action against Chinese kudzu and Asian long-horned beetles, which he described as colonizing new habitats at exponential rates. Elton established and led the Oxford University Bureau of Animal Population to carry out this line of thinking and culling. Not until the 1990s and early 2000s did its prominence grow as a biological term for ecologists. Rhetoric matters. Twentieth-century words should not necessarily determine twenty-first-century environmental policy. Mongoose infiltration has wrought an estimated $79 billion (USD) worth of damage on the United States in Puerto Rico, and mongoose management is a politicized matter for taxpayers. Storytelling, lore, and fables determine policy, none of which have much to do with the reality of the mongoose itself.

Indigenous scientists and theorists in North America, on the other hand, have explored other non-extractive modes and relational methods based on Native philosophies of ecosystems.[36] Revering the natural environment does not mean that Native people value only endemic species. That would be a perverse logic. Again, the segregation of life and the notion of property govern colonial ways of thinking. Following the mother mongoose's journey has revealed how racist thought shaped climate history and continues to threaten the viability of future ecosystems. The mongoose survived beyond the lab. Indian laborers may have been surprised to see not only the familiar sugarcane transplanted from India to the Caribbean, but also the mongoose.

As will be explored in the next chapter, over one million Indians who became Indo-Caribbean over the generations were also agents of ecological change. They stowed away plant life and brought it to the Caribbean, giving birth to new botanical worlds. The zoological worlds are continuing to thrive as well. On my most recent trip to Jamaica, I saw a mongoose darting across the street. Mongooses are adaptive by nature, known to leave their home colonies to join others, especially where there may be a more dominant position to possibly assume. As for themselves, mongooses are skeptical of newcomers who seek to join the colony, but after a process of integration they are slowly welcomed and trusted in the night watch. There are so many lessons to learn from the mother mongoose about survival. She has made living an art form. It can be heard in the patter of her furtive footsteps. She mothers for the multiple time lines beyond apocalypse.

8

Pedagogies of Smoke

W eeds are teachers. With a patient pedagogy, their method of growing defies expectations. Far from the horticultural failures they are maligned to be, weeds thrive against eradication. Where life is supposed to be impossible, these marvelous plants offer many lessons on survival against the odds of annihilation. What more ambitious a weed is there than weed? Through smoke, marijuana's lessons for humanity have been many. The flowering plant has co-mutated and arguably adapted human life and society. Cannabis complicates Darwinist notions of survival and competition. Who, we must ask, is cultivating whom? Couldn't it be argued that marijuana has cultivated human societies for its survival?

Marijuana's propagation is a form of gardening and experimentation in the dark. Often grown in dark grow rooms, the plant is demonized merely for flourishing. There is a dark colonial imagination about marijuana's psychotropic properties, but less is imagined of its healing and spiritual uses. How, I ask, could there be strategies to learn from marijuana's journey to and survival in the Caribbean? Could there even be potential to reduce carbon emissions in the propagation of the herb? Hemp is two times more effective than trees as a source of carbon capture.[1] In the present day, the illicit cash crop could have an important role to play in climate action as laws banning its sale and usage are relaxed and overturned. First, we begin with its Caribbean history. The shadow economy of the ganja trade blossomed in the Americas during the nineteenth century. The plant traveled in the darkened hulls of ships from British India to Jamaica with indentured Indian laborers. Merchant ships, as we have seen in the previous chapters, were veritable laboratories beyond colonial regulation where a myriad of biological transformations and African, Asian, and European exchanges took place as the ships made their transatlantic voyages.

From the Ganges River we can hear its moniker in ganja. Marijuana traveled unsanctioned by the regulations of the British colonial authority when indentured Indians from Uttar Pradesh and Bihar were brought to the Caribbean in the 1830s. Since then, weed has captured the global imagination with its contested economy and its own soundtrack pushing for legalization. Within its tendrils exists the foundation of a Black philosophy that is an environmental philosophy of liberation. Jamaican reggae singer and producer Lee Scratch Perry put it poignantly when he said, "The Earth is the Lord. Everybody walks on the earth. And nobody respects the Earth. Everybody who walks on the earth, shits on the Earth. Spits on the Earth. Don't respect the Earth. So the Earth didn't like it. [T]he Earth [is] fighting back. The Earth call for a revolution. The Earth call for justice. And the Earth get justice. 'Cause the Earth release ganja. The Earth release herbs."[2] These sentiments characterize Rastafarian botanical and religious attitudes toward the plant. Climate justice as imagined by Perry is manifest in the power of ganja to cause a revolution against pollution.

Tracing the colonial story of ganja unravels an environmental history in the shadow of the trade of Africans who were enslaved, and subsequently of Asians who were transported to the Caribbean through exploitative debt bondage labor schemes. Considering that Africans and Asians were forced to cultivate the soil of the U.S., Caribbean, and Latin America, what did it mean for them to have choices in cultivating their own plants, such as marijuana, secreted away? Counter to these bounded and confined spaces, "provision

grounds" emerged across the diaspora. These plots of land functioned as alternative spaces of agriculture that opposed the logic
of the plantation. On provision grounds, allotments of fertile earth
are nurtured by Black people. Healing through botanical knowledge becomes possible in these radical spaces of transformation.
Cultivators of color have long been exploring marijuana's potential
for healing and have created sustainable economies, sanctioned or
unsanctioned. Governments are beginning to incentivize taxes gen-
erated from the legal sale of marijuana in the United States. Certain land rights now include usage of the soil for self-determined
purposes. Many traditional practices of agricultural cultivation are
banned because of the psychedelic effects of Native science, with
peyote and ayahuasca being two prime examples of plant-based substances that have been sanctioned and criminalized. Much about the
future of regulations, legislation, and taxation, the fate of cannabis
and those whose lives it has altered forever remains to be determined in the U.S. and Jamaica in the twenty-first century.

The traffic of ganja is responsible for the incarcerations of millions, especially Black people and people of color. Ultimately, the
history of weed and how it arrived in the Americas in the nineteenth century illuminates the legal construction of race and the
racializing currents of the global drug trade. As it becomes legalized, climate questions as well as racial questions emerge about justice. Will criminal records for using or selling ganja be expunged
retroactively? Each puff, toke, and vape is politicized as new laws
shape the carceral landscape. Behind the debate lies a history of
Indigenous healing through smoke that stretches across continents. Growing in basements and warehouses under UV lamps, the
shadow economy thrives regardless, including in the forms of hemp
fibers, oils, and other products.

Smoke has taught us many lessons, and we can look to parallels about the tobacco industry for a pertinent contrast with marijuana's legalized potential. Tobacco was first cultivated in the first
century BCE by Native peoples across the Americas, from the Maya
to the Mississippi Valley.[3] It is recorded that Christopher Columbus
received dried tobacco leaves as a gift from Amerindians in 1492.

Botanical drawing
of cannabis

The monocrop agriculture of tobacco shaped the terroir of North Carolina, Kentucky, Tennessee, and Virginia. Native communities are seizing the opportunity to reconnect with pedagogies of smoke and traditional botanical practices. The ritual of smoking herbs, including marijuana and tobacco, is a genre of Indigenous medical knowledge, discredited as primitive by Europeans who later profited from the wholesale trade. Much was learned from the ceremonial practice of Native peoples in the Americas about how to use tobacco. It became an industry beyond these sacred practices. The farmlands of the East Coast were forever altered by the nightshade plant of the genus *Nicotiana*.

Today, tobacco harvesting is seasonal work in the United States that brings laborers from the Caribbean to locations like Connecticut, where they live in labor camps. The leaves are cured and left to hang and dry in sheds for months. Workers from Jamaica on H-2A visas are permitted to stay only from August to December for this intensive work. The temporary agricultural worker visa program was established in the 1980s to help U.S. farmers fill employment gaps. Such immigrant labor schemes encouraged by the government have been common throughout U.S. history. The word "tobacco" is believed to be derived from the Arawak language. Its name is found across the islands, and specifically in the Spanish translation of Tobago, the twin island of Trinidad, which translates as "tobacco." Global tobacco circuits come full circle in the Caribbean and form a contrast to the regulation of the ganja traffic.

GANJA, THE GANGES, AND HOW INDIANS BROUGHT MARIJUANA TO JAMAICA

In Hindu traditions, marijuana was part of rituals of transcendence. Plants that could have healed the soil were adopted en masse and instead exhausted the earth. From smoke there are ashes, which fertilize; the cycle of renewal should be one of nourishment and replantation. Today, ganja, unlike tobacco, remains demonized as a gateway drug, but few know its history of healing on Caribbean sugarcane plantations during the nineteenth century. The experiment was not a white British one, but rather an unsanctioned one that Indian migrants undertook in the nineteenth century. In the hold of a ship, clippings and seeds of *Cannabis sativa* made their way from British India to British Jamaica. The herb came to shape Jamaican life when a recorded 36,412 South Asians were brought to Jamaica through debt bondage labor schemes from 1845 to 1917. In contrast to the story of the mongoose's arrival, weed was a stowaway.

Victorian-era Irish physician William Brooke O'Shaughnessy is famous for bringing *Cannabis sativa* to Britain as a treatment in 1841, overlapping with the era of Asian indentured labor in the Caribbean.[4] He was a member of the Medical and Physical Society of Calcutta, where he learned about the therapeutic uses of cannabis. Working in the botanical garden at Calcutta, O'Shaughnessy brought Western forms of anesthesia to India. This is the conventional telling of the botanical history of marijuana in the British Empire. But we rarely acknowledge that when South Asians secreted away marijuana to Jamaica, they shifted the world order for their own spiritual purposes. The plant's psychoactive properties facilitated a type of ritual transcendence important across Asian cosmologies. Marijuana was known to treat asthma, glaucoma, diarrhea. Described as a holy herb, it was used to channel transcendence to another plane through its psychoactive properties. Many wars have been waged over what are determined to be controlled substances. From the narcotics trade and the Opium Wars of 1839–1860, the matter of dependency was systematized to design commercial imbalance.

Some refer to these laborers as "indentured servants," but there are key distinctions between the white history of "indentured servitude" and this new era of racializing plantation labor in the shadow of chattel slavery. The labors Indian people were forced to perform in the Caribbean were far from acts of mere servitude. European colonizers forced them to live in abhorrent conditions in plantation barracks, leaving behind the plains of Bhojpur without speaking English or understanding the fate that awaited them. Abolition led to a scramble for new agricultural solutions for cheap labor. From racializing institutions that advanced after the abolition of racial slavery such as indenture and sharecropping that bound people to the land, debt was fashioned in perpetuity.

Laborers practiced sadhana, a method for attaining God-realization through intoxication. Shiva became the god of cannabis, emulating his consciousness, created from his body to purify the elixir of life from the churning of the ocean by demons and gods. The goddess Ganga is also worshipped as a representative of the Ganges River. One potential root of ganja is the Sanskrit *ganjaka,* with *gan* meaning "to tend to," and *jaka* meaning "plant." These spiritual tenets of botanical worship were translated into the broader Jamaican culture at the birth of the Rastafarian religion in 1930s Jamaica. The Rastafari religion leader Leonard Percival Howell cultivated a plant-based philosophy and diet as an act of protest against British colonialism.[5] The Ital way of eating, which centers raw, unprocessed, vegan methods of cooking, is in tune with Hindu dietary prescripts of devotion. Similar genealogies can be identified with dreadlocks, which have origins in Sadhu aesthetics. The asceticism of stylized matted hair alludes to the worship of Shiva. Howell traveled to the Panama Canal, Costa Rica, and New York City as a laborer before returning to Jamaica in 1932.

Howell used an Indian pen name, the Gong or G.G. (Gangun Guru) Maragh, when he began publishing pamphlets in the 1930s. Maragh is a Jamaican-adapted form of the Indian name Maraj. Howell based his writing on the Hebrew Bible and rejected slavery, imperialism, and capitalism; Rastafari quickly became a powerful movement. At the core of the religion is a belief in a living Black God, the emperor of Ethiopia Haile Selassie. In Jamaica, Afro-Asian botanical knowledge and scientific exchange occurred through the sacrament of ganja. In a documentary on the Hindu legacy in Rastafari, commentator Ras Moqapi Selassie says of the genesis, "There are Africans taken out of Africa and brought down to the Caribbean, and there are Indians taken out of India and brought down to the Caribbean, and they corresponded, and out of that correspondence has come something new."[6] Jamaican historian Verene Shepherd, who specializes in histories of Indo-Jamaican labor, also speculates on the cultural, philosophical, and religious transfer between South Asians and Africans on Jamaican plantations during the early twentieth century.[7]

Marijuana is often racialized as Black and is most readily associated with Jamaica because of reggae music and attendant Rastafarian religious practices. The British Ganja Law of 1913 outlawed the consumption and sale of the herb in colonial Jamaica.[8] Across the Caribbean, Rastafarians and especially those wearing dreadlocks were targeted and ostracized by police violence and societal ridicule because of associations with weed. The British authorities could never have anticipated these social interactions between South Asian and Afro-Caribbean people, let alone predicted that they would bear fruit for new religions and new horticultural practices.

THE GLASSHOUSE EFFECT:
CHINA, SCOTLAND, THE CARIBBEAN

The technology of the glasshouse transformed global botanical knowledge and colonialism. Hothouses and conservatories in the metropole were populated by stolen flowers. The politics of gardening carry many lessons to adapt to the climate crisis. Exchanges of botanical knowledge take place inside and outside of formal institutions, from botánicas to grow houses to botanical gardens. There are sanctioned and unsanctioned forms of agriculture under colonial rule. Gardens are never neutral. Scottish plant hunters were famous for their expeditions and the buds, clippings, and seeds they would bring back to be grown in Britain from the seventeenth to the twentieth centuries.[9]

Many noted Scottish plant hunters were the agents of theft across the British empire, working in an administrative capacity at the bidding of the Crown.[10] While this book is not about them, it is worth noting the time line of their exploits as follows. Author of *A General System of Gardening and Botany* George Don (1798–1856) went to Madeira, Tenerife, Cape Verde, Gambia, Guinea, Sierra Leone, Brazil, Cuba, and the United States to collect specimens for the Royal Horticultural Society. Botanist David Douglas (1799–1834), namesake of the Douglas fir tree, traveled to the United States, Portugal, Hawai'i, and the Galapagos. Robert Fortune (1812–1880) introduced 250 ornamental plants from China, Indonesia, Japan, Hong Kong, and the Philippines into Britain over his lifetime. Commissioned by the Russian monarchy, botanist John Fraser (1750–1811), who went to the United States, Cuba, and Russia, sold his herbarium to the Linnean Society. Namesake of the plant genus *Lyallia*, David Lyall (1817–1895) explored Antarctica, the Arctic, New Zealand, and the United States. Francis Masson (1741–1805) traveled to South Africa, Spain, Portugal, Tenerife, the Azores, the West Indies, Canada, and the United States. Archibald Menzies (1754–1842), a surgeon, botanist, and naturalist, traveled to the United States, New Zealand, Tahiti, Hawai'i, and Chile. George Sherriff (1898–1967) journeyed across India, Tibet, and Bhutan. These eight men are only a handful of the most famous Scotsmen: there were many others who ventured to the colonies because botanical theft was highly incentivized by the glory and renaming of plants.

The field of ethnobotany is a colonial discipline and is chiefly to blame for the beginnings of Big Pharma. Indigenous knowledge has been exploited, and plants are stolen from those who have nurtured them, sometimes as part of their sacred origin stories. Imagine if global climate justice were incentivized in this fashion today. Perhaps the harm these men (amateur botanists at best) did to the natural environment would be reversible. Only since the 2010 Nagoya Protocol has it been internationally mandated that permission must be obtained to use plant knowledge and ensure profit sharing.[11]

Collectively, Scottish plant hunters disrupted numerous ecosystems based on the whims of what they chose to collect and were able to transport successfully overseas. Sometimes the reasons were aesthetic; in other cases, they learned from Indigenous science and Native communities about the healing properties of certain plants. They could not retrieve everything they desired, but the plant hunters' actions over three centuries led to the globalization of modern horticulture as we know it. From what they gathered, these hunters' conquests led to the development of crossbreeding and cross-pollinating and formed hybrids. The effect of biological transplantation on human life is untold. Colonization threatens biodiversity and replaces it with monocultures. The legacy is the fiction of the botanic garden in the United Kingdom as English, when it is really a stolen Eden.

George Forrest, among the most renowned of the plant hunters, worked in the herbarium at the Royal Garden of Edinburgh. He left Scotland for Yunnan Province in 1904. Many of the ornamental flowers and plants associated with English and British gardening are from China, Nepal, and other parts of South Asia. Forrest returned to Scotland with over 5,000 rhododendrons comprising more than 300 new species that were native to Asia.[12] The flowering shrub, with its bright pink flower and hardy stem, is known as *Lali gurans* and is Nepal's national flower. After six more expeditions to Yunnan, Forrest died in 1934 in China of massive heart failure and was buried there.

The European tradition of plant hunting soon extended to American and Canadian forms of capturing botanical life as part of an unspoken colonial mission for amassing scientific intellectual property. In issues of *National Geographic* from 1954, the green and militaristic visions of empire are placed side by side. While one article, "Rhododendron Glories of Southwest Scotland," delights in the verdant floral beauty of these flowers at the Royal Horticultural Show in Scotland, the next article, "War and Quiet on the Laos Frontier," depicts plumes of smoke rising above the trees in the town of Thakhek.[13] Following the imprint of French colonialism in Southeast Asia, it would not be long before Americans would arrive and take over the mantle of regional environmental devastation. The ecological impacts of war are often last thought of, but they may be the underlying incentive, as they exemplify theft of natural resources. *National Geographic* was an engine of environmentalist, imperialist propaganda. In its pages was a battle plan that invited the average reader to participate in a colonial way of looking that paired botanical and military conquest. The magazine's board of directors, headquartered in Washington, D.C., reflects the "military industrial complex," U.S. president Dwight D. Eisenhower's term for what was evolving as the shape of U.S. power in the postwar period. Plant life has long been a part of this complex.

GENRES OF GARDENS

There are botanic gardens, pleasure gardens, physic gardens, the English country garden, provision grounds. The violence of botanic gardens contains many subgenres of green space where beauty is defined through the fascist forms of control disguised as elegant gardening. Many forms of morality are mapped onto Western gardens, which were modeled on the Garden of Eden as a genre of fantasy. Botanic gardens such as Kew in southwest London showcase the perverse colonial landscaping design as a capitalist project of extractivism. The colonial logic of collecting in botanical gardens like Kew is cold and exploitative, rarely designed for sustenance. At the heart of the modern climate crisis, botanic gardens must be understood as engines that fueled the economy of empire. Such "ornamental" gardens fueled deforestation. Many tropical plants exist only in private collections, like so many rare works of art. Naming is part of the branding and marketing process. What is Kew's core mission as a royal institution?

One garden became a template for utopia, structuring the Abrahamic world. Medieval physic gardens traditionally have an emphasis on herbal and medicinal propagation of plant life. The first physic garden of the United Kingdom, the Oxford Botanic Garden, was established in 1621. Founded on a former medieval thirteenth-century Jewish cemetery after the expulsion of Jewish people from England in 1290, the grounds were already contested. The modern Western idea of the pharmaceutical was first fashioned here, associated with medical schools and training until the twentieth century. Botanic gardens developed out of physic gardens, where cataloging in Western Europe in the 1500s took place. Other genres of gardens began to evolve, including pleasure gardens, landscaped for the public's entertainment and recreation.

The second botanic garden was established in 1673: the Chelsea Physic Garden. I visited it in its 350th year and was greeted by a marble statue of patron saint Hans Sloane at the focal point of the garden. Sloane's 1707 book, *A Voyage to the Islands Madera, Barbados, Nieves, S. Christophers and Jamaica*, shows how the British viewed islands as laboratories of human and ecological experimentation. A man of his times, Sloane's notes document his abuse of people of color and people who were enslaved.

Modeled on gardens like the Chelsea Physic Garden, the Castleton Botanical Garden and the Hope Botanical Gardens are on high in Kingston and could not be further from Tivoli Gardens, one of the more notorious neighborhoods on the island. While the global guano wars were taking place, as detailed in chapter 5, the British were also sending expeditions to the Peruvian Amazon to steal precious cinchona trees and plant them in Jamaica. The prized cinchona tree is depicted on the flag of Peru. The British military required quinine to combat malaria, the invisible enemy of the tropics. Also located high in St. Andrew's, Cinchona Gardens was established in

1868 by Sir John Peter Grant in the mountains of Kingston. These environmental entanglements show how the botanical and the militaristic were entwined projects of colonial greed.

In 1847, the world's first Museum of Economic Botany was founded for commercial purposes at Kew. From this legacy, today Kew has a living collection of 27,000 taxa. Reflecting on their checkered history of exploiting Indigenous science, on March 2, 2021, Kew published their "Manifesto for Change 2021–2030." They describe it as a ten-year strategy to end extinctions and protect nature. Written in a moment of post-Brexit colonial reckoning, the representatives of the gardens declared the sins of the past but did not admit or acknowledge how these practices are ongoing. The British podcast "Dirt on Our Hands" connects the dots between the global pandemic and the role of the gardens in systemic racism. It is not surprising that a person of color, British Asian botanist James Wong, was chosen as the spokesperson to talk about the botanical connections to the murder of George Floyd. He called for reevaluation of the garden's legacy when it comes to the fact that racial slavery was based on botanical cultivation of sugarcane and other crops.[14] Are there parallels to draw between the biodiversity crisis and the diversity crisis in prominent cultural institutions?

The Kew podcast outlines a five-point plan as follows: 1) science-based solutions; 2) access to knowledge; 3) training students from the U.K. and around the world; 4) extending their reach; and 5) influencing global policy and leading by example with net-zero carbon emissions by 2030.[15] In the accompanying glossy corporate report probably designed by a public relations firm, Kew pledged to expand their colonial practices abroad. They emphasize their work at the Kew Madagascar Conservation Centre, showcasing their colonial satellites. The garden's mission is a far cry from repairing the ecological damage produced or restitution to the communities who have been harmed by the stealing of plants. Instead, Kew Gardens commits to further contributing to the brain drain by recruiting talent away from the Global South.

Richard Deverell, director of Kew Gardens, did not take responsibility for a 260-year history of ecological extraction. Instead, he said: "All life depends on plants and fungi, but natural resources are being degraded and destroyed at a rate unprecedented in human history. We stand at a crossroads—the next decade will be critical if we are to reverse this environmental devastation. Royal Botanical Gardens Kew is perfectly placed as a globally revered plant science institute to lead efforts in creating a world where nature is protected, for the benefit of humanity and our planet. As *Our manifesto for change* demonstrates, we will no longer be silent."[16] Hollow statements treat scientific inquiry as if there is only one tradition and it is neutral. The research at Kew incentivized by the British Crown and Western capitalist growth is precisely what built the demand for these extractive economies that destroy natural resources.

A similar attitude of neutrality is taken by many universities who own research botanical gardens. I noticed this at the Leiden Garden (Hortus Botanicus Leiden), among the oldest in Europe, when I was a research fellow in the Netherlands. The signage in the garden celebrates the pilgrims who departed from Leiden to the Thirteen Colonies without acknowledging the ecological impact of what was uprooted. Oxford Botanic Garden, founded in 1621, is the oldest in England. Its arboretum was acquired in 1947.

In the United States, the Cornell Plantations was renamed Cornell Botanic Gardens because of controversy in 2016 over what the name may conjure even though the university was founded after the Civil War.[17] Harvard owns not only a botanic garden and herbaria in Cambridge, and an arboretum in Jamaica Plain (Boston) but also, as of 1988, a 4,000-acre forest.[18] The Harvard Forest is touted as a laboratory and classroom, bought after the school lost grounds on Long Island. The forest is a department of the Faculty of Arts and Sciences and a member of the U.S. Long Term Ecological Research Network. An alternative asset, it is a special class of nontraditional investment that benefits and diversifies the university holdings, which total an endowment of $50.7 billion USD. Forestry funds are an emerging form of greenwashing endowments and investment portfolios for major universities.

PROVISION GROUNDS AND BOTÁNICAS:
BEYOND THE ENGLISH COUNTRY GARDEN

There is a stark contrast between the botanical knowledge of botanic gardens and the invisible network of botánicas across diasporic cities. I should not have been surprised to see spiritual botanical shops in Brixton because gardening is a profoundly Caribbean act, as Antiguan author Jamaica Kincaid puts it in her *My Garden (Book).*[19] She found she was organizing her gardens in the shape of the Caribbean, the arc, without realizing it. In Britain, Caribbean gardens are arranged this way too. In her Vermont garden, Kincaid shows us the relationship between gardening and conquest. She meditates on how Columbus named the Caribbean as well as how the Linnaean hierarchy dissects the world with its Latin names.

Provision grounds are the plots of land where Africans who were enslaved cultivated produce. Whether permitted or not, these communal spaces became ones of guerrilla agriculture and botanical healing. In the soil of diaspora gardens from Africa, Asia, the Caribbean, the Americas, and Europe, the jigsaw pieces of Pangaea fit together again, rejoined in cities that I know, such as New York and London. The geography of the Caribbean forms a mirror of place-names with Great Britain, yet the shared topography has more in common with the red earth of the West African coastlands. Ripped away during Pangaea, the African agrarian tradition somehow lives on, carried out by peasant farmers on opposite sides of the Atlantic. Diasporic connection between the Caribbean and Africa converges in Brixton Market.

If England is only for the English, then they will have to uproot those gardens. I have learned more from urban communal garden plots in Martinique, Jamaica, Trinidad and Tobago, Antigua, the Bahamas, Cuba, Suriname, Grenada, and Aruba than from any of the perfectly manicured gardens of England. Immigrants stowed away clippings, Africans and Asians planted strange fruits in cold soil, and somehow both managed to cultivate the future. In South London, my paternal grandmother planted mint and other herbs during the 1950s. White English landlords refused to rent to anyone who looked like her. I wish I knew her London gardens, which must have been a triumph for her to own as a Black woman in a cold place. The land is a pact, and her father, Allan, was a Maroon born within the community and traditions. I will never know what the intimacy of the soil meant to my grandmother, but I know she was from a small village called Vanity Fair in St. Catherine Parish, Jamaica. I was four when she died and remember her funeral in South East London.

When I hear the tune of the Jamaican folk song "Linstead Market," I think of her town Vanity Fair, which is not far from there.[20] The lyrics are tragic, like many Jamaican and English ditties, with an upbeat tone:

> *Mi carry mi ackee go a Linstead Market.*
> *Not a quattie worth sell.*

A mother and marketwoman has failed to sell her ackees, and therefore her children will starve. The macabre song speaks to the delayed and difficult transactions of intimacy in diasporic motherhood. Today barrels with presents and supplies are sent between the Caribbean and its diaspora in lieu of being able to spend the holidays together for these split-household families. Half the children live on one continent, while the other siblings patiently wait for their visas to clear, if they ever do.

Gardening and farming are gendered work. Women are responsible for 60 to 80 percent of food production in the Global South and rural farming communities. My grandmother, like many Afro-Caribbean migrants, pooled capital and other resources to find collective means of ownership through systems of *sou sou,* also called *pawdna*. She also worked as a seamstress, and for a time as a bus conductor in London. Despite restrictive covenants and racial segregation, mutual aid and systems of loan rotation led many Black migrants to become property owners of Victorian houses because they could not rent. Calling at Clapham, Streatham, Herne Hill, Thornton Heath, Lambeth, and across London, my grandmother cultivated her plants and cared for local children who did not always have enough to eat. Cultivating gardens across oceans was painful work, and the maternal labor of sustaining required patience and planning in these grounds.

The botanical glory and heritage of the tropics thrive in small plots of land in urban centers of the North. Caribbean knowledge practices and ways of tending to the land have evolved in London and New York. West Indians of African, South Asian, Indigenous, and East Asian heritage shaped the metropolitan terroir when they cultivated their native herbs on British soil. Once they were able to own property, yards made homes out of Victorian houses. The freedom to cultivate botanical worlds and terrariums abroad helped them withstand the cold. Black and Asian gardens in the U.K. were sanctuaries of survival, tree grafting, plant healing, pruning, and brewing. Migrants plowed the grounds for the sustenance of future generations. Fante, Akan, Aja, Fon, Fula, Koramantee, Ìgbò, Ewe, Ga-Adangbe, Ashanti, and Yorùbá, among other Western African ways of tending the soil, live on in multiple entangled and microbial forms in those gardens.

In Brixton Market, shoppers haggle for produce from local grocers along Electric Avenue and Coldharbour Lane. The flavor of the Caribbean, Africa, Asia, and Latin America is replanted at home and

Brixton Market

cooked in kitchens. As you walk in between clouds of marijuana smoke and impromptu sound clashes, music plays from different speaker systems. The heady fragrant scent of the Rastafarian sacrament is part of the biological experimentation and cross-pollination derived from Indian horticulture and Hindu rituals. The aroma of home-cooked food wafted through the kitchens of Victorian houses on Sundays, ackee and callaloo, part of a ritual Ital nourishment. We claim land back through the nutrition provided intergenerationally by these garden plots.

MUTINY IN THE GARDEN:
TWINNING TAHITI AND JAMAICA

The Garden of Eden was mapped onto Tahiti in Oceania as it was in the Caribbean Sea. The islands were treated as laboratories through which botanical exchange, cross-pollination, and transplantation were taking place both for nutrition and for the purposes of cultivation. The fruits native to other shores are core to the Jamaican palette—the Otaheite apple, East Indian mango (black mango), breadfruit (from Tahiti). While visiting Tahiti in 2023, I learned how the gardens of our paradise islands are interconnected through centuries of botanical exchange. On April 28, 1789, Acting Lieutenant Fletcher Christian led a mutiny against Lieutenant William Bligh. Dramatized in the film *Mutiny on the Bounty* starring Marlon Brando (1962), it is one of the most infamous mutinies, and behind it lies a history of colonial island gardening practices that transformed the diets of Black people. Echoes of dissent haunt this botanical chapter of the climate history of how islands have been used as laboratories.

The task of the British seamen aboard the *Bounty* in 1789 was to transport breadfruit plants to Jamaica to be propagated as a cheap food source for Africans who were enslaved. Chief botanist David Nelson had the responsibility of collecting the plants at the right moment of dormancy, building on prior experience of traveling with Captain Cook. Bligh had sailed with Cook, too, and so the colonial circuits were ones of ecological extraction and exploitation of Native peoples. The crew of the *Bounty* were ill-treated by the sadistic Bligh, who decided that water should be prioritized for the plants over the men.[21] These inhumane decisions in valuation illuminate the politics of personhood, race, and planthood under British colonialism.

I learned another botanical lesson from the significant Hakka Chinese community across the islands of Tahiti. During the U.S. Civil War, whites feared that Black American emancipation would end global cotton production. Scottish planter William Stewart petitioned a British company in 1866 to devise a labor scheme to coerce Chinese indebted laborers to French Polynesia.[22] The labor experiment of the Atimaono plantation was short-lived and failed by 1873, because not enough cotton was produced to compete with the United States. By 1870 after the Civil War, the oppressive system of sharecropping by African Americans had produced more cotton than before the Civil War. But the Chinese community remained in Tahiti, and transformed the Polynesian diet with the addition of rice, soy sauce, and scallions to the local flavor palate of coconut and limes. Stewart also brought Chinese laborers to the German colony of New Guinea from 1891 to 1903—yet another island lab, complete with bioengineering dreams of European racial labor substitution schemes.

Otaheite apples thrived in Jamaica and are a staple of the diet. Their name denotes the origins of Tahiti. The Native Polynesians I met informed me that they were grateful for the botanical exchange of marijuana to the Tahitian islands. Under French law, marijuana is regulated, but as part of local philosophies and cosmologies, its uses are multiple. It is enjoyed across the island and there are anti-colonial affinities between Caribbean and Polynesian peoples based on Rastafari tenets and the music of Bob Marley. Fletcher Christian perished, and Bligh succeeded in transplanting breadfruit, which grows plentifully on trees across the archipelago. Bligh was also responsible for ackee's presence in the West Indies, having shipped it there from the Gold Coast in 1788. Under the Linnaean binomial nomenclature, ackee was named after the brutal captain: *Blighia sapida*. Though it is not eaten there, the plant of the soapberry family became a mainstay of the Jamaican diet. In 1793, William Bligh took samples of the ackee fruit to Kew Gardens, the metropolis and heart of the operation, continuing the island laboratory experimentation of British empire. In doing so, he transformed the Jamaican microbiome and the gastronomy of the islands. Is there a way for Tahitians and Jamaicans to commune on this colonial history? Together, could we build a plan for future food sovereignty and nourishment for islands that would enrich the diets and livelihood of Black and Indigenous peoples?

COMMERCE, CRIMINALIZATION, CLIMATE, AND LEGALIZATION

The gradual strides in the legalization of marijuana are opening potential botanical and economic pathways for climate repair. In some Indigenous communities in Jamaica, New York, and New Jersey, the legal ganja trade is helping to bolster movements for sovereignty and self-determination. Perversely, the U.S. government has deployed marijuana raids on Native lands as a ploy to undermine sovereignty and challenge tribal authority—the raids on the Flandreau Santee Sioux are one example.[23] Sometimes tribes—exempt from state tax—will tax at the same level as the state and put these monies toward the tribe's revenue and monthly payments to members, as is the case with casinos. Indigenous rights are constantly at risk of losing federal recognition by the United States depending on the will of the administration in power. For those who have pursued Bureau of Indian Affairs (BIA) recognition successfully, the path forward is cultivation; however, it puts them at risk with the law and subject to more surveillance. In Jamaica, too, state power is tested by the burgeoning recreational and medicinal marijuana trade.

In New Jersey, the Munsee Three Sisters Medicinal Farm is one example of a federally recognized Native tribe going into business for themselves as part of their mission of food sovereignty. Their lands are not far from the pristine and protected New Jersey Botanical Garden, with their manicured lawns, in nearby Ringwood. The Munsee Three Sisters Medicinal Farm was founded in 2019 as the vision of Chief Vincent Mann and Michaeline Picaro, who work to heal the members of the Turtle Clan who suffer the long-term carcinogenic effects of the Superfund site known as Ringwood Mines, on which their tribal land sits.[24] They cultivate local produce and sell smokeable flowers, CBD, edibles, pipes, Native crafts, incense, sage, sweetgrass, and botanical bundles. This for-profit venture blends with the nonprofit activities of the tribe for education on Native science and history. The Three Sisters Farm is an important source of employment for the tribe. Supporters of the farm forage for medicinal plants and plant new crops. It is critical to the Turtle Clan's agricultural philosophy that there is a balance on the farm, not a monocrop marijuana approach to farming. The name "Three Sisters" refers to the cosmological and intercropping principles of Native farming: planting maize, beans, and squash together.

The hope is to protect the Wanaque Reservoir from any more toxic waste and dumping by the State of New Jersey, and to contend with the poisoned groundwater from the Ford Motor Company's dumping of toxic paint sludge down mines on the clan's land in the 1980s.[25] Through food justice, the Ramapough Lunaape Nation hope to find a sovereign niche within the local economy in which marijuana can be central. They helm New Jersey's environmental justice organizing, collaborating with local universities to create initiatives for repair and regeneration of the Munsee lands. The tribe has published environmental reports and documents such as *Our Land, Our Stories: Excavating Subterranean Histories of Ringwood Mines and the Ramapough Lunaape Nation* as resources for all those battling toxicity in New Jersey.[26]

In New York, less than a hundred miles away from Lunaape territory, the Shinnecocks have opened Little Beach Harvest, a cannabis dispensary in Southampton. It opened in November 2023 near the tribe's sovereign land. Also located off a highway, the venture is eight years in the making. Since marijuana was in the process of being legalized in New York State, it has the potential to transform the Indian nation's economy. The Marijuana Regulation and Taxation Act was passed in 2021, legalizing recreational use of the controlled substance. Momentous as these developments are, the legislation is bittersweet, considering the perverse ways in which the carceral landscape for people of color has been shaped by this one plant. The first tribally owned dispensary on Long Island, Little Beach Harvest is the result of a partnership with Power Fund Partners in Boston, who have invested $1.4 million to help develop the enterprise after another investment partner named TILT left the venture. With the magnitude of luxury tourism that comes to the expensive holiday location of the Hamptons each summer, there is a strong potential customer base beyond tribe members. The legalization of marijuana in New York State has major possible reparative impacts for the Shinnecock and other tribes.

The hope is that Little Beach dispensary will lead to a different space of encounter and commerce between the nation and other Hamptons residents. The inventory features Indigenous brands, and profits will go toward the tribe. They'll sell edibles, pre-rolls, vaporizers, smoking accessories, concentrates, and marijuana flowers. With brand names like Tribe Tokes, Shinnecock Hemp, and Sovereign AK, they are marketing based on Indigenous pride and taking business into their own hands for the tribe's profit. They are gearing Little Beach toward wellness for customers twenty-one and older. With the grand opening taking place on the Shinnecock harvest holiday Nunnowa during Native Heritage Month, November is an auspicious time of year for the tribe.[27] In opposition to Thanksgiving, their celebration is an Indian Thanksgiving in which a bounty of local delicacies includes oyster stew, succotash, turkey, and clam chowder. The logo of Little Beach is a whale, representing Shinnecock Nation's reverence for the sea and marine life. Whaling economies have sustained the community for millennia and have taken members of Shinnecock Nation to ports across the world such as the South Pacific where they have sometimes settled.

In Jamaica, we see parallels to what is taking place for Native nations in New Jersey and New York in the marijuana trade. Contrary to what many non-Jamaican tourists likely assume, marijuana has long been illegal in Jamaica. The island's laws may be among the most punishing and draconian historically. As in the United States, marijuana sale, possession, and usage have led to many becoming incarcerated. This is part of why countless reggae anthems

campaign to legalize the herb. The laws are highly racialized, dating back to the colonial period. The nation is taking steps toward
reform. Rastafarians have been singled out, facing discrimination
and state-sanctioned violence in Jamaican and other Caribbean societies. Despite Rastafari being the global face of Jamaica in many
ways, marijuana laws have been used to criminalize and persecute
the religious community.

Decriminalization of possession and cultivation began on the
island in 2015 with the Dangerous Drugs (Amendment) Act.[28] Medical practitioners were permitted to import and export marijuana
for the treatment of chronic illnesses, which impacts Maroon livelihoods, as herbalism is core to their religious beliefs and to the
healing services they could offer to the broader Jamaican public.
Jamaican attorney, lecturer, and consultant to the Charles Town
Maroons Dr. Marcus Goffe has written about the emerging ganja
industry in the Maroon territories.[29] Just as with Native nations in
the United States, there is hope being placed in the botanical cultivation of weed in grow houses under Maroon control. Maroons have
historically grown ganja for medicinal and commercial purposes.
With their sovereignty backed by the eighteenth-century treaty
with the British, they have historically tested the limits of Jamaican
law when it comes to the marijuana trade. In the 1970s, ganja possession was deployed by the Jamaican government, as we saw with
the U.S. government, as a pretense to raid the Accompong Maroon
territory. Chief Richard Currie has been a proponent of the Accompong Cannabis Trust on social media, showcasing footage of ganja
plants cultivated in the hills of St. Elizabeth. He captions the images
with "Indigenous Self Determination #420."

The protected status of Indigenous people throws into contention whether Jamaican Maroons are deemed Indigenous by the
Jamaican government. Jamaica has not ratified or acknowledged
that Indigenous peoples should be protected by the International
Labour Organization (ILO) convention, purposely leaving Maroon
status ambiguous. Debates about genetic testing become part of
the argumentation to make cases for or against them as being
of Arawak heritage. Archaeological evidence in Nanny Town also

reveals Maroon and Arawak coexistence and coalition, but no amount of proof will matter. The right to self-government and self-determination, including commerce, to sustain the nation was first secured in the 1739 British Treaty. In contrasting and neighboring examples for Afro-Indigenous communities, the Maroons of Suriname and Belizeans have been granted these rights by the state, though they are indigenous to West Africa. Marcus Goffe uses these legal case studies to push for an expansive definition of indigeneity and sovereignty through the cultivation of ganja. The bud could be a loophole to enshrine Indigenous protections, permits, and licenses to allow the Maroons to be recognized for their sovereignty unambiguously. Why should it be only multinational corporations and wealthy established Jamaican brands such as the Marley Natural Dispensary of Jamaica who play a part in the botanical heritage of the island nation? What is homegrown could possibly be the answer to climate sovereignty.

9

Affective Plate Tectonics

> Linear time is a western invention; time is not linear,
> it is a marvelous entanglement, where at any
> moment, points can be chosen and solutions invented,
> without beginning or end.[1]
>
> —LINA BO BARDI

Touching down at Martinique Aimé Césaire International Airport in the winter of 2019, I began to understand more fully the perverse reality that is the European Union in the Caribbean—Martinique, Guadeloupe, and French Guiana—all départements of France. A powder keg of dissent, some mountains are dormant volcanoes, and Martinicans know this well. The euro is the common currency and you can buy champagne at the French grocery store Monoprix. When Columbus first arrived in the Caribbean, he noted seeing the island and that the Taíno called it Madiana. He left pigs and goats on a subsequent voyage in 1502, when he called the island the most beautiful place on earth. The French agree. The Caribbean in many ways functions as a time-share of European, U.S., and Canadian empires, each with their convenient vacation homes where locals speak the imperial language, and you don't even have to exchange currency at the cambio. That is until disaster strikes, as sulfur is emitted from mountain peaks across the archipelago and the name La Soufrière, which means "sulfur mine," resounds. French and former French Caribbean territories understand how the volcano is a timekeeper. The sulfurous fumes and pyroclastic flows have claimed tens of thousands of Black lives from Martinique to Guadeloupe to St. Vincent to Montserrat, thanks to European colonial mismanagement.

Just as tectonic plates continuously shift underfoot, a harmonic dissonance of global lava flows seemingly without beginning or end. This is the course of geologic time, determined by the shifting plates that divided Pangaea into the continents. Across divergent and convergent boundaries of geopolitics, we need an "affective plate tectonics" as a transnational climate policy and a way to feel crisis, to have coordinated global responses for nations not our own. Haiti and Jamaica are intimately connected, as both straddle the Enriquillo–Plantain Garden fault zone, a major left-lateral, strike-slip fault system. This means when natural disaster strikes one island it will likely soon strike the other. Isolated islands do not experience pain in isolation.

For instance, in the eastern part of Cuba there are people known as "los palestinos" who live in destitute conditions. Palestinian suffering is affectively invoked for these Cubans, and it is no accident that the impoverished palestinos are Black Cubans. Deeper than this turn of phrase are the histories of Middle Eastern migration to the Caribbean, where a significant Palestinian, Syrian, and Lebanese diaspora has found home via merchant intimacies over the past two centuries. In Jamaica, Gaza is a popular slogan and almost a catchphrase for the controversial dancehall singer Vybz Kartel, who until recently was incarcerated on charges of murder. An area of the singer's native Portmore, Jamaica, called the Borderline is referred to as Gaza because of the intensity of gun violence in this strip of land. While these parts of the world, the Caribbean and Palestine, may not be in direct communication today, they are deeply kindred through colonial intimacies and the violence of dispossession.

Is there a global emergency system that can be coordinated, such that dispossessed nations are tethered in response to one another? How will the planet survive if we cannot feel for one another beyond the fault lines of national disaster? Seismic hazard should not only be felt in the alignment these zones. A single tremor along one tectonic plate should be felt across continents because it impacts us all.

Geological time or time as recorded by the earth's rock record far precedes the borders of nation-states. A relatively recent construct of the Westphalian order (1648), this notion of nationalist division was enshrined in the United Nations charter. Since the international governing body's establishment in 1945, we have been pushed to feel allegiance and empathy along national lines; meanwhile, the well-being of our planet requires more expansive understandings of where the state begins and ends. Volcanic eruptions, tsunamis, and earthquakes form a cascade of interconnected disasters, one leading to the next. When the Afro-Martinican anti-colonial thinker Aimé Césaire wrote *Discourse on Colonialism* in 1950, he critiqued the impact of capitalism and colonialism on the global majority. The book was inherently a climate critique written from his unsovereign island of Martinique. We are all sinking because of decisions on international governance made in 1945 to address genocide that refused to reconcile the ongoing genocidal paradox of colonialism by European powers.

The Yellow Vests Protests of 2018 (Gilets Jaunes) are ongoing because dissent against colonialism has never ended within France. In 2017, suffering colonial abandonment and neglect like many Caribbean nations, Guianans blocked roadways in protest of the contrast between the standard of living in Paris and their capital city of Cayenne. The French Yellow Vests Protests were catalyzed by rising crude oil and fossil fuel prices and were said to be the most violent French movement since 1968. Reliance on diesel fuel for French car manufacturing and on heavy taxation of the working and middle classes led citizens to the streets. French Guianese protesters won this battle due to their disruption of the peace and damage to state buildings. The French government has since agreed to devote $3.4 billion USD to social and civic services across the Caribbean nation.

Even if countries successfully tackle carbon emissions, extreme weather events show us how entangled the crisis is. In 2021, Tonga and St. Vincent both survived volcanoes. Island nations, "small places," have big lessons for the globe. Why not let the islands lead? Continental rule has led to our current impasse, where we are unable to imagine or invent a way out of crisis. Shared experiences of colonial rule and neglect attune islands to local strategies for sustainable recovery from crisis. Distant archipelagoes could act in unison to form a geodynamic policy of climate federation. For the far future, this legislation is urgent based on the political power of fragmented unity. If we were able to move people beyond nation-state boundaries, climate grief could spur a potential affective plate tectonics, of which we are in dire need.

As noted in chapter 4 on underwater ecologies, Charles Darwin did not anticipate the theory of plate tectonics when he wrote his theory of corals and rising land as balanced by subsidence in ocean areas. For that matter, geologists did not fully understand or accept the theory of the seafloor spreading until 1977. The concept of continental drift was hypothesized but not fully theorized until fairly recently. New seafloor rises and rolls out of underwater vents shift the earth's tectonic plates like so many conveyor belts in a continuous flow of magma and circulation. The earth's surface is continually moving and expanding into new islands. Land that belongs to one country is not property at all; it should be beyond this status or sanction. Darwin's preoccupation with coral reefs, underwater volcanoes, and geological formations eventually led him to his grand unifying theory of evolution. Darwin's expeditions cannot be decoupled from Victorian science and how it divided and dehumanized people of color across the planet. These scientists mapped racist ideas onto geology as a justification for exhausting islands and their peoples.

Accelerating eruptions of volcanoes form a ring of fire around the globe that is resetting according to the longue durée of colonial

time and the accelerating pressure of hypercapitalism. The Pacific
and Caribbean islands experience this danger first, and always have.
Colonial historiography produces a cycle of forgetting the impact
of natural disasters and the lack of crisis forecasting, management,
or relief. Historical events erupt as part of a cycle; molten lava hard-
ens to form glassy igneous rock. Sedimented, a new layer of bedrock entombs fossils. Will we be among them? The rock cycle is
one full of magnificent transformations, constant metamorphosis
under pressure and over time. Climate insurgency and insurrection
foment in such conditions where survival for those who protect the
natural environment becomes an art form.

DECENTRALIZED ISLAND FUTURES: CLIMATE FEDERATION

Beyond governing bodies like the UN, the definition of sovereignty
exists in fragmented formations across the Caribbean. The geomor-
phology of the archipelago has led to geopolitical configuration of
how to exist with difference and sameness. The vexing matter of
island federation in the region—its horizon of hope and its cycle
of failure—has many valuable lessons for the world on coexistence
and new political modes that need to be imagined. If we rely on
the United Nations for our official definitions to script the future,
we can dream only within the limits of bureaucracy. While diplomacy is necessary as we address the climate crisis, so is insurgency
against the transnational corporations and forces that dispossess
sovereignty and extract from the earth's mantle. The seismic power
of climate insurgency could send shock waves of inspiration as protest swells. The climate crisis intensifies for the most vulnerable and
impoverished. Innovation and emergency planning without those
most affected in the room is unethical. New climate philosophies
born of collective dreaming are waiting to emerge if we can connect across tectonic plates and beyond the friction of fault lines.

There have been various attempts to consolidate power across
the chain of islands, at times led by European powers and at times
led internally by independent Caribbean countries. The cycles of

failure in the Caribbean federation (1890s, 1920s, 1960s) carry critical lessons regarding statecraft, policymaking, and fragmented governance for the planet. The short-lived West Indies Federation (1958–1962) is perhaps mourned most of the political unions. Each island became a laboratory of policymaking. Today, leaders of sovereign island nations, such as Barbados's prime minister Mia Mottley, are at the helm of the fight for global climate justice that addresses the colonial realities of the Caribbean. At COP 26, the United Nations Climate Change Conference held in Glasgow in 2021, Mottley took to the world stage to amplify the message of small island nations against the extractive ills of ongoing colonialism. She called a two-degree-Celsius rise in global temperature a "death sentence." She described how "life and livelihoods" in the Caribbean are the measure of climate change.[2] Mottley wanted major financial investments that would put the Caribbean on the map for sustainable development without relying on the U.K., the U.S., or Canada.

As part of these efforts, Mottley declared popstar Rihanna a national hero and recognized the role she could play in advocacy for the hearts and minds of the climate future of the region. Jamaica has yet to bestow this honor on Bob Marley because of debates about his lifestyle choices. The site-specific impact of oil economies and sand mining transforms these matters into crises that directly affect the everyday life of Rihanna's family members, who live in the Caribbean and are of both Bajan and Guyanese heritage. So far, the singer's nonprofit organization, the Clara Lionel Foundation, has donated $15 million USD to environmental groups, including the Black Feminist Fund, the Indigenous Environmental Network, and the Caribbean Climate Justice Project. In June 2023, Rihanna tweeted at U.S. Treasury Secretary Janet Yellen and World Bank chief Ajay Banga asking them to join Mia Mottley and "step up for communities hit hardest by climate emergencies." Using the power of her social media platforms, Rihanna asked for "bold commitments to finance and debt reforms" in a rare political action for a celebrity today.

Decentralized autonomous organizations, or DAOs, are another possible financial strategy for reorganizing sustainability efforts and fundraising for global climate disasters in the future. DAOs were first developed as a system of financing in 2016 by those invested in cryptocurrency and blockchain, but there are other ways to consider the homegrown potential for this form of banking in the Caribbean. Intentionally fragmented collectives are taking action for the climate from the bottom up. These financial technology banking structures may be needed more and more because, as of 2019, the Caribbean has suffered what is called de-risking by Canadian banks. Scotiabank, CIBC, and the Royal Bank of Canada are the main banks across the region for colonial reasons, and they have begun to abandon the region because of a lack of population growth. Canada's sudden choice to divest from the Caribbean shatters the colonial infrastructure they caused average citizens to become dependent on over the decades. A Commonwealth nation itself, Canada is not exempt from colonial attitudes and actions. However, grassroots efforts of traditional mutual aid have space to flourish, and there is potential for remittances from the diaspora to play a large role. Systems of loan rotation and microfinance have thrived across the Caribbean region, unrecorded and unknown to colonial authorities. Networks of care and local funding emerge in times of crisis and mourning to pay for burial and more. While many may be part of the unbanked and untaxed—especially those who are Black and/or Indigenous—they more likely function in nimble ways that are not always legible to state surveillance and financial institutions.

Prime Minister Mottley has been outspoken in calling for institutional overhaul that could lead to a regional Caribbean movement. Her policy recommendations include economic demands for global financing pacts and reparations paid by colonizing nations. She has said, "Global problems like the Climate Crisis show us that we simply can't address modern issues with institutions, which were created for a very different world nearly 80 years ago." In declaring Barbados free of the Commonwealth, Mottley established a major break, inspiring other small island nations who bear the brunt of increased frequency and severity of natural disasters like hurricanes and volcanoes. Independence from the United Kingdom is a holiday celebrated at least every six days across the planet. When I heard that fact and let it sink in, I heard the lyrics to "Rule, Britannia!" in my head: "Britannia, rule the waves! / Britons never, never, never shall be slaves." The United States is one of these nations, but there is a cycle of forgetting sovereignty and Commonwealth status here, too.

The forthright manner of Mottley's rhetoric draws on the language of Walter Rodney's Pan-Africanism by evoking the Caribbean and Latin America as part of the extended process of Africa's underdevelopment, because African people and people of African descent live in these regions. She is aware that Guyana is one of the fastest-growing economies due to its mining of oil underwater, and we must wonder what Rodney would have made of the windfall had he not been assassinated at the age of thirty-eight in 1980. Will the discovery of oil and deep-sea mining lead to further racial divisions in Guyana? Cycles of contested rule and turmoil due to U.S. and other external interference have bitterly splintered the nation of Guyana since it became independent in 1966.

Until recently, Guyana rivaled Haiti as the poorest country in the Western Hemisphere, according to its gross domestic product (GDP). To tackle the climate crisis, we need new metrics to truly measure wealth and perhaps an index calculating the value of a nation *not* mining its natural resources. Pollution is expensive. Eviscerating local agriculture is ultimately very expensive, creating dependencies on food imports. Africa's wealth should not be measured by gold, cocoa, bauxite, ivory, or lithium; rather, despite the bounty of these materials, we ought to see the wealth beyond what can be excavated from the earth. The price of development is that it depends on the devaluation of other nations. For this reason, fields such as development economics often participate in shoring up colonial infrastructure, especially through the proliferation of nongovernmental organizations. Rodney understood this, and he paid the price with his life. If the stakeholders at COP (the major yearly UN climate conference) do not include representation and space for vibrant participation from countries with small populations and minority communities—including Black people and Indigenous peoples—climate negotiations will end before they have begun. And we are already out of time.

When I most recently visited Jamaica, I was given a tour of the Mona Geoinformatics Institute in Kingston, and it gave me hope and inspiration. I had met the director, Dr. Ava Maxam, in Dominica, where she gave a demo on drone technologies in mapping climate crisis damage since Hurricane Maria. Dr. Maxam explained to me the practical ways in which data and mapping have the power to directly affect policy among MPs in Kingston. From car crash data to mapping violent crimes to coastal erosion decisions, so many of these systems of information are determined based on digital maps.

Dr. Maxam and her colleague Gabrielle Abraham used drone footage to survey river sedimentation rates from flooding since Hurricane Maria. From the aerial photography and the use of LIDAR technology, which uses light to record accurate aerial geospatial data, the software could track a hurricane's impact on buildings. Recently constructed structures built by Chinese contractors did not fare well when the 174-mile-per-hour winds of Maria struck the Eastern Caribbean. At Mona GIS, Dr. Maxam showed me more about the drone footage and how GIS was used to observe Jamaican mangroves. Tragically, the coastal trees with their underwater roots are being strangled by garbage that flows downstream daily from Kingston gullies. From the ridge to the reef, the water leads through the antiquated British colonial drainage architecture, and the vertical city of aqueducts overflows with refuse. With no way to recycle and no incentive for limits on the accumulation of plastic, the trash piles up.

Working with geographical information systems and in partnership with corporations such as Garmin, Jamaicans have been able to create their own independent satellite navigation databases. JAM-NAV is Jamaica's first GPS: a navigation system that gives real-time, street-level directions narrated by voice assistance, first developed over fifteen years ago. In spite of brain drain and competition with

U.S. and U.K. universities for talent and funding, the University of the West Indies is taking initiative to invent nimble homegrown technologies and cooperatives. While these systems must also contend with the perils of surveillance, Mona Geoinformatics deploys the Garmin platform for local uses, and, of course, data, as I learned from a tour from GIS experts, could be used to save lives in the future. This data can be used for emergency planning to prevent subsequent catastrophic climate losses. Without an ethical road map to navigate surveillance and especially predictive technologies, these tools could be deployed for further extraction.

＊

When was the genesis of modern capitalism?

＊

It is so easy to confuse capitalism with commerce, but they are not the same. Still, some Marxists do not fully connect the dots between climate and labor politics, because they do not attend to the rubric of race as relevant. Multiple global systems of commerce existed before modern capitalism, and for a sustainable future, we will need a new, reimagined order of commerce. Race existed before capitalism, and scholars look to the medieval period for its origins, but capitalism became refined as a colonial sorting tool only after 1492. Modern capitalism's origin depended on the trading of humans as chattel. Contrary to the philosophy of trickle-down economics, the global economic system of capitalism impoverishes us all. The true cost has been the natural environment, of which we are a part. Capitalism has impoverished our imaginations by discrediting the traditions of the global majority. It is insistent and seductive and convinces us that there can be no other system of commerce, trade, or exchange beyond itself. Whether new and more stable financial systems will develop using the banking platforms of Web3 or whether they will be entirely analog, different organizations of power are a requirement for a sustainable green future.

Competitive markets have long existed beyond Europe, but capitalism works as an ongoing dynamic that consistently bolsters the stock portfolios of the ruling elite. The colonial alibi is always the same for those who profit, most often described as "men of their time." Every generation, the future is foreclosed by such men. It is a reflex to shut down conversations with accusations of presentism, and in so doing, society lowers its collective moral standards. Innovation also slows when "men of their time" are revered without accountability. We must hold these leaders responsible for the damage they have done, or a more sustainable and equitable future will never be able to take its first breath. Climate repair will require rebuilding societies from the ruins of capitalism. While islands will become submerged, new ones will form. New markets based on new values and definitions of wealth are necessary. A reevaluation of the existing forms of currency is necessary. Why aren't the nations that protect the most biodiversity the wealthiest? Every appraisal of GDP leads to the devaluation of other sovereign and Indigenous nations and their economies.

The Western canon's approaches to science and philosophy, as systems of hierarchy and devaluation, have led us to this cannibalizing present. Because theories of the Enlightenment must be supreme in order to be true, they have bolstered themselves by devaluing the philosophies and ways of life of other societies. The Light and Truth is blinding. My book is a call to embrace the dark, which is to say the vastness of the interior unknown. The dark is also the vastness of galaxies mysterious to us. There are other sciences, methods, and schools of thought that do not require the destruction of other worlds to be correct. The existential doom of our environmental crisis is far beyond the terms of German philosopher Walter Benjamin's oft-quoted Angel of History, an unending cycle of despair.[3] The figure Benjamin describes in a 1940 article witnesses wreckage piling into a heap in a continuous catastrophe called history. With the cycle of hurricane season, Caribbean people cannot afford to live in this despair. The Antilles witness the wreckage in a different way. The repetition and timescale is oriented toward nonlinear transcendence beyond fixed Eurocentric definitions of catastrophe.

THE PYROCLASTIC TRAGEDY OF MOUNT PELÉE, MARTINIQUE, 1902

Much of the field of modern vulcanology is based on a tragic case study of Caribbean disaster research, the lessons of the tragedy that befell Martinique in 1902. Pyroclastic flows billowing avalanches of steam claimed the lives of up to 40,000 Martinicans. The city of Saint-Pierre became engulfed and was later christened the West Indian Pompeii. The French still claim this Eastern Caribbean island but offer only lip service in terms of their negligence in managing environmental crises and racism. Martinique is a part of the European Union, and yet there is no restitution for how the French have poisoned the earth with pesticides. French officials say it was too long ago to determine who is to blame. When it comes to the mass death of Black people due to environmental disasters in the Caribbean, there is no justice. Today, up to 95 percent of people in Martinique and Guadeloupe have been found with chlordecone in their blood due to the chemical (similar to DDT) that farmers have sprayed across banana fields for two decades in spite of its documented toxicity to humans. People suffer side effects from the carcinogen including slurred speech, low sperm count, and nervous tremors. The use of the chemical was authorized in 1972 by then agricultural minister of France Jacques Chirac, who would become president. It is a bitter irony that Martinique is not yet sovereign, though it remains a vibrant locus of fomenting anti-colonial philosophy.

In 1902, Saint-Pierre's estimated 30,000 to 40,000 Caribbean people were killed by a devastating eruption of Mount Pelée. Pyroclastic flows are fast-moving currents of hot gas that can reach temperatures of 1,000 degrees Celsius. In French, the term *nuée ardente,* or "burning cloud," described what islanders witnessed in Martinique. Grief tends to be segregated along racial and national fault lines; however, this volcano impacted the globe in the way that nations reckon with climate crisis today. Communist dissident Rosa Luxemburg, who was white and of a Polish and German background, wrote a speech about the French Caribbean island that demonstrated the magnitude of natural disasters.[4] Though she had never visited the faraway tropical island, Luxemburg empathized with the islanders and was an ardent student of ethnology across the Americas. Before the Richter scale was developed in 1934, Luxemburg, who would later be martyred for the Marxist cause, described Mount Pelée as a giant. She personified the island of Martinique, the volcano, and the shock waves it sent across the globe, especially for colonized territories. She threaded together the colonial plight of the Philippines, South Africa, and Mozambique, showing how these interconnected catastrophes led to similar death tolls over longer periods of time.

The tragedy of the volcano is documented at the Natural History Museum in London. Display cases contain glass bottles and other artifacts that melted in the Caribbean in 1902. How these morbid objects ended up in England is anyone's guess. The macabre fascination with the deaths of Black people became part of the luster of the Barnum & Bailey Circus, often under the guise of scientific inquiry. One man, Ludger Sylbaris, walked away from the eruption, and the circus transformed him into one of the world's first Black celebrities.[5] His prison cell kept him safe from the volcano, but Barnum & Bailey dehumanized and exploited him. Was Sylbaris the first Black climate "refugee" on the global stage? Circus posters describe him as "the only living object that survived in the 'silent city of death' where 40,000 human beings were suffocated, burned, or buried by one belching blast of Mount Pelée's volcanic eruption." Sylbaris is depicted holding the signature Martinican bakoua straw hat, made from the leaves of the bakoua tree plaited into a wide brim. What was his fate as the lone survivor of a violent regime?

Sylbaris, who was born on a plantation, became known as "the man who lived through Doomsday." Reportage of the time embellished and further racialized him. "Not one of these was saved, but a coarse negro, jet black and stupid, with little more mind than an animal, was strangely preserved to continue a worthless life."[6] The ugly language and its anti-black dehumanization show how it was possible for 40,000 Martinicans to die due to colonial negligence in the first place, because Black lives don't matter to France. The use of "jet black" here is both anti-black in how it devalues Sylbaris based

Saint-Pierre on Martinique was destroyed in 1902 with the eruption of Mount Pelée. The remains of the town's bank in the wake of the eruption.

on an epidermal calculus and capitalistic in its emphasis on the minable resource of obsidian and the igneous temporality of Mount Pelée. Jet is a gemstone, the lowest rank of coal, derived from petrified wood. Is there a blacker black than jet? The cataclysm of the Antilles became violently inscribed in the scars that would form on Sylbaris's skin as proof. But of which crime? The carceral system? European colonialism? The aftermath of enslavement? The volcano from which 40,000 Black people should have been evacuated by the white governor? The images are similar to photographs circulated of enslaved Africans with scars from the whip as testimony, brutality inscribed on the body. Yet again, there is *too much evidence* of brutality, which has been consumed by the West as a macabre curiosity.

The disaster of Mount Pelée is a textbook example of state and imperial negligence when it comes to Black life mattering as part of the climate in crisis. There are many state-sanctioned examples in the present that are best described as, in the words of Black studies scholar Clyde Woods, "planned abandonment."[7] While he describes the devastation of New Orleans, the parallels globally when it comes to Black life and climate risks are clear. Ruth Wilson Gilmore, too, writes of the consequences of this systemic abandonment of Black people and the need for abolition. Race determined the laissez-faire French governmental response to the fates of tens of thousands of Afro-Caribbean subjects. Many warning signals—none clearer than clouds of ash falling from the sky—went unheeded. This refusal to acknowledge disaster was typical of the imperial state's crisis management. Instead, Saint-Pierre became prey to scavengers who gathered for many years to gaze at the wreckage from the port. They took morbid souvenirs and transported them around the globe, as became Sylbaris's fate.

Even in 2025, earthquakes, volcanoes, and tsunamis are difficult to predict, despite vulcanologists' extensive research. Pyroclastic flows not far from Martinique are in motion on the much smaller island of Montserrat, a British territory where an active stratovolcano lies. Neither Martinique nor Montserrat are sovereign nations, which determines what climate action and aid will look like in their futures. Before 1995, the Montserrat volcano experienced a 400-year period of dormancy. Now vulcanologists predict that it will erupt again within two years. With half of the island's residents currently living in climate exile, the population of Montserratians went from 8,000 to 4,000. The islanders await the impending disaster in a British Overseas Territory with a majority Black population. What will this mean post-Brexit? Now the volcano of the Soufrière Hills is expected to erupt in thirty-year intervals. Vulcanologists do not know for certain what causes this increased frequency. However, increased global seismic activity due to sudden climate shifts, such as glacial melts, mountain erosion, and aquifer depletion, is one answer. Emitting carbon dioxide, erupting volcanoes add to global warming.

I do not remember the volcanic exodus of Montserrat's eruption as a televised catastrophe amid the other racializing events of the 1990s. I was a child, but I do remember the O. J. Simpson trial and the meaning of the race card. I vaguely remember news broadcasts reporting on Apartheid and the victories of Nelson Mandela as a political prisoner who survived. I remember Oprah, but I do not remember Montserrat. Digging for memories, even prosthetic ones, of Montserrat's tragedy in the 1990s, I cannot find them. Many islanders still remember the destruction and have never been able to return home properly, even if they can afford to visit.

The British Empire is relinquishing territories and the Commonwealth because of the upkeep and the expenditures for countries it ravaged and exploited for centuries. As the sun sets, an everything-must-go fire sale is part of the United Kingdom's downsizing and its cycles of forgetting. There are some prized territories it will never release. The imperial amnesia must deny any part of its role in inciting atrocities. Meanwhile, ongoing crimes against humanity appear as arbitrary divisions on a map of grief. How can these nations translate to aid one another across oceans impacted by British violence? Has the U.K. earmarked its domestic budget for Montserrat's next crisis, or will it shift responsibility and blame like France did with Mount Pelée in 1902? The U.K.'s share of carbon emissions during the Industrial Revolution must be tallied as part of the reparative actions it should undertake as a debt to Asia, Africa, the Middle East, Polynesia, and the Caribbean.

In Montserrat, vulcanologists fly helicopters daily over the steamy peak, observing billowing clouds of sulfur that obscure the view. During my 2023 visit I heard the sound of choppers, an ominous callback to those who could not be medevacked off the island in 1995. Yet no one knows *when* the eruption will happen, only that it *will*. What is U.K. prime minister Keir Starmer's climate relief plan and budget for the inevitable disaster of the Soufrière? Does King Charles III have a canned speech pre-written by his team for the inevitability? The monarch who claims to have advocated for climate protection since 1970 should have an environmental plan over his dominion. Charles III took a vow that he will not be a political leader, an easy speech to make to his Montserratian subjects in the coming years. The 1995 volcano showed the world how easy it is to forget Black tragedy under the long durée of British colonialism. The humanitarian crisis also demonstrated how easily the U.K. rewrites the rules of citizenship and asylum.

BLACK IRISH FREEDOM AND THE FIRE THIS TIME

Ireland raised almost $2 million (USD) in resources to help Navajo and Hopi Nations in the beginning of the Covid-19 pandemic, an example of aid as a gesture of anti-colonial struggle and the affective plate tectonics I am calling for. In 1847, the Choctaw had sent $170 to the Irish people as a gesture during the Irish Potato Famine. Both nations are owed reparations by England, I would argue.[8] Irish planters lived and defied England's reign on the island of Montserrat over the centuries. They welcomed the French, who ruled for a short time after the French Revolution. Irish planters were also the slavers on Montserratian estates, operating the whip against African people to cultivate wealth from sugar, limes, coffee, cacao, and arrowroot. Montserrat is known as the Emerald Isle; every March, the island's majority-Black population celebrates St. Patrick's Day wearing green-plaid kilts. It is the only country aside from Ireland to celebrate the patron saint's day as a bank holiday, and the weeklong festival that attracts revelers from neighboring islands is a profound statement against the British Crown. St. Patrick's Day is a yearly act of defiance masked by merrymaking. In its costumes, it resembles many African masquerade traditions in the Caribbean and Latin America, from Junkanoo in the Bahamas to Canboulay (*cannes brulées*) and Carnival in Trinidad and Tobago. What if these celebrations are missed opportunities for climate action and fundraising? Afro-diasporic carnival festivals unite the fragmented islands. The calendar extends from country to country throughout the year, from Boxing Day to New Year's Eve to pre-Lent. On St. Patrick's Day, it's Montserrat's turn, adorned in green. The weeklong celebration in March disguises a plot for Black freedom that took place centuries ago on the island. In Montserrat, there were coordinated rebellions between mountain peaks and valleys.[9]

To echo C. L. R. James's sentiment again on how power determines the way official histories are written, it is only in the pages of capitalist historians that Black struggles of revolution and emancipation did not take place.[10] Black people and others who have been historically oppressed by European colonialism have always fought and staged mutinies. Plots of social and labor upheaval were a given, whether they were thwarted or not. We heard these echoes in the coordinated labor movements of the nineteenth-century guano trade. Africans who were enslaved masterminded rebellions, determined to overturn their confinement, torture, and the very terms of chattel slavery.

While many Black people on Montserrat are likely of both African and Irish heritage, the festivities ultimately celebrate the deferred dream of emancipation. Though slavers foiled the African Montserratians' rebellion in 1768, the act still inspired hope across generations. The leaders of the revolt chose St. Patrick's Day knowing that the Irish overseers and plantation owners would be inebriated and incapacitated at Government House in Plymouth. It is said that a white woman overheard the plot and reported it to the authorities. Had it succeeded, would Montserrat have taken the sovereign path of Haiti—the first successful revolt by enslaved Africans? If so, would the island ever have become the beloved vacation sanctuary of '80s musicians such as Elton John, Mick Jagger, Stevie Wonder, and Paul McCartney? They all recorded at the famed Sir George Martin's AIR Studios. Looking at the accumulated debt and climate catastrophe half of Hispaniola suffers under the weight of France's colonial tax, the price of freedom weighs heavily.

The saccharine soundtrack of luxury Edenic tourism erases Black life and historical resistance across the Caribbean. At all-inclusive resorts like Hedonism in Negril, Jamaica, a new erotic Edenism is the governing order for nudists and self-described "swinger-friendly" guests. Eden often carries a connotation of white innocence and abandon. Eve becomes a sex object in the colonizer's fantasy. Many Westerners travel to the Caribbean and Latin America to escape and to seek a specific realm of sexual abandon. Hawaiian shirts, mai tais, and orientalist Polynesian tiki kitsch have become conflated with the Caribbean, part of a mellow tropical island aesthetic. It masks the trans-island anthem of anti-colonial rebellion.

The 1988 song "Kokomo" by the Beach Boys typifies the way the West flattens the West Indies as a vacation paradise through fantastical island-hopping. Kokomo is not a real place. The song creates a dreamy space of romantic escape for wealthy Americans.

> Aruba, Jamaica, ooh I wanna take ya
> Bermuda, Bahama, come on pretty mama[11]

As we drove with our tour guide, I noticed a row of expensive houses high on a mountainous peak on a Montserrat street called Kokomo Drive. North Americans and Europeans build their dream houses based on the catchy song. Even the Beach Boys, it turns out, didn't like that song, though it brought them a lot of money.

By contrast, I turn to the poetry of Kamau Brathwaite, the late Bajan philosopher, for a Caribbean soundtrack that celebrates the geometry and the geology of the islands. His words give shape to the arc with the delicate lilt of Black poetic intonation and intention. Brathwaite's island-hopping is entirely different from the way the Beach Boys advertise the islands for Western consumption. We can begin to find fragments of a potential future climate theory the world desperately needs in Brathwaite's mapping of decentralized unity. In the poem "Calypso," he writes:

> The stone had skidded arc'd and bloomed into islands:
> Cuba and San Domingo
> Jamaica and Puerto Rico
> Grenada Guadeloupe Bonaire[12]

Brathwaite does not fear the coral teeth, the geology that gave rise to the islands. The reef represents the living genesis of what is yet to be born across the shorelines of the Caribbean. The difference in cadence between the Beach Boys and Brathwaite is liberatory hope. He activates the archipelago as part of what was perhaps, for him, a deeper Pan-Caribbean political dream. Not only was he a bard, but an architect for those sovereignty movements. It has always been my contention that we desperately need more poets influencing policymakers. With poets at the international table to develop climate policy, what new horizons are possible? They invent and distill the language needed for an optimistic future. In 1958, Brathwaite hoped for the success of the Federation of the West Indies and expressed heartbreak as Jamaica—the largest member due to its bauxite economy—withdrew in 1962. Despite Jamaican premier Norman Manley's opposition, 54.1 percent of Jamaicans voted "No" to the federation by referendum. Among the reasons may have been not wanting to tie their fate to others, as the larger of the economies in the region. Brathwaite believed the union failed because "there was no Kingdom of the Word." He said that "a magic wand over our shattered islands" was not enough because it enthroned "our own kings and queens (disguised as presidents and premiers)."[13] The problem, as Brathwaite saw it, was the monarchical thinking and aspirations of Jamaicans. These are impediments to climate sovereignty, too. Many seek to overthrow the governing order with only the small imagination to reproduce violent regimes and economies. Newly anointed overlords who might even be people of color govern as the European monarchs did, as authoritarians.

Returning to Kokomo, driving just outside the exclusion zone, we arrived at the entrance to the fabled ruins of the Montserrat Springs Hotel. Nature reclaimed the edifice after just thirty years, and now it is time for the volcano to erupt again. Thousands of termites work dutifully to form faint lines embedded in the walls, foreshadowing the fate of the hotel. They are hollowing out the wooden structure through a choreography of interconnected tunnels. Their passageways of decomposition will eventually make way for new forests and new life, though not in our time. We must honor cycles of decay as much as those of harvest to truly reckon with the climate crisis. The ritzy hotel was named for its thermal springs, from which a spa was carved on the premises. Bands such as the Rolling Stones enjoyed the heat when they frequented the exclusive establishment in the 1980s. Standing just by the edge of the former pool at the Hotel Montserrat Springs, I peered into a region now overgrown with unruly, unrulable plant life. To my left, far in the distance, there is an unobscured view of the ruins of the buried capital city, Plymouth, which was destroyed instantaneously in 1995. Nineteen people died. Soon, the celebrity hotel will collapse into the mountainside and aging British rock stars will perhaps raise a glass to the good Caribbean times of the 1970s and '80s. They will rhapsodize about the magic of the recording studio but will never understand the cycles of foiled Montserratian plots of Black emancipation. How well does Sting—who flew into an airport that now sits in the exclusion zone—remember those years? The tarmac is melted and it has been left abandoned in Plymouth. Now, the only route onto the island is to book a charter flight on a six-seater aircraft. Twenty minutes away from Antigua, one can look down and see a stunning aerial view of curved stone and hissing coral reefs through the sapphire-blue water.

The more affordable interisland ferry transport for locals of neighboring islands was out of service for a few years since the Covid-19 pandemic began and is only now back up and running. The transportation needs of Western tourists are valued more than those of people of the diaspora or other parts of the Caribbean. As is common for most Caribbean nations, more Montserratians live outside the country than within it. Many Montserratian climate refugees have settled in London or New York and do not even hope to return, nor is it easy to afford the flight. The vista of the active steaming peak, paired with geologists flying helicopters overhead, portends the next expulsion of volcanic matter. Yet old stone sugar mills, military outposts of the brutal period of racial slavery, stand eternal.

When we exited the exclusion zone, dump trucks barreled by, blowing plumes of dust into the air. Laborers, clad in luminescent hard hats, wore masks or scarfs loosely over their noses and mouths. They looked like henchmen from the 1962 James Bond film *Dr. No*. Our guide glibly related that Montserratians had been ready for Covid from 1995, having worn N95 and respirator masks since then. Sand mining, a new economy heralded as the economic savior, has blurred the distinction between debris and the sedimented ash. We already know the consequences of this toxic inhalation. Some of the workers loosely cover their mouths with bandanas; others have decided that it might not make a difference, considering how much dust they inhale daily.

The mining, whose products are shipped to other countries, recalls the operations of the nineteenth-century guano trade in chapter 5. Are there no hemispheric lessons here about the climate consequences of vacating and exporting the geology of an island? Abandoned vacation homes sit empty, some with automatic lights still coming on and shutting off each day. On the opposite side of the island, far from the exclusion zone, mansions are kept in working order by staffs of Montserratian housekeepers and gardeners. The parallel economies of cyclic extraction are clear for the island underclass. They have no choice but to rebuild their houses anew, on different island slopes. Black communities are thriving in these other locations, and hopefully they will be able to avoid the path of pyroclastic flows next time.

The earth's crust is a dynamic system, a conveyor belt of geological instability and the political memory of fomenting rebellions. Each impossible and unthinkable revolution could change the direction of destiny, and Black and Indigenous people know this well, having lived through multiple time lines of apocalypse. Montserratians, Vincentians, and Martinicans continue to show the world how to live in the after. They have had no choice, but survival is an art, and their coordinated mobilizations carry hope in decentralization. Still, island sovereignty is a prerequisite to avert further preventable crises. Dependencies form in the aftermath of external aid.

The island of St. Vincent, whose eponymous peak bears a French name, saw the tragic eruption of La Soufrière in 2021 transpire without much notice in the nearby United States. When I taught a graduate class at Cornell University's Architecture School on Black and Indigenous Metropolitan Ecologies, none of my students had heard about the devastating eruption. We talked about this knowledge gap and the boundaries of climate aid. BBC global coverage showed the nation—its former colony—blanketed in ash. The carcinogenic impact is untold, and, tragically, the epigenetic effect of the sulfur and noxious fumes will continue to unveil itself. Home to one of the largest Afro-Indigenous communities, the twin island nation St. Vincent and the Grenadines has long been a sovereign haven. But in majority Black countries such as these, the sirens fail to trigger a global affective or aid response.

Legally, what is classified as an act of God? Insurance clauses, refunds, and rebates have a definitive answer. It's an act of nature that couldn't have been predicted or prevented, and for which no human is to blame. Beyond the future failures of FEMA and the Red Cross, island nations have depended on one another even when they have nothing. NGOs do not have the capacity to save us, even if they are approved as fiscal sponsors, because they are not incentivized to. Many across the world profit from forming dependencies of bureaucracy and money laundering under the guise of philanthropy and what Canadian author and climate activist Naomi Klein would call disaster capitalism. Decentralized networks and informal economies have been invisible systems of recovery. But these modes of self-organization are discredited, as if there are not Indigenous forms of policymaking and mutual aid in place. Neighboring islands have most swiftly provided safe harbor and supplies to climate refugees, but nothing about this intervention is evenly distributed.

Another soundtrack, the hit song "Hot, Hot, Hot" by Montserratian-Calypsonian musician Arrow, takes on a haunting resonance. It is the extant soundtrack of all-inclusive resorts across the Caribbean. "How you feeling?" Tourists in wet T-shirts mock Caribbean culture and accents in a call-and-response of bottomless pitchers of rum punch at poolside bars. They move through the limbo alongside dance instructors but know nothing of the dance's choreography of purgatory during the Middle Passage. Though the game and dance became popular in the 1950s in Trinidad, it relates to the funereal Legba dances. Papa Legba is the guardian of the crossroads between life and death. Stick dances and fire dances are common on other tropical resorts too. In 2021, after attending a friend's wedding on Big Island, part of another colonized island chain, I saw Native Hawaiian men perform the same nightly masquerade of their island culture on former coastal sugar plantation sites. They run to light tiki torches at sunset, leading a procession of gleeful tourist children who have just finished their hula lessons.

TV programs like *The White Lotus* are an inadequate means to show the perversion of colonialism because they are themselves part of a visual colonizing agenda.

"NOT IN OUR TIME":
THE VOLCANO AS TIMEKEEPER

I had asked my guide if the new parts of Montserrat made by lava flows spilling into the sea would be habitable and fertile land because of the volcanic ash.

✳

"Not in our time," he said.

✳

Hadn't the Caribbean archipelago begun this way? Thousands of small islands rising out of the sea before they were deemed America's backyard with the Monroe Doctrine?

"Not in our time." The tour guide's response gave me a sense of the timescale and poetics of relation, to use Édouard Glissant's term. The volcano is a cosmological timekeeper, and Black and Indigenous people across the Caribbean islands and Central America understand this intimately. Big Island understands too. Laguna Pueblo writer Leslie Marmon Silko gave us a codex in her novel *Almanac of the Dead* in 1991, and it included the Black Indians in the mountains of Haiti.[14] She drew her own map that encompassed the hemisphere, understanding the dark laboratory of stolen land and stolen life converging.

Islands everywhere understand this timescale of climate peril as conjoined with militaristic peril. How do we define war in the twenty-first century? Today the United States describes islands in the South China Sea—contested territories it hopes to claim— as "unsinkable aircraft carriers."[15] This imperial language refuses to acknowledge the lands as islands, demonstrating their necessity to China and the United States. If the U.S. does not have to acknowledge the islands, then their military seizure is not an act of war or

colonization. New citizens are not naturalized, just left in limbo—
taxation without representation across each new territory. Remote
islands are out of sight, out of mind until they become of strategic
importance for the military or for offshore banking. It's no coincidence that the drone's-eye view of tourism provides the same
vantage as the one the U.S. military's technology yields. Battleship
cruiser or cruise ship? Many have made these parallels as harbors
need to be widened for both types of vessels, careening into the
delicate reef. In extractivist capitalism's ongoing war on the earth's
climate, the Caribbean and the Pacific remain pivotal, albeit small,
places of reinvention contesting Western sovereignty daily.

What is a desert island? Desertion is a myth. Nonhuman ecologies are always present. The alibi of virgin land is the colonial plot
of militarized seizure. Columbus arrived in the Bahamas with the
sword and the backing of the cross behind him for this reason. From
Lost to *Love Island* to Ja Rule's Fyre Festival to *Treasure Island* to *Desert Island Discs,* the fantasy island is a genre of Edenic storytelling.
It is so often a mirage and a trap. Its plot is anything but innocent:
nature's witnesses—the singing corals, the dancing blue-footed
boobies, the furtive mother mongooses, and the blossoming buds
of marijuana—have shown us that.

Big places have always depended on small places. How can we
expect meaningful climate policy without a federation of the global
majority? The crisis is one of governance. Capitalist greed only accelerates it, but it was also not inevitable.

Beneath the water's surface in the Caribbean, Kick 'Em Jenny is
an active underwater volcano near Grenada preparing to rise out
of the sea. Vulcanologists and oceanographers keep watch over
her eruptions, measuring each centimeter of underwater activity.
The designated maritime exclusion zone is monitored by the Seismic Research Centre of the University of the West Indies. A new
seamount, Kamaʻehuakanaloa (previously known as Lōʻihi), grows
twenty miles off the south coast of Hawaiʻi's Big Island. Indigenous
and vernacular names connect the cinder cones and hot spots across
the two archipelagos of Hawaiian and Caribbean. Mythologies of
the volcanoes and coral as origin stories connect them, too. The

Menehune are a race of mischievous miniature mythic protectors and artisans in Hawai'i, fabled to be invisible guardians of the molten landscape. There are many similar beings in the Caribbean and Latin America.

In the South Pacific, as 40 percent of Tuvalu's capital district sinks below sea level at high tide, the Maldives, located in the Indian Ocean and the Arabian Sea, have already foreshadowed what underwater governance means for the planet. The Maldives' president Mohamed Nasheed agitated for the urgency of the crisis with his Underwater Cabinet in 2009.[16] As sea levels rise, the suboceanic thermal vents increase the temperature beneath the island of Montserrat. Who will the climate crisis exile next? Without an affective plate tectonics—without caring beyond the fault lines of race and nation—the planet is doomed.

Some see the Covid-19 pandemic as creating what they call "species solidarity" through a shock that pushes human evolution forward. But this line of argumentation is dangerous, as it leads to a eugenics logic. Should we not always attempt to prevent the death of society's most vulnerable? The CDC does not believe so. Scholars such as Dipesh Chakrabarty and Timothy Morton purport that the Anthropocene way of thinking pushes for "species thinking."[17] However, I contend the notion of the species is specious, and never as objective as it may seem. Scientifically, this notion has long been disproven and debated, and scientists consistently propose new definitions of species.[18] The alternative definitions show the flaw in the concept. Species were defined by interbreeding potential and intercommunicating gene pools. But this has never been as neat as biologists have wanted it to be, especially as more advances are made in molecular biology. Biological hybrids complicate notions and neat assumptions of species. Linnaeus's follies teach us the error of species thinking, because the very concept of species is a construct. Early biologists followed racist precepts to hypothesize that Africans and Europeans were different species. A human-only species-centered approach will lead the planet further into crisis because it is myopic and dangerously universalist. It silences the history of social problems that race constructed. It is convenient to jettison "race" as a concept once it has served its purpose of division. Climate universalism that erases racial difference is as much a problem as racism, and it is a common principle to hear in white-majority environmentalist circles. Many would rather forget history to focus on the crisis of the present, supposedly putting our differences aside. Reckoning with and embracing our differences could save us, if we would acknowledge how we arrived in the current climate crisis.

GHOST CHICKENS COME HOME TO ROOST

While in Montserrat, I wanted to learn about the geology of the island as much as I wanted to learn the songs of the birds that call that rock home. I wanted to listen to the echoes of Black vitality amid the narratives of death and suffering. A friend who is an ornithologist from Jamaica had told me that Montserrat was one of the best places in the world to see orioles. I was surprised to know the Baltimore bird made it that far, but I should not have been. The Montserrat oriole (*Icterus oberi*), a medium-sized black-and-yellow bird, is listed as Critically Endangered due to habitat loss. What does this say of the index for Montserratian life and who determines it? The bird is named after Frederick Albion Ober, a white British naturalist. Even in this nomenclature, Afro-Montserratian ornithology is erased. Black ornithology reveals other names, usually by the sound the bird makes as it flits through the tall grass. Or the name mimics the bird's mating call.

The biggest birding surprise for me was that the free-roaming red junglefowl—called "ghost chickens" locally—turned out to be a more fascinating bird than the treasured oriole. The island junglefowl stand as a spectral reminder, haunting the igneous black isle. Roosting cockerels are a common sight and incessant soundtrack across the Caribbean and other islands, but in Montserrat, the bounty of these chickens symbolizes something more morbid. They are the birds left behind by those fleeing the volcanic eruption. From 1995, generations of the birds have found roosts in abandoned buildings and have also populated the rainforest. The climate consequences for domesticated animals cannot be underestimated. Families of wild chickens cross the streets across the mountainous island, emerging from the forest brush at sunset.

On a wildlife tour with James "Scriber" Daley, the island's foremost bird-watching expert and nature guide, I asked him why there were so many chickens. "Does anyone own them? Could anyone who wished to eat them?" Scriber explained that as people fled the island in 1995, they had to leave everything behind, including livestock and poultry. Anyone can catch today's population of wild birds, and occasionally they are cooked up for a nice feast. There is a bounty such that chicken pilau, a local specialty cooked with curry, tomatoes, onions, and garlic, is bountiful. The reminder of the eruption looms in the grazing of the roaming fowl, untended and untamed. The capital city of Plymouth remains covered in ash. The errant junglefowl inhale the dust clouds that circulate with every gust of wind, a haunting reminder of the Commonwealth failure to protect islanders who were forced to become climate refugees or were killed.

The proverbial chickens coming home to roost seek vengeance against the colonial order. Malcolm X's famous words echoed as I watched the birds. He spoke to the reparations due from colonizing nations of the West. Climate reparations are part of this toll. He said in 1964, "It was, as I saw it, a case of 'the chickens coming home to roost.' I said that the hate in white men had not stopped with the killing of defenseless Black people, but that hate, allowed to spread unchecked, had finally struck down this country's chief magistrate." In response to the assassination of John F. Kennedy, Malcolm X continued, "Being an old farm boy myself, chickens coming home to roost never did make me sad; they always made me glad."[19] For Malcom X, there is a rural philosophy of living that embraces the rhythm of the natural world.

In his autobiography, he recounted how his words were taken out of context. Malcolm X wrote, "I said it was the same thing as had happened with Medgar Evers, with Patrice Lumumba, with Madame Nhu's husband." These words echoed across the world and still do across time. Later, in a 1964 TV interview, Malcolm X said,

"If you stick a knife in my back nine inches and pull it out six inches, there's no progress. If you pull it all the way out that's not progress. Progress is healing the wound that the blow made. And they haven't even pulled the knife out, much less healed the wound. They won't even admit the knife is there." While he may not have had environmental crisis in mind, when Malcolm X said these words, they would speak to the injury of climate ruin for generations to come. Climate degradation is a flesh wound left to fester, and so many in the West refuse to admit the knife is there. The impact of this climate denial is felt most severely by island nations like Montserrat, Jamaica, St. Vincent, and an island of Malcolm X's lineage: Grenada. Did he know much of the seamount Kick 'Em Jenny?

Malcolm X began his life in Nebraska and spoke of his rural upbringing as the fourth of seven children in a family that moved to Milwaukee in 1926 before settling in Michigan. In Lansing, his family lived on a farm, and he knew well the labor of the land and the habits of the domesticated animals. "Not only did we have our big garden, but we raised an appetite for chickens. My father would buy some baby chicks and my mother would raise them. We all loved chicken. That was one dish there was no argument with my father about." Malcolm X had fond memories of agrarian life as a child, expressed in communion with the birds that sustained them. While a sense of shame has been associated with an appetite for chicken and Black people, there is an avian philosophy of coexistence and nourishment that X identifies. As demonstrated in food historian Psyche A. Williams-Forson's *Building Houses out of Chicken Legs,* there is a rich history of poultry husbandry and cultivation in Black American socioeconomic mobility. At the same time, shame and stereotypes abound surrounding perceptions of sophisticated eating habits.[20] Malcolm X's mentions of eating chicken render a Black culinary tradition, signifying nostalgic comfort food. Black foodways and sustenance are entangled with Black climate issues and the risks of environmental racism. Continuing with the metaphor of the chicken, Malcolm X later described his time dealing drugs as if he were a vulture. He wrote, "I'd drop my stuff, and they would be on it like a chicken on corn. When you become an animal, a vulture, in the ghetto, as I had become, you enter a world of animals and vultures. It becomes truly the survival of only the fittest." As we saw in chapter 6, vultures have been unfairly maligned, but his point about scavenging is well taken.

Further back, before the United States, Malcolm's mother, Louise, was from the Spice Isle Grenada in the Caribbean. Two disparate geographies—Nebraska and Grenada—connect the affective plate tectonics through Malcolm X's genealogy. Louise Little, born Helen Louise Langdon in the 1890s in Grenada, migrated to Montreal in 1917, where an uncle of hers introduced her to Marcus Garvey's Black nationalist movement. She soon became the chapter secretary for the Universal Negro Improvement Association (UNIA). In his autobiography, Malcolm X mentions Grenada only once, saying his mother was born in the British West Indies. He notes that she looked like a white woman and that her father was white. While these details are debated, Malcolm X wrote, "She had straight black hair, and her accent did not sound like a Negro's. Of this white father of hers, I know nothing except her shame about it."[21] She was also a reporter for the UNIA newspaper, *The Negro World*. In Montreal, she met Malcolm X's father, Earl Little, a Baptist minister from Georgia. Earl died in what Louise believed to be a Ku Klux Klan plot, a motor vehicle collision in Lansing, Michigan.

While he says little of his mother in the autobiography beyond her trauma, Malcolm X remembers gardening with her, and in these lines, I hear Grenada and the arc of Caribbean islands connected across the affective plate tectonics toward Michigan and Nebraska. He writes:

> One thing in particular that I remember made me feel grateful toward my mother was that one day I went and asked her for my own garden, and she did let me have my own little plot. I loved it and took care of it well. I loved especially to grow peas. I was proud when we had them on our table. I would pull out the grass in my garden by hand when the first little blades came up. I would patrol the rows on my hands and knees for any worms and bugs, and I would kill and bury them. And sometimes when I had everything

straight and clean for my things to grow, I would lie down on my back between two rows, and I would gaze up in the blue sky at the clouds moving and think all kinds of things.[22]

The Michigan soil meant a lot to young Malcolm, such that he would recount this relationship to the plot as one of connection and chosen cultivation, as we saw with provision grounds in the last chapter. The terroir of Georgia and Grenada, the soils of which were tilled by his African ancestors, were part of Malcolm X's auto-biography. Today biographers are doing important work to illu-minate the story of Louise Little, which is remarkable in its own right.[23]

Malcolm X returned to the avian metaphors in other geopolitical statements. He wrote rather presciently:

Red China after World War II closed its doors to the Western white world. Massive Chinese agricultural, scientific, and indus-trial efforts are described in a book that *Life* magazine recently published. Some observers inside Red China have reported that the world never has known such a hate-white campaign as is now going on in this non-white country where, present birth-rates con-tinuing, in fifty more years Chinese will be half the earth's popu-lation. And it seems that some Chinese chickens will soon come home to roost, with China's recent successful nuclear tests.[24]

The memory of chickens is such that they return to the same spot reliably to sleep. Malcolm X extends the chickens-roosting meta-phor to China in recognition of the global majority and the escala-tion of the Cold War. What would he have made of China's role today when it comes to agriculture, science, and industry? Chinese birth rates have slowed, but perhaps in 1965, Malcolm X was predict-ing the economic shift that took place for China and continues to threaten U.S. dominance. Could he have imagined the combined pollution and climate impact of both superpowers and how it would affect a small island like Grenada?

When I asked a Grenadian tour guide about Louise Little, he told me he did not know that Malcolm X had anything to do with the Spice Isle. He could, however, rattle off quaint facts about '80s singer Billy Ocean, who sang "Caribbean Queen," and could speak at length about the British race car driver Lewis Hamilton, both of Grenadian roots. The guide also regaled our tour group about the tense jockeying between China and Taiwan to build stadiums on the island. But he failed to speak of the United States' invasion of Grenada beginning on October 25, 1983. Yet the global resonances of Black Power in the 1970s across the island include the legacy of the revolutionary Maurice Bishop, for whom the island's international airport is named. Bishop was martyred at age thirty-nine, similar to or the same age at which Malcolm X, Martin Luther King Jr., and Walter Rodney died. What does it mean for models of leadership in our world amid crisis if martyrdom is the only fate for the true radical?

Though he was assassinated in 1965, Malcolm X continued to shape the foundational politics of the Black Power movement, which included the Leeward Island of Grenada. Travel and a Pan-African global vision of liberation were core to his philosophy; the countries he visited included Syria, Egypt, Iran, Ghana, and Nigeria. Though Malcolm X did not travel to Grenada, he was born into an Afro-Caribbean Garveyite tradition of the UNIA, which echoes in his rhetoric. Theirs was a doctrine of Black nationalism and self-determination led by the teachings of Afro-Jamaican political figure Marcus Garvey. Malcolm X's origin story as a farm boy shows how he understood the pecking order of race and color in environmental terms. The clockwork homing instinct of the chickens coming home to roost has been invoked since his death to describe the colonial present. The rationale applies to the climate crisis, though Malcolm X would not have articulated crisis through this lens in his time. Retribution is due.

What grew in Grenada was radical because of how Black people resisted French and British rule. From 1838, South Asians brought their spices to this island, as described in the botanical and zoologi-

cal transplants of the previous two chapters, as part of the history
of racial indenture. Among the spices, nutmeg, with its lucrative
cultivation, has become a national emblem; nine pods of the fragrant seed are depicted on the Grenadian flag. Curiously, nutmeg
plays a significant role in *The Autobiography of Malcolm X* in 1965. He
describes the way it became a valued currency in prison because
of its hallucinogenic properties when consumed in large quantities.
Grinding the pods into a powder that is ingested can lead to temporary feelings of euphoria, unreality, and delirium: an escape.

From the nutmeg to the proverbial chickens, both botanical and
nonhuman animal ecologies were important to Malcolm X's articulation of karmic vengeance against white hatred and violence. Since
1977, when seafloor spreading was first identified as the way the continents formed, a reckoning with the breakup of Pangaea has been
overdue. Race formed, segmented by the continents; it became a
hierarchy based on the arbitrary trajectory of geological fractures.
Worlds separated by oceans for millennia found one another again
as globalization and modern capital took hold. Over the ages, sediments accumulate, stalactites form from calcium deposits, and glaciers erode rock to carve out canyons and salt lakes from Utah to
Arizona. Tributaries flow to the ocean, moving cloudy water full of
sediment and breaking down mountains over time.

As I waited to leave from the tiny Montserrat airport, I watched
as commercial flights were delayed to make way for the impromptu
landings of private jets. I overheard U.S. or Canadian tourists advis-
ing one another about how to adjust to what they called "island
time." "The drivers are always late," they said. Little acts of resistance and slowing down. I noticed what I could call a *deliberate* pace.
Little did the tourists know that the volcano is the timekeeper for
that island time. Everyone must wait for island time. The volcano is
anchored by a subduction zone of Black Atlantic histories waiting
to erupt to the surface. Witnesses—the corals, the mongooses, the
cormorants, the plants—have recorded the myriad crises of fallen
ash upon hardened lava. What is the planet's fate?

The nine chapters of this book are the tiny sediments, accumulating into the strata of the bedrock of a future island. Can we imag-

ine sovereign islands beyond coloniality? How impossible is a world beyond capitalism? If we do not dare to ask, then climate ruin is the only possible answer. A return to Eden would be violent for many, just as a future without race does not mean a future without racism. Whiteness did not exist in the story of Adam and Eve, and yet supremacists have projected race onto it. The other gods beyond the garden inform untold scientific modes that hold technological promise. Other galaxies of thought, endless worlds, with potential other climate strategies.

Did mountains become gods or did gods become mountains? Orogeny, or mountain building, describes the action of creative friction on a tectonic level. Two plates collide at the margins of continents, where ecologies merge and form new ecosystems. Along fault lines, islands emerged, born out of the sea, promising new possibility. The grounds of freedom shifted as formerly enslaved Africans captured land and stole back time. Were the mountains gods who had once been venerated ancestors? The bitter irony is that European planters did not plant at all. They reaped what they stole: the land and the people. They exhausted the soil while becoming masters over men, frustrated that they could never become gods.

Mountains and valleys hold the echoes of myths, lullabies, and dirges for the overtaxed soil. Ballads preserve histories and rituals of cultivation. In these songs, we could listen to the story of your four grandparents too, shaped by an ecology and ecosystem, whether known to you or not. But we should also listen across tectonic boundaries to one another's stories beyond biological relation, extending our awareness to the maps of chosen family. At the scale of island thinking, the magnitude of what climate justice needs to encompass becomes precise. Colonialism, whether of the settler or of the absentee variety, is a rigid structure that lacks imagination. It violently forecloses futures by reproducing old world orders. The system has discredited and tried to erase the very scientific knowledge that could promise climate salvation across the fault lines that divide us.

Coda

Burial at Sea

تقبرني

"May you bury me."

At the age of ten, I was pulled underwater by the currents of the stormy Atlantic. I nearly drowned.

My family had recently migrated from London to the United States, where I was already becoming a bold little New Yorker. The five boroughs embraced and initiated us, immigrants, in a way that London had always refused. Britishness, with its deceptive polite exterior, dismisses foreigners and uses unspoken codes that immigrants can never become fluent in. My little sister prayed at night for an American accent, hoping to trade in our British ones.

When I was pulled beneath the waves, I held on to my sister and my cousin and gasped for air, able to breathe only water into my lungs. We were being swept away by the cold and unforgiving gray tide. We were dragged beneath by the strengthening force of the undertow. Each wave buoyed us and crashed us down deeper and farther from the cold, sandy shore of Montauk. The rhythm of the waves gave us hope that sank with each cycle as we spun farther and farther away from the coast. I have been writing this coda since I was ten.

Montauk is the end of the line, the end of the island—a town nicknamed "the End"; only when I was older did I learn that it was named for a people continuously denied their sovereignty.[1] The Montaukett, also known as Manitowoc, were stripped of their legal title as a federally recognized tribe in 1910 and have been fighting to get it back ever since. It should come as no surprise that Robert Moses is part of this New York story of dispossession.[2]

When I was in fourth grade, we read textbooks about the design of Iroquois longhouses, architectures suspended in the past, as a mandated part of New York State's curriculum. I did not know then that the Iroquois are the Haudenosaunee, or the People of the Long-houses. They were named not only for an architecture but for what those single-room communal dwellings structured about life for several generations of an extended family living together. At the age of ten, I was not taught about the survival of Algonquian-speaking tribes, much less Indigenous futurisms of the Montaukett—or that North America will always be part of Turtle Island.[3] In art class, we made models of longhouses by gluing together stacks of Popsicle sticks. Schooling suggests these placed the Haudenosaunee in the past, frozen in time as if the tribes were not expected to engage in the present. What of futurist Indigenous confederacies?

I did not know about the historically Black community of Free-town in East Hampton, not far from Montauk.[4] Or that the Mon-taukett were forcibly displaced to Wisconsin in 1784, or that Sag Harbor was a historically Black town, or that Shinnecock Nation's territory is in the resort town of Southampton. Afro-Indigenous coalitions were storied and strong here, as they were in the Carib-bean. The Dutch and the English were responsible for both colonial incursions. Native people from what is now the United States were also conscripted into indentured labor and forced to go to the Carib-bean during King Philip's War (1675–1676).[5] Ports of call for inden-tured Native Americans included Barbados, Bermuda, and Jamaica.[6]

Their fate is unknown to us, but their descendants could be many across the Caribbean islands.

Our six small feet reached for something that was no longer there: the sandbar. It was growing cold as summer turned to fall. What had been beneath us just moments ago was gone. Our parents chatted on the beach, not noticing we were in danger until it was too late. I looked upward and began to accept our fate. With each scream, the possibility that we could get back to shore felt more and more distant.

Montauk had not yet become the Hamptons brand, the tourist destination further erasing the Native tribe's name by using it. During the New England Hurricane of 1938, Montauk became an island due to extreme flooding. Black feminist meteorology forecasts the future based on cycles of memory and experience over deep time, which is Black geologic time. The Category 5 hurricane originated from the coast of West Africa on September 9. It reached the Sargasso Sea and the Bahamas on September 16. New York was not ready. It arrived on Long Island as a Category 3 hurricane on September 21. Harvard and Yale owned forests as part of their forestry school on Long Island that were ravaged by the 1938 hurricane. Again, we see the colonizing mission of U.S. institutions that seek to dispossess the sovereignty of Native peoples into the twentieth century, amid the shifting climate. Such efforts continue across all levels of U.S. governance, and Native peoples and lawyers continue to fight, sometimes buying back land privately.

In the water, I felt calm and cold, resigned to my fate. I loved the water but had never learned how to swim despite many beginner lessons. My sister and I had never completed the courses, as our lessons were interrupted when we moved to the U.S. The last swim teacher I remember taught in a rec center in Forest Hills, Queens: a white coach who scolded me. How could a Black girl have no rhythm, he demanded to know. I thought freestyle meant I could choose to breathe when I wanted. All I could seem to perfect were the butterfly and back float—the dead man's float. (I later learned it is also called the survival float.)

I understand the rules but am never good at following them exactly. Black people are required to know the rules better than anyone, as a matter of life or death; we are always reinventing and re-creating them. We subvert styles and invent new forms. Survival is our rule. I never learned to swim, and whether that is related to being a so-called Black inner-city urban youth, I will never know. My parents were city dwellers who also never learned to swim. My mother, who was born in Kingston, grew up not even owning a swimsuit. This was not uncommon. My father never learned throughout his childhood in London. That day in Montauk, they desperately tried to swim anyway, wading fully clothed and yelling out from the shore once they realized the three of us children were being swept away.

When I returned to Long Island's shores in 2021, it was because an experimental film was commissioned by the subscription streaming service Hulu to be made about how my life and research bring African and Asian worlds together. The director had somehow chosen this location based on reading my work. For the voiceover of the film, I recited the words, "I am the sedimented sum of four islands—the Caribbean, Hong Kong, the British Isles, and New York—archipelagoes fragmented." I must have said the words four hundred times, and they helped me realize the ethos of this book as an island story of climate crisis and climate hope. None of the takes I recorded were good enough, according to the director. "Say it like Robert Redford looking out over a river," they told me. I am not an actor, and try as I might, I couldn't get the words out the way they wanted them performed. A crew member took my place, and she nailed the voiceover recording with one wistful take.

As in Montauk, each shoreline has its forgiving and unforgiving lessons. We need only remain ready to hear them. We must be open to learning from nature's contradictions. I conversed with the director of the Hulu production about the fact that, although we were not related by blood, our shared heritage of African and

Asian, Atlantic and Pacific currents made it feel like we had met many times before. We knew, from piecing our family histories together and filling in each other's gaps, that Africa and Asia had met many times before over the centuries. Filming on that shore, New York became our collective island inheritance, an archipelagic story beyond laboratories of experimentation.

The beachfront movie set became a ritual space that taught me just how interconnected the geologies of those four islands are. Everyone has a climate story like this within them, known or unknown. We are bound to protect these climates because the land is a pact. From New York, where there are vibrant Black, Indigenous, Asian, and Caribbean communities, I have inherited the philosophies of neglected urban wetlands and swamp territories. I have learned from the patterns of white flight in Queens. From the coastlines of Shenzhen in southeastern China and the Gold Coast of West Africa, I have learned of the polluting effects that attend booming commerce and unbridled corporate greed. The contested contact zones of hypercapitalism, from Guangzhou to Lagos, have been deeply shaped between river deltas, too, where the Pearl River meets the Niger. Ecological histories of cultivation in these sedimented riverbeds carry futurist technologies of climate repair.

What the philosophy of the watercourse way, the *Tao Te Ching*, has to do with me baffles most Chinese people from China. They are puzzled, assuming this Black woman's confused parents might have been Taoists to bestow her with such a name. For me, the name Tao guides an intergenerational climate philosophy of reckoning with my own disinheritance from native lands that I will never know. Over the long span of time, the *wushu* of the river has the power to move mountains and sculpt boulders. If we zoom out to reckon with geological time in the climate crisis, all hope is not lost. My home of New York is made up of forty-plus islands and continues to be shaped by the surrounding estuaries and flood zones. Irene, Sandy, and Ida have taught us this. Each day, just outside of New York Harbor, at the mouth of the Hudson, the archipelagic patterns instruct us about coastal repair and water's memory. From the Dark Lab offices on the Upper East Side, we venture out on field research trips, sailing outward to Manhattan's Pier 11 at Wall Street. We depart for whale watching expeditions and together we learn about the restorative science of oyster reefs and how sightings of dolphins in the Hudson and humpback whales in the Atlantic mean hope is not lost.

But I have learned even more by listening to the Shinnecock summer dance and to philosophies of aquaculture: whaling, clam digging, and scalloping in the unceded territory of Shinnecock Nation. The Montaukett are also still here on Long Island, fighting against their disenfranchisement and lack of federal recognition. Golf courses are a sacrilegious and fabricated landscape of deforestation, marring burial mounds of Native peoples and using Native peoples as mascots and logos. The decadence of the Hamptons is haunted by unrelenting Shinnecock activists who protest outside country clubs.[7] Many tribe members are people of African and Native heritage, and they fight for Land Back and the repatriation of their ancestors' bones. The American Museum of Natural History has only just returned skeletal remains to Shinnecock Nation as of 2023.

When we embrace nondualism, the binary between the land and the sea becomes so clearly invented. Like the sandbar beneath my feet as a ten-year-old in Montauk, structures that feel so fixed can radically shift underfoot and surprise us. But if they can disappear, they can also reemerge unexpectedly. Climate storytelling requires tuning in to heed the lessons of corals before we lose them all. Through coral, we comprehend how newer islands are alive with polyp activity. Corals are the missing link between the inanimate geology and the animacy of life. They are an ancestor. The waterways of the island of Manhattan, since built over by Europeans, show us where life gathered in Lenapehoking, or the homelands of the Lenape tribe. Collect Pond, the freshwater pond in what is now China-town, was a bountiful space of genesis; then a jail nicknamed the Tombs in 1838, it is now a park in front of the criminal and supreme courts. What justice is there without climate justice, without racial justice? Pearl Street is named for the riches of oysters that thrived downtown. Water Street downtown is named for where the water used to be before the island was expanded. When the memory of this water floods the southernmost point of Manhattan, there is a climate story of the shorelines occupied by the Dutch and then the British. The economy and exchange of oysters and clams connected the Munsee bands to Shinnecock Nation, where these bivalves were a valuable part of the ecosystem and filtered water of impurities as part of the delicate balance of the coastal waters.

As we filmed on the shorelines of Long Island in 2021, the setting of the Atlantic seaside was an intimate homegoing. The camera allowed us to time travel through multiple time lines. The actors ended up dancing in the unplanned formation of what began to look like a ring shout, a dance enslaved Africans traditionally performed in a circular formation of hand-holding and ecstatic chanting. All of this took place not far from where I had almost drowned at the age of ten. Coincidences on set between the cast and crew felt like the names of ancestors, unknown to us and revealed to guide our way. Using experimental filmmaking has become a core method for my art practice and research at Dark Lab. The shoreline ecologies hold many unexpected lessons, powerful as riptides. The ebbing tide at Shinnecock Inlet extends more than 980 feet offshore.

That day in my childhood, a young off-duty lifeguard happened to be running by along the beach, probably exercising, when he spotted us, our small arms breaking through the water and reaching for sky. He dove into the ocean and pulled the three of us from the undertow. Had he not been there, the Atlantic would have kept us forever. Luck does not begin to describe the timing. Salvation seemed foreclosed, and the fear and respect I gained that day for the power of the ocean has shaped my research. Mami Wata is vengeful. I've devoted my life to studying the currents of consequence for peoples whose destiny was changed forever by the Atlantic, and I have done this work as a university educator since 2011. I have studied the haunted Atlantic crossings as rip currents of history that have claimed so many people of color, entire histories. The riptide of climate crisis will claim us all if we continue to skirt addressing race as a critical part of the crisis.

Conservationism and industrialism work hand in hand to omit race as a relevant factor to progress, often in ways that perpetuate racism. In my archival research through the personal effects of New England conservationists, I have found shares and receipts of stock options for mining companies. The decision-making of our shared futures depends on embracing the horizon of limitless possibility that will come with restoring sovereignty over the land and oceans to those who have been their protectors. The only other option is to continue to dispossess future generations, spending their inheritance before they are born. The failure to see how labor history is intimately a part of climate history shows us another fallacy. Committed radicals must not consider environmental debates as secondary or sentimental. If we cannot breathe, nothing else is possible. What could be more political than the air itself?

Nothing is green about the monocrop violence of pastoral Long Island farmlands. They are part of an extended and fabricated bucolic imaginary of homogeneity. And we all pay a devastatingly high climate price for the manicured lawns of suburbia. California, with its Manifest Destiny innocence, knows this well. Like the sugarcane plantations of the Caribbean and South America, these plots of suburban land had been dense woodlands and forests before the European incursion. Will the rich biodiversity ever return to Long Island? Country clubs and golf clubs are not inevitable; nor is it incidental how racism structures membership and the landscape architecture of such coded places.

Shinnecocks—people of the stony shore—have been pushed to the edge of their land, which builds at the easternmost point of Long Island, a knowing finger that juts out into the Atlantic. It is estimated that sea level rise will claim the land, with between 2.1 and 4.4 feet of rise by the end of the century.[8] Shinnecock scientists and climate experts are producing their own reports and meteorological models for what is to come. Still, the tribe stewards island survival by facilitating the repair of Long Island shorelines for everyone. They intend to plant flood-tolerant trees such as black gum and bald cypress to stabilize the shoreline and control its erosion. Amid the profligate wealth of the Hamptons, which functions under the same absentee logic of Caribbean vacation homes, Shinnecock Nation has held on to the sovereignty of its ancestral lands.

The Shinnecock know Long Island's true name as Sewanhacky, the isle of shells, because of the abundant quahog, or hard clam. These bivalves play an important role in water filtration for the marine ecosystem. They also protect the island as a barrier and foundation of seawall that guard coastlines against storm surges. As ecological extraction accelerates, the failure to recognize race as a factor in systemic disenfranchisement puts the climate more and more at risk. Environmental racism continues to show us whom society regards as expendable.

Time traveling from 1492 to the far future has been necessary for the scale of imagination of this book and will be necessary to face the climate crisis. Poets and policymakers will be critical to the scale of empathy we need. Everyone must take accountability as a first responder, like the lifeguard on that fateful day, in the crisis instead of waiting for governments and corporations to solve a problem they created. Would we turn to men like Columbus to solve the climate crisis?

*

May we live to unlearn the American kind of love, which is a dangerous kind of love. It is a possessive love. It is a colonizing love, a greedy love. "I love your shoes" has meant "I want your shoes to be my shoes and I will take them from your feet while you wear them." "I like your hair" has meant "I will scalp you." American love is dangerous to those on whom it fixates. It is dissection. It is violence and possession.

*

May you bury me.

While on a studio visit, I learned this Arabic phrase from an MFA art student whose work I was evaluating. Sharing the meaning with me in Lebanese, the student, who is of Irish and Lebanese heritage, said, "Ya'aburnee." This, it occurred to me, is a different type of love. They were exploring expressions of love in Gaelic and the tradition of street art in solidarity by Irish people with the Palestinian cause through this phrase. I found the student's insight so poignant. An idiomatic declaration of love that wishes not to live to see the other die. May you bury me. It would be unbearable to continue living without you. It is said that no parent should have to bury their child. It is not the natural order of things. In the climate crisis we bury the future every day. Palestinian parents are burying their children every day, as the world watches. The only kind of commitment that can save us must be global and interdependent against the regime of stolen land and stolen life.

All of us or none of us will survive the crisis before exhaling the last breath.

5

Acknowledgments

There are many experts, scholars, researchers, and enthusiasts without whose guidance, teaching, and intellectual generosity this work would not exist. In particular, I would like to thank radical intellectual companions in parallel struggle to produce art that articulates the urgency of the multiple time lines of apocalypse. I knew I would not be ready to write this book until my characters arrived to me, and they have emerged in the form of budding corals, the mother mongoose, and lava flows. I am also forever indebted to the rainforest guides who entrusted me with their knowledge, from those in the Maroon territories of Jamaica to the Kalinago Amerindian reserve of Dominica. The blueprint of ongoing Black Indigenous struggle for over 400 years forms the deep optimism with which this book was written.

For the final push to the top of the mountain, I will always be grateful to my editor, Thomas Gebremedhin, at Doubleday for guiding this process as one that is urgent, vital, and about the beauty of healing. For her assistance as part of the New York team, I am thankful to assistant editor Johanna Zwirner, especially as we decided on images of coral and birds together. I'm grateful to Kristen Bearse,

Milena Brown, Sara Hayet, and the rest of the Doubleday team who worked to bring this book to life. To my agent Ian Bonaparte, the interpreter of my dreams of Eden whose wish for me to spread my wings beyond the coop has come true. To Emily Bonaparte for her generous advice during the writing of this book. To the U.K. team, Hamish Hamilton editor Simon Prosser and assistant editor Ruby Fatimilehin for her keen intellect and attention to historical details. The welcome in London, as we watched the ebb of the Thames, was so meaningful to my Drexciya dreaming. All of you have provided needed encouragement, clear vision, and wisdom, for which I'm immensely grateful. To my agent Leslie Shipman and her team, including Katie McDonough, I am ever appreciative. Thanks also to editor Hannah Chukwu for first hearing the rainforest symphony and for the evening at the Royal Society of Literature soiree. I have had an excellent transatlantic team at Janklow and Nesbit UK as well, with the guidance of agent Rebecca Carter. A dream team of support.

Thanks are due to Oliver Munday for seeing the texture and color of my climate vision in designing the book cover. I am grateful to David Lindroth for attempting to encompass the submerged unity of the Caribbean islands. Thanks to production editor Melissa Yoon for her careful and thoughtful eye. At Stanley Nelson's Firelight Media, thanks to Justin Sherwood for his strategic insight. I would like to thank the members of the Dark Lab advisors board, who are the leaders in Hollywood, Big Tech, and academia. They form a constellation of Black and Indigenous philosophy, including Henry Louis Gates Jr., Fred Moten, Claudia Rankine, Tracy Rector, Kamal Sinclair, Tracy K. Smith, Eve Tuck, and many more. To the Dark Lab and Afro-Asia Group teams, which encompass interns, technicians, and theorists worldwide, I would like to extend special gratitude to Hashem Abushama, Amber Starks, Sara Salem, Joshua Bennett, David Gonzalez, Jesse Sgambati, Matt Hooley, Aree Worawongwasu, Mónica Ramírez Bernal, Rewa Phansalkar, and Lydia Macklin Camel. To summer intern research assistants from Cornell, Amherst, Princeton, Yale, and Fordham: Leanna Humphrey, Kendall Greene, Koby Chen, Humyra Karim, Nour Darragi, Arianna Qianru James, Ariana White, Tatyana Tandanpolie, and Maria Aguirre. To the hundreds of students over the years who have enrolled in my courses, too many to count: please know the imprint of each of your lessons of call-and-response is collectively here.

I am enormously grateful to those who read part or all of the book at various stages of its growth: I could not have done without their insight and sensitivity. Thank you to Kiese Laymon for reminding me about the truth of why we keep going. Invitations to share work in progress and to present keynote lectures helped me crystallize the vision for Dark Laboratory as a book. Thanks to Rachel Bernard, Benjamin Keisling, and Raquel Bryant for the invitation to give the Geosciences keynote in Washington, D.C. Thanks to Stanford University in Global Studies for the invitation to present a keynote on mangroves as a Caribbean method. Thanks to Tom Western and Tariq Jazeel for the invitation to share my coral art research at University College London. It was an honor in 2023 to deliver a UCLA keynote in Performance Studies on what it means to "be led." Thanks also for the long-term support of my writing by editors Elizabeth Ault and Ken Wissoker.

For many kinds of inspiration and generosity along the way I thank Shannon Gleeson, Shannon Mattern, Ashmita Lama, Ana Paulina Lee, Courtney Sato, Heather Hart, Yasmin Elayat, Pamela Phatsimo Sunstrum, Mary Sabbatino, Imani Perry, Anibal Luque, Cajetan Iheka, Martina Abraham-Ilunga, Sim Chi Yin, Amber Musser, Alex Moulton, Jane Debevoise, Emma Dabiri, Kandis Williams, X Zhu-Nowell, Julie Dash, Tuan Andrew Nguyen, Andrea Chung, John Tain, Denise Ryner, Merv Espina, Ruha Benjamin, Simone Browne, Kaysha Corinealdi, Felicia Chang and Zaake DeConnick of Plantain, Keolu Fox, Lisa Kim, Xiaowei Wang, Courtney Bryan, Uli Baer, Eric Zinner, Furen Dai, Kevon Rhiney, Maaza Mengiste, Marco Navarro, Sonya Posmentier, Monica Kim, Grace Hong, Ariella Azoulay, Dan-el Padilla Peralta, Adam Banks, Mireya Loza, Kayoung Lee, Maria Koenigs, Chloë Bass, Kris Manjapra, Erna Brodber,

Gloria Wekker, David Scott, Tim Barringer, Jon Goff, Iracel Rivera, Chitra Ramalingan, Ashley James, Annie DeSaussure, A. Kristin Okoli, Carmen LoBue, Courtney Cogburn, Jen Tomassi at Hulu, Monique Truong, Joan Kee, Orlando Jones, Nari Ward, Anne Ishii, Courttia Newland, Cecile Chong, Mpho Matsipa, Vivian Crawford, Carolyn Cooper, O'Neil Lawrence, Ben Mylius, Alisa Petrosova, Schuyler Esprit, Ava Maxam, Anthony Browne, Arlene Torres, Jillian Baez, Alhaji Conteh, and Deborah Thomas.

I am thankful to Hunter College, CUNY, my new academic home, for celebrating the book and to Dean Andrew Polsky and Eric Chito Childs for offering a sabbatical for the research. I am thankful to Ruth Wilson Gilmore at CUNY Graduate Center for her ground-breaking work, and for the invitation to be in community next year on anti-capitalism and environmentalism. I am indebted to many wonderful teachers and mentors who have guided me over the years, in particular Michael Denning, Wai Chee Dimock, Anthony Reed, Hazel Carby, Stéphane Gerson, and Annemarie Luijendijk. Without Toni Morrison as a professor, this book would not exist, and neither would the lab. For her writing and tutelage, I am eternally grateful and lucky. The support of the Boys Club of New York, especially Stephen Tosh, Avita Bansee, and M. Shadhee Malaklou at the bell hooks center, has been very meaningful as we build new Dark Lab initiatives in the outdoors.

I am thankful to the sport of fencing. Since I first picked up the sword at the age of seven, it has enabled a philosophical and embod-ied process of martial arts. Thanks to the Peter Westbrook Foun-dation, a Black club in an elite white space where I learned more lessons on and off the strip than I can fully comprehend about the footwork of navigating institutions designed to exclude us based on the color of our skin and so-called pedigree.

Thanks to the Caribbean naturalists, hikers, bird-watchers, and orni-thologists I have met along the way: Scriber in Montserrat, Colonel

Sterling in Jamaica, Dr. Birdy in Dominica, Sean Dilrosun and Fred Pansa in Suriname, and Leo Douglas in New York.

I am grateful for the support of several institutions: Yale University, Princeton University, where I spent over a decade in study and apprenticeship. At Leiden University, Phillipe Peycam, the director at the Institute for Asian Studies, was a wonderful mentor for Afro-Asia research and decentering the West. I am grateful also for the fellowships I was awarded to perform research in Shanghai in 2017 and London in 2018. The writing I was able to do at Senate House, the British Museum, and the British Library was invaluable. Being artist in residence at Columbia University's Columbia Climate School over the past two years has truly supported the research for this book and my creative practices. Art residencies with Andrea Chung in California and at the University of Connecticut were also an education in how to navigate between institutions. With thanks to the Guggenheim for the invitation to participate in the Sixth Asian Arts Council in Kingston, Jamaica, a life-changing experience. I have also been fortunate to be a part of NEW INC at the New Museum, led by director Salome Asega and deputy director Raul Zbengchi, who supported these dreams with physical space, mentorship, and the community of a bustling co-working space. I would also like to thank the Harvard Carr Center at the Kennedy School for their support of the Dark Laboratory for our work on racial justice and coastal ecologies. Any omissions of institutions I may have been affiliated with in the past are deliberate, because not all are supportive of the type of research and art practice I advocate for in this book. It is an uphill battle, which is why I am evermore grateful to those that financially helped with Dark Lab's vision for climate and racial justice.

The archives and respective custodians of the Geological Society, the Royal Academy, the Linnean Society, Cold Harbor Spring, the Beinecke, Kew Gardens, the New York Botanical Garden, the New York Society Library, the New York Public Library, the Schomburg

Center, Asia Art Archive in Hong Kong and America, the Museum of Food and Drink, the Asian Art Museum, the Museum of the African Diaspora, Wave Hill, the National Gallery of Jamaica, and the Museum of Contemporary African Diasporan Arts. I am grateful to the CV Starr Fellowship that enabled me to travel to Suriname, where I met my long-lost family members. Gratitude is owed to my co-PI Eddie Bruce Jones for being an intellectual partner in our transatlantic National Endowment for the Humanities and AHRC UKRI grant. Thanks to the Dark Lab artists in residence who I have been able to sponsor over the years with financial support from grant funding I have raised: Nadia Huggins, Abigail Hadeed, and Maya Cozier. Thanks to chef-in-residence Jenny Lau and terroir specialist Patricia Powell.

Countless hours of music have helped me to think and feel my way through the scale and register of this book. Of particular importance have been the sounds of Drexciya, Maroon drumming, Jacob Miller, Dolly Parton, Prince, Audioslave, Rage Against the Machine, Waka Flocka Flame, Jeff Buckley, Sade, the Revolutionaires, King Tubby, Ennio Morricone, Kiddus I, and Little Roy.

The two main places that have guided the emergence of this book are Jamaica and Hong Kong. To these islands, and to all those who inhabit them, I owe more than I can express. For their inspiration, love, wisdom, and endless patience I am grateful to clan and kin: Daksha, Dinesh, and Sahil in West Bridgford to Streatham to Aotearoa to the five boroughs to Shenzhen to Paramaribo. Thank you especially to my cousin from Suriname, Natasha. To my parents and sister in New Jersey with all the love that is possible to give. Love also to my brother-in-law and nephew. Thanks to Tim and Sarah for the inspiring visit to Brantwood, the historic house of John Ruskin in Cumbria, and to see peregrine falcons at the Limestone Pavement in 2008. I have learned as much from the dales, the moors, and the peaks of England as I have from visits to stately homes like Harewood House. Thanks to Louis Goffe and Kayoung Goffe, the kids, and the bumblebees for the encouragement along this journey. Most of all, with love always, I thank Romil, who joined me in the rainforests of Jamaica and Tahiti and read all the drafts of this book as it gestated over the years. Without him I could not have had the courage to photograph and film the strange underwater magic of corals, stingrays, and reef sharks. From this eternity to the next, I know we will continue our island adventures together. In this year of the dragon, our love continues to grow as we celebrate the meaning of "life after life" together.

Notes

INTRODUCTION

1. Ivan Penn and Eric Lipton, "The Lithium Gold Rush: Inside the Race to Power Electric Vehicles," *New York Times*, May 6, 2021.
2. James Baldwin, *Go Tell It on the Mountain* (New York: Vintage International, 2013).
3. Martin Luther King Jr., *Where Do We Go from Here: Chaos or Community?* (Boston: Beacon Press, 1967).
4. "Uncovering Tribal Connections to the Underground Railroad," National Park Service, https://www.nps.gov/articles/000/tribal-ugrr-connections.htm.
5. Grace L. Dillon, ed., *Walking the Clouds: An Anthology of Indigenous Science Fiction* (Tucson: University of Arizona Press, 2012).
6. Institute for Economics & Peace, "Over One Billion People at Threat of Being Displaced by 2050 Due to Environmental Change, Conflict and Civil Unrest," September 9, 2020. Intergovernmental Panel on Climate Change IPCC, 2022: Climate Change 2022: Impacts, Adaptation and Vulnerability. Contribution of Working Group II to the Sixth Assessment Report of the Intergovernmental Panel on Climate Change [H. Otto-Pörtner, D. C. Roberts, M. Tignor, E. S. Poloczanska, K. Mintenbeck, A. Alegría, M. Craig, S. Langsdorf, S. Löschke, V. Möller, A. Okem, B. Rama (eds.)]. Cambridge University Press, Cambridge, U.K., and New York, NY, USA, 2023, 3,056 pp., doi:10.1017/9781009325844.
7. "Mountain Areas Crucial for 'Survival of the Planet,'" *UN News*, May 6, 2016.
8. bell hooks, "Appalachian Elegy," in *Appalachian Elegy: Poetry and Place* (Lexington: University Press of Kentucky, 2012).
9. Steven Field used the term "acoustemology" in 1992 to foreground the sonic

experience as a mode of knowing. He derived the definition from fieldwork in Papua New Guinea with the Kaluli.

10. Peter Tosh and the Wailers, "Go Tell It on the Mountain," 1970.

11. From Auckland to Oakland, the Polynesian Panther Party (PPP) formed in 1971, influenced by the Black Panther Party. The Dalit Panthers, inspired by the Black Panthers, formed from 1972 to 1977 to fight against caste discrimination.

12. David A. Biggs, *Footprints of War: Militarized Landscapes in Vietnam* (University of Washington Press, 2018).

13. Ruth Iyob, *The Eritrean Struggle for Independence: Dominion, Resistance, Nationalism* (Cambridge University Press, 1995).

14. "Sagarmatha" translates from Nepali as "forehead or head in the sky."

15. "Planetary History: Growth in the Anthropocene Research Project 2018–2021," University of Chicago.

16. Other scholars such as Donna Haraway and Anna Tsing use terms such as "Cluthocene" and "Plantationocene." While these in some way attend to the flaws on the Anthropocene vantage, they still conform to this rigid view of geologic time.

17. Laurel Wamsley, "Supreme Court Rules that About Half of Oklahoma Is Native American Land," NPR, July 9, 2020.

18. Thomas W. Church and Robert T. Nakamura, *Cleaning Up the Mess: Implementation Strategies in Superfund* (Washington, D.C.: Brookings Institution, 1993).

19. Gaia Goffe, *South China Morning Post*, 2016.

20. Diane Bernard, "The Creator of Mount Rushmore's Forgotten Ties to White Supremacy," *Washington Post*, July 2, 2020.

I. ISLAND LABORATORIES

1. Kenneth Bilby, *True-Born Maroons* (University Press of Florida, 2005).

2. For more on this articulation of warfare, see Vincent Brown, *Tacky's Revolt: The Story of an Atlantic Slave War* (Harvard University Press, 2020).

3. Jen Kinney, "The Fight Against Mining in Jamaica's Rainforest," *Vice*, September 24, 2021.

4. British Treaty of 1739 with the Jamaica Maroons. https://cyber.harvard.edu/eon/marroon/treaty.html.

5. John Kuo Wei Tchen and Dylan Yeats, eds., *Yellow Peril!: An Archive of Anti-Asian Fear* (Verso, 2014).

6. "PM Welcomes $300-million Investment by Huawei," *Jamaica Observer*, October 21, 2022.

7. Niko Block, "Toronto's Buried History: The Dark Story of How Mining Built a City," *Guardian*, March 3, 2017, https://www.theguardian.com/cities/2017/mar/03/toronto-hidden-history-how-city-built-mining.

8. Emma Jinhua Teng, *Eurasian: Mixed Identities in the United States, China, and Hong Kong 1842–1943* (University of California Press, 2013).

9. Ibid.

10. Yoshiko Shibata, "Revisiting Chinese Hybridity: Negotiating Categories and Re-constructing Ethnicity in Contemporary Jamaica—a Preliminary Report," *Caribbean Quarterly*, 51, no. 1 (March 2005): 53–75.

11. Small House Policy of 1972, Legislative Council of the Hong Kong Special Administrative Region of the People's Republic of China. https://www .legco.gov.hk/research-publications/english/essentials-1516ise10-small -house-policy.htm.

12. Lucy Kwan and Alexis Wong, "How Hong Kong's Artificial Island Project Can Tell the City's Story Well," *South China Morning Post*, May 10, 2023.

13. Sandra Laville, "Beijing Highway: $600m Road Just the Start of China's Investments in Caribbean," *Guardian*, December 24, 2015.

14. Olive Lewin, *Rock It Come Over: The Folk Music of Jamaica* (The University of the West Indies Press, 2000).

15. Vere T. Daly, *A Short History of the Guyanese People* (Macmillan, 1975).

16. Audre Lorde, "Poems Are Not Luxuries," *Chrysalis: A Magazine of Women's Culture*, vol. 3 (1977).

17. Hans Sloane, *A Voyage to the Islands Madera, Barbados, Nieves, S. Christophers and Jamaica* (1707).

18. Patrick Browne, *The Civil and Natural History of Jamaica* (1756).

19. Ed Yong, *An Immense World: How Animal Senses Reveal the Hidden Realms Around Us* (Random House, 2022).

20. Thomas Nagel, "What Is It Like to Be a Bat?," *Philosophical Review* 83, no. 4 (October 1974): 435–50.

21. "Jamaica Secures First Limestone Shipment to the US," *Jamaica Gleaner*, October 28, 2022.

22. L. R. Gallant et al., "A 4,300-Year History of Dietary Changes in a Bat Roost Determined from a Tropical Guano Deposit," *JGR Biogeosciences*, March 23, 2021.

23. Aimee Nezhukumatathil, *World of Wonders: In Praise of Fireflies, Whale Sharks, and Other Astonishments* (Minneapolis: Milkweed Editions, 2020).

24. Jamaica Environment Trust, https://jamentrust.org/.

25. Brown, *Tacky's Revolt*.

26. Philip Sherlock and Hazel Bennett, *The Story of the Jamaican People* (Ian Randle Publishers, 1998).

27. "Challenge Against Leader of Moore Town Maroons Fizzles," *Jamaica Gleaner*, May 8, 2022.

28. Robert Lee et al., Land-Grab Universities: A *High Country News* Investigation, https://www.landgrabu.org/.

29. Cassandra Khan, "New $61 M Sand-Mining Operation for Canal #2," *Guyana Chronicle*, January 23, 2023.

30. Christopher Serju, "A Beloved Jamaican Beach Is Succumbing to Climate Change. It Won't Be the Last," *Guardian*, October 27, 2020.

31. Charles Benedict Davenport and Morris Steggerda, *Race Crossing in Jamaica* (Negro Universities Press, 1970).

2. CLIMATE CRISIS, GENESIS 1492

1. Nicolás Wey Gómez, *The Tropics of Empire: Why Columbus Sailed South to the Indies* (MIT Press, 2008).

2. Christopher Columbus, *Personal Narrative of the First Voyage of Columbus to*

America: From a Manuscript Recently Discovered in Spain, ed. Bartolomé de las Casas, trans. Samuel Kittle (1827), https://www.loc .gov/item/02008207/.

3. Carib Reserve Act of 1978. https://kalinagoaffairs.gov.dm/.

4. Emma Gaalaas Mullaney, "Carib Territory: Indigenous Access to Land in the Commonwealth of Dominica," *Journal of Latin American Geography* 8, no. 2 (2009): 71–96.

5. Gerald Vizenor, *Survivance: Narratives of Native Presence* (University of Nebraska Press, 2008).

6. Édouard Glissant, *Caribbean Discourse,* trans. J. Michael Dash (University of Virginia Press, 1989).

7. Elizabeth Kolbert, "Talk to Me: Can Artificial Intelligence Allow Us to Speak to Another Species?," *New Yorker,* September 4, 2023. https://www.new yorker.com/magazine/2023/09/11/can-we-talk-to-whales.

8. MS 7294 Hester Merwin Carib Indian drawings, Smithsonian Institution, 1940.

9. North-East Women in Agriculture Movement Dominica.

10. Daphne Ewing-Chow, "Caribbean Islands Are the Biggest Plastic Polluters Per Capita in the World," *Forbes,* September 20, 2019, https://www.forbes .com/sites/daphneewingchow/2019/09/20/caribbean-islands-are-the -biggest-plastic-polluters-per-capita-in-the-world.

11. Ibid.

12. Leander Raes, Damien Mittempergher, and Aanchal Jain, "The Economic Impact of Marine Plastic Pollution in Saint Lucia." IUCN. 2022. https:// www.iucn.org/sites/default/files/2023-01/report-economic-assessment -st.-lucia.pdf.

13. "Seagrass—Secret Weapon in the Fight Against Global Heating," United Nations Environment Programme, November 2019, https://www.unep .org/news-and-stories/story/seagrass-secret-weapon-fight-against-global -heating.

14. Tao Leigh Goffe, "Bigger than the Sound: The Jamaican Chinese Infrastructures of Reggae," *Small Axe,* 24, no. 3 (November 2020): 97–127.

15. Jacques Derrida, *Archive Fever: A Freudian Impression,* trans. Eric Prenowitz (University of Chicago Press, 1996).

16. Columbus, *Personal Narrative.*

17. Jean Rhys, *Wide Sargasso Sea* (W. W. Norton, 1966).

18. Cassandra Garrison, "Besieged by Seaweed, Caribbean Scrambles to Make Use of the Stuff," *Reuters,* September 29, 2021.

19. Ada Ferrer, *Cuba: An American History* (Scribner, 2021).

20. Bartolomé de las Casas, *A Short Account of the Destruction of the Indies,* 1552.

21. Columbus, *Personal Narrative.*

22. Frantz Fanon, *The Wretched of the Earth,* trans. Constance Farrington (Grove Press, 1965).

23. Toni Morrison, *Sula* (Alfred A. Knopf, 1973).

24. Sylvia Wynter, "Black Metamorphosis: New Natives in a New World," unpublished manuscript, n.d. Eric Williams, *Capitalism & Slavery* (University of North Carolina Press, 1944). Jamaica Kincaid, *A Small Place* (Farrar, Straus & Giroux, 1988).

25. *Atlas of the Bible: Exploring the Holy Lands, National Geographic,* June 29, 2018.

26. W. E. B. Du Bois, "To the Nations of the World," 1901.

27. Walter Rodney et al., *How Europe Underdeveloped Africa* (Verso, 2018).

28. Chronixx, "Capture Land," *Dread & Terrible*, 2014.

29. Tao Leigh Goffe, "Not Another Lab: The Forever Crisis of Incentivizing Collaboration in the Humanities," *ASAP/Journal* 8, no. 2 (May 2023): 248–54.

30. Stuart Hall, *Familiar Stranger: A Life Between Two Islands* (Duke University Press, 2017).

31. Toni Morrison, *Playing in the Dark: Whiteness and the Literary Imagination* (Harvard University Press, 1992).

32. UN Factsheet on Indigenous Peoples, Indigenous Voices.

33. William Bligh, *The Dangerous Voyage Performed by Captain Bligh: with a Part of the Crew of His Majesty's Ship Bounty, in an Open Boat, Over Twelve Hundred Leagues of the Ocean; in the Year 1789* (1824).

34. Audre Lorde, "Litany for Survival," in *The Collected Poems of Audre Lorde* (W. W. Norton, 1997), 141.

3. NATURAL HISTORY MUSEUMS, SHRINES TO EXPLORERS

1. Natalie Diaz, "American Arithmetic," in *Postcolonial Love Poem* (Graywolf Press, 2020).

2. Will of Fulke Rose, https://www.ucl.ac.uk/lbs/person/view/2146650539.

3. Zachary Small, "Facing Scrutiny, a Museum That Holds 12,000 Human Remains Changes Course," *New York Times*, October 15, 2023.

4. Robert Krulwich, "What (Not?) to Do When You Meet the Last Great Wild Buffalo," NPR, November 11, 2011.

5. Toni Morrison, *Unspeakable Things Unspoken*. https://sites.lsa.umich.edu/mqr/2019/08/unspeakable-things-unspoken-the-afro-american-presence-in-american-literature/.

6. Jacques Derrida, "Archive Fever: A Freudian Impression," *Diacritics* 25, no. 2 (Summer 1995): 9–63, https://doi.org/10.2307/465144.

7. Max Weber, *The Protestant Ethic and the Spirit of Capitalism*, trans. Talcott Parsons (G. Allen & Unwin, 1930).

4. BREATHING UNDERWATER

1. William Shakespeare, *The Tempest*, ed. Alden T. Vaughan and Virginia Mason Vaughan (London: Bloomsbury, 2011), 1.1.20–26. References are to act, scene, and line. https://doi.org/10.5040/9781408160183.00000045.

2. Greg Tate on Drexciya from Mike Rubin, "Infinite Journey to Inner Space: The Legacy of Drexciya," June 29, 2017. https://daily.redbullmusicacademy.com/2017/06/drexciya-infinite-journey-to-inner-space.

3. "Africa's Disappearing Lake Chad," NASA, https://earthobservatory.nasa.gov/images/1240/africas-disappearing-lake-chad#:~:text=Once%20a%20great%20lake%20close,it%20was%2035%20years%20ago.

4. Peter Tosh, "Downpressor Man," 1977.

5. Nina Simone, "Sinnerman," 1965.

6. Virginia Hamilton, *The People Could Fly: American Black Folktales* (Alfred A. Knopf, 1985).

7. Margaret Washington Creel, *A Peculiar People: Slave Religion and Community-Culture Among the Gullahs* (New York University Press, 1988).

8. Walter D. Mignolo and Catherine E. Walsh, eds., *On Decoloniality: Concepts, Analytics, Praxis* (Duke University Press, 2018).

9. Edouard Glissant, *Caribbean Discourse: Selected Essays,* trans. J. Michael Dash (University of Virginia Press, 1989).

10. Kevin Dawson, *Undercurrents of Power: Aquatic Culture in the African Diaspora* (University of Pennsylvania Press, 2018).

11. Ibid.

12. Shavonne Smith quoted in Somini Sengupta, "The Original Long Islanders Fight to Save Their Land from a Rising Sea," *New York Times,* March 5, 2020.

13. Sengupta, "Original Long Islanders Fight to Save Their Land."

14. "The Myth of Drexciya," *Terraforma Journal #2,* February 2022, https:// greyoverblue.net/2022/02/08/the-myth-of-drexciya/.

15. Ellen Gallagher, *Watery Ecstatic.* https://www.maxhetzler.com/exhibitions /ellen-gallagher-drawings-series-watery-ecstatic-zimmerstrasse-9091-berlin -mitte-june-06-july-26-2003-murmur-animation-2003.

16. Paul Gilroy, *The Black Atlantic: Modernity and Double Consciousness* (Harvard University Press, 1993).

17. "2022 Sea Level Rise Technical Report," https://oceanservice.noaa.gov /hazards/sealevelrise/sealevelrise-tech-report.html.

18. Ian Fleming, *Dr. No* (Jonathan Cape, 1958).

19. Dow V. Baxter, "Photograhic Foray," *Quarterly Review: A Journal of University Perspectives* 48–49 (1941).

20. Rachel Carson, *The Sea Around Us,* 3rd ed. (1950; repr., University Press, 2018).

21. Annual Report of the Board of Regents of the Smithsonian, Smithsonian Institution, 1921.

22. Mohsen Manutchehr-Danai, *Dictionary of Gems and Gemology* (Springer, 2013), 333.

23. LaTasha Lee, Kim Smith-Whitley, Sonja Banks, and Gary Puckrein, "Reducing Health Care Disparities in Sickle Cell Disease: A Review," Public Health Reports, 134, no. 6 (November/December 2019) 599–607.

24. Rita Rubin, "Tackling the Misconception that Cystic Fibrosis Is a 'White People's Disease,'" *JAMA,* 325, no. 23 (June 15, 2021): 2330–32. doi:10.1001/jama.2021 .5086, https://jamanetwork.com/journals/jama/article-abstract/2780564.

25. Danielle Hall, "Joan Murrell Owens and Her Button Corals," Smithsonian, September 2022, https://ocean.si.edu/human-connections/careers/joan -murrell-owens-and-her-button-corals.

26. Frederick Douglass, [Letter], Edinburgh, Scotland, July 30, 1846. To William A. White, from Frederick Douglass Manuscripts, Douglass Memorial Home, Anacostia, D.C.; transcribed in Philip Foner, ed., *The Life and Writings of Frederick Douglass,* vol. 1 (New York: International Publishers, 1950), p. 181.

27. Charles Darwin, *The Structure and Distribution of Coral Reefs* (1842).

28. Charles Darwin, *The Life and Letters of Charles Darwin* (1887).

29. Charles Darwin, *Descent of Man, and Selection in Relation to Sex* (1871).

30. André W. Droxler and Stéphan J. Jorry, "The Origin of Modern Atolls: Challenging Darwin's Deeply Ingrained Theory," *Annual Review of Marine Science* 13 (January 2021): 537–73.

31. Trinidadian French Mauritian poet Khal Torabully—himself a descendent of indenture—says of coining the term "coolitude," riffing off "négritude."

See Khal Torabully letter to Amitav Ghosh, "Coolitude and Khal Torabully," Amitav Ghosh blog, October 1, 2011, http://amitavghosh.com/blog/?p=1210. See Marina Carter and Khal Torabully, eds., *Coolitude: An Anthology of the Indian Labour Diaspora* (London: Anthem, 2002).

32. While some reports say Darwin was fifteen or sixteen, others say he was seventeen.

33. R. B. Freeman, "Darwin's Negro Bird-Stuffer," *Notes and Records of the Royal Society of London* 33, no. 1 (August 1978): 83–86.

34. Darwin, *Life and Letters of Charles Darwin*.

35. Marlon James, *A Brief History of Seven Killings: A Novel* (Riverhead Books, 2014).

36. *The Silent World,* directed by Jacques-Yves Cousteau and Louise Malle (Columbia Pictures, 1953).

37. Timothy Lamont et al., "The Sound of Recovery: Coral Reef Restoration Success Is Detectable in the Soundscape," *Journal of Applied Ecology* 59, no. 3 (December 2021): 742–56.

38. *Conscience Point,* directed by Treva Wurmfeld (Conscience Point Film, LLC, 2019).

39. Julian Granberry, *The Calusa: Linguistic and Cultural Origins and Relationships* (University of Alabama Press, 2011).

5. GUANO DESTINIES

1. Alan Fram and Jonathan Lemire, "Trump: Why Allow Immigrants from 'Shithole Countries'?," *AP News,* January 12, 2018.

2. Ralph Waldo Emerson, "Fate," in *The Conduct of Life* (1860).

3. C. L. R. James, "Revolution and the Negro," Marxists Internet Archive, https://www.marxists.org/archive/james-clr/works/1939/12/negro-revolution.htm. Originally published under the name J. R. Johnson as "The Revolution and the Negro" in *New International* 5 (December 1939): 339–43.

4. "An interview with Gregory T. Cushman," Cambridge University Press, 2015. https://www.youtube.com/watch?v=5NegNnPM91k.

5. Filmore was born into poverty; his parents had been tenant farmers in the Finger Lakes area of upstate New York. A bootstraps model of American advancement through meritocracy, the connection to the soil was important to his upbringing. The contested soil of the Underground Railroad bordering Canada was also the terrain of historic vaunted battles won during the Revolutionary War, massacres of Native peoples to claim the land.

6. Aimé Césaire, *A Tempest* (1969).

7. Treaty of Rijswijk, 1697.

8. Andrew C. Revkin, "New York Tries to Clean Up Ash Heap in the Caribbean," *New York Times,* January 15, 1998.

9. See Christina Duffy Burnett, "The Edges of Empire and Sovereignty: American Guano Islands," *American Quarterly* 57, no. 3 (September 2007): 779–803. Jennifer James, "'Buried in Guano': Race, Labor, and Sustainability," *American Literary History* 24, no. 1 (Spring 2012): 115–42. Jimmy M. Skaggs, *The Great Guano Rush: Entrepreneurs and American Overseas Expansion* (St. Martin's, 1994).

10. Trista Talton, "Part of Former Navassa Superfund Site Up for Highest Bidder," *Coastal Review Online*, August 31, 2023.
11. https://www.nybooks.com/articles/1984/11/22/mr-america/.
12. Alicia A. Caldwell, Chad Day, and Jake Pearson, "Melania Trump Modeled in US Prior to Getting Work Visa," November 4, 2016, https://apnews.com/article/lifestyle-travel-immigration-migration-election-2020-37dc7aef0ce440 77930b7436be7bfdod.
13. George Jackson's use of fertilizer has been linked to Emerson's guano by literary critic Eduardo Cadava, "The Guano of History," in *Cities Without Citizens*, ed. Eduardo Cadava and Aaron Levy (Philadelphia: Slought Foundation, 2003).
14. George Jackson, *Soledad Brother: The Prison Letters of George Jackson* (Coward-McCann, 1970).
15. Sonia Elks, "Haitians Say Underaged Girls Were Abused by U.N. Peacekeepers," *Reuters*, December 18, 2019.
16. Benjamin Franklin letter to Peter Collinson, May 9, 1753. New York Public Library; also copies: Public Record Office, American Philosophical Society, and (part only) British Museum.
17. Tao Leigh Goffe, "'Guano in Their Destiny': Race, Geology, and a Philosophy of Indenture," *Amerasia Journal* 45, no. 1 (2019): 27–49.
18. September 22, 1900. https://www.nytimes.com/1900/09/22/archives/island -sold-at-auction-navassa-in-caribbean-sea-goes-to-only-bidder.html.
19. Humphry Davy, *Elements of Agricultural Chemistry* (1813).
20. Susan E. Bergh, *Wari: Lords of the Ancient Andes*, Cleveland Museum of Art, 2012.
21. Pedro Rodrigues and Joana Micael, "The importance of guano birds to the Inca Empire and the first conservation measures implemented by humans," *International Journal of Avian Science*, August 18, 2020.
22. Rosa Luxemburg, *The Complete Works of Rosa Luxemburg, Volume 1, Economic Writings 1*, ed. Peter Hudis, trans. David Fernbach, Joseph Fracchia, and George Shriver (Verso, 2013).
23. Ibid.
24. George W. Bush, *Decision Points* (Crown, 2010).
25. Frederick Douglass, "Cheap Labour," *New National Era*, August 17, 1871.
26. Ibid.
27. Daniel Immerwahr, *How to Hide an Empire: A History of the Greater United States* (Picador, 2019).
28. A. J. Duffield, *Peru in the Guano Age: Being a Short Account of a Recent Visit to the Guano Deposits* (London, 1877), 45.
29. Joseph Victor von Scheffel, "Guano Song," in *Gaudeamus! Humorous Poems* (1872), 24–5. http://www.public-domain-poetry.com/joseph-victor-von -scheffel/guano-song-36522. In the poem, which for the most part is an ode romanticizing guano work, von Scheffel mocks Hegel, saying the birds make better waste than him.
30. William Plomer letter to Ian Fleming, in *The Man with the Golden Typewriter: Ian Fleming's James Bond Letters*, ed. Fergus Fleming (London: Bloomsbury, 2015).
31. Ian Fleming, *Dr. No* (Jonathan Cape, 1958).
32. Percival Everett, *Dr. No* (Graywolf Press, 2022).

33. Edward Wybergh Docker, *The Blackbirders: The Recruiting of South Seas Labour for Queensland, 1863–1907* (Angus and Robertson, 1970).

6. COLONIALISM, THE BIRDER'S COMPANION GUIDE

1. Christian Cooper, *Better Living Through Birding: Notes from a Black Man in the Natural World* (Random House, 2023).
2. Annette Gordon-Reed, *The Hemingses of Monticello* (W. W. Norton, 2008).
3. W. E .B. Du Bois, *The Souls of Black Folk* (A. C. McClurg & Co., 1903).
4. Cooper, *Better Living Through Birding.*
5. Yale Land Acknowledgement Statements. Office of the Secretary and Vice President for University Life. https://secretary.yale.edu/services-resources/land-acknowledgment-statements
6. "Meet Yale Dishwasher Corey Menafee Who Smashed Racist Stained-Glass Window," *Democracy Now!,* July 15, 2016.
7. CIA, The World Factbook, October 20, 2022.
8. Charles Darwin, Letter to J. D. Hooker, August 1, 1857. Darwin Correspondence Project. "It is good to have hair-splitters & lumpers."
9. See Daina Ramey Berry, *The Price for Their Pound of Flesh: The Value of the Enslaved from Womb to Grave in the Building of a Nation* (Beacon Press, 2017).
10. See Lennox Honychurch, *In the Forests of Freedom: The Fighting Maroons of Dominica* (Papillote Press, 2017).
11. Charles Darwin, *The Descent of Man, and Selection in Relation to Sex* (1871).
12. Ibid.
13. Calvin Tomkins, "The Epic Style of Kerry James Marshall, *New Yorker,* August 2, 2021, https://www.newyorker.com/magazine/2021/08/09/the-epic-style-of-kerry-james-marshall.
14. Andrew Maas, "NYC Bird Alliance to Change Name to Better Reflect its Values, Mission, and Work," *Syrinx* (blog), March 22, 2023, https://nycbirdalliance.org/blog/audubon-name-blog. Christian Cooper also directly commented on the change: https://www.nytimes.com/2024/06/11/nyregion/why-nyc-audubonv-name.html#:~:text=Last%20week%20a%20well%2Dknown,He%20also%20enslaved%20people.
15. Jean Rhys, *Wide Sargasso Sea* (W. W. Norton, 1966).
16. Scott Cummings, *Left Behind in Rosedale: Race Relations and the Collapse of Community Institutions* (Perseus, 1998).
17. Clyde Rigney, "A Story from Community," Emerging Minds podcast, https://emergingminds.com.au/resources/podcast/a-story-from-community/transcript/.
18. adrienne maree browne, *Emergent Strategy: Shaping Change, Changing Worlds* (AK Press, 2017).

7. THE CURIOUS CASE OF THE CALCUTTA MONGOOSE IN JAMAICA

1. Rudyard Kipling, "Rikki-Tikki-Tavi," in *The Jungle Book* (1894).
2. W. B. Espeut, "On the Acclimatization of the Indian Mungoos in Jamaica," *Proceedings of the Zoological Society of London,* 1882: 712–14.
3. Ibid.
4. Louis Kirk McAuley, " 'Calcutta Still Haunts My Fancy,' or the Confusion of

Old and New World Ecologies in Early Caribbean Literature," in *The Edinburgh Companion to Atlantic Literary Studies,* ed. Leslie Eckel (Edinburgh University Press, 2016).

5. Lucy Cooke, *Bitch: On the Female of the Species* (Basic Books, 2022).
6. Anthony Trollope, *The West Indies and the Spanish Main* (1860), 15.
7. W. B. Espeut, "The Timbers of Jamaica," speech delivered at the Town Hall, Kingston, on March 8, 1881.
8. "Abortion Is Murder! Deacon Says Women Have No Right to Terminate Pregnancies," *Jamaica Gleaner,* May 17, 2018. http://jamaica-gleaner.com/article/lead-stories/20180517/abortion-murder-deacon-says-women-have-no-right-terminate-pregnancies.
9. H. H. Marshall et al., "A veil of ignorance can promote fairness in a mammal society," *Nature Communications* 12, no. 1 (2021): 3717. https://doi.org/10.1038/s41467-021-23910-6.
10. Ibid.
11. See Amrita Pande, *Wombs in Labor: Transnational Commercial Surrogacy in India* (Columbia University Press, 2014).
12. Espeut, "Timbers of Jamaica," 15.
13. Caitlin Rosenthal, "Slavery's Scientific Management: Masters and Managers," in *Slavery's Capitalism: A New History of American Economic Development,* eds. Sven Bekert and Seth Rockman (University of Pennsylvania Press, 2016).
14. Paul H. Baldwin, Charles W. Schwartz, and Elizabeth Reeder Schwartz, "Life History and Economic Status of the Mongoose in Hawaii," *Journal of Mammology* 33, no. 3 (August 1952): 335–56.
15. Ann Laura Stoler, *Carnal Knowledge and Imperial Power: Race and the Intimate in Colonial Rule, with a New Preface* (University of California Press, 2010), 13.
16. Jared Sexton, *Amalgamation Schemes: Antiblackness and the Critique of Multiculturalism* (Minneapolis: University of Minnesota Press, 2008).
17. Espeut, "On the Acclimatization of the Indian Mungoos in Jamaica," 15.
18. Ibid.
19. Ibid.
20. Ibid.
21. Ibid., 10.
22. Trollope, *West Indies and the Spanish Main,* 91.
23. Ibid., 95.
24. There are records of an Espeut property in Cap Dame Marie in Saint-Domingue, but no deeds of ownership in Jamaica. In addition, though it had been stated that the father of Peter Espeut's wife, Dorcas, was an admiral in the British navy, he was more likely a merchant or plantation owner. After her husband, Peter, died in 1802 in St. Domingue from an old wound, Dorcas left for London, where she married James Whyte in 1804. It is believed that Dorcas made a false declaration of relation to the family of "Parish of Standing Hall in the County of Hertford" that gave rise to the apocryphal provenance of the family. What historians can corroborate is that the Espeut family moved from the Caribbean and settled in England for the first time in the 1880s. The Jamaican Espeuts more than likely invented a distinguished English heritage and told the story to Burke, who published it. The family's

origins were probably in France, where the Espeut name is found in Auriac and Mouthoumet via Saint-Domingue.

25. The CIA named a Cuban mission to remove Fidel Castro from power Operation Mongoose.

26. Derek Walcott, "The Mongoose." See "Rhyme and Punishment for Naipaul," *Guardian*, May 31, 2008.

27. Jahan Ramazani, "The Wound of History: Walcott's *Omeros* and the Postcolonial Poetics of Affliction." PMLA/Publications of the Modern Language Association of America 112, no. 3 (1997): 406–7. https://doi.org/10.2307/462949.

28. Ibid.

29. G. Llewellyn Watson, *Jamaican Sayings: With Notes on Folklore, Aesthetics, and Social Control* (University of Florida Press, 1991).

30. Sam Manning, "Sly Mongoose," *Volume 1: Recorded in New York, 1924–1927*, 1925.

31. See Cindy Hahamovitch, *No Man's Land: Jamaican Guestworkers in America and the Global History of Deportable Labor* (Princeton University Press, 2011).

32. Felix (Gregory Felix) and his Internationals on the clarinet, trumpet, fiddle, and bass.

33. Lord Invader, "Sly Mongoose," 1946.

34. William C. White, "The Calypso Singers," *Folk Review of Peoples' Music*, ed. Max Jones (Jazz Music Books, 1945), 46.

35. Charles S. Elton, *The Ecology of Invasions by Animals and Plants* (University of Chicago Press, 2000).

36. Priscilla M. Wehi et al., "Contribution of Indigenous Peoples' Understandings and Relational Frameworks to Invasive Alien Species Management," *People and Nature*, no. 5 (October 2023): 1403–14. https://doi.org/10.1002/pan3.10508.

8. PEDAGOGIES OF SMOKE

1. Jeremy Plester, "Could Hemp Be a Key Tool in Fight Against Climate Change?" *Guardian*, November 24, 2022. https://www.theguardian.com/environment/2022/nov/24/could-hemp-be-a-key-tool-in-fight-against-climate-change.

2. Stephen Cooper, "Scratch in LA: An Interview with Lee 'Scratch' Perry," *CounterPunch*, November 7, 2017.

3. Iaian Gately, *Tobacco: A Cultural History of How an Exotic Plant Seduced Civilization* (Grove Press, 2002).

4. Natalie R. Lenggenhager-Krakoski et al., "William Brooke O'Shaughnessy, an Irishman Who Introduced Anaesthesia to India and Brought Eastern Medicine to Britain," *British Journal of Anaesthesia*, March 13, 2023, https://www.bjanaesthesia.org/.

5. Clinton A. Hutton, Michael A. Barnett, and Daive A. Dunkley, eds., *Leonard Percival Howell and the Genesis of Rastafari* (University of the West Indies Press, 2015).

6. Ras Moqapi Selassie quoted in "Erased from Collective Memory: *Dreadlocks Story* Documentary Untangles the Hindu Legacy of Rastafari," *Afro-Asian Connections in Latin America and the Caribbean*, ed. Luisa Marcela Ossa and Debbie Lee-Distefano (Lexington Books, 2019).

7. Verene A. Shepherd, *Maharani's Misery: Narratives of a Passage from India to the Caribbean*, (University of the West Indies Press, 2002).

8. The 1913 Ganja Law of the Legislative Council was later developed into the Dangerous Drugs Law (1924). See Louis Moyston, "The Ganja Law of 1913: 100 Years of Oppressive Injustice," *Jamaica Observer,* December 1, 2013.

9. Ann Lindsay, *Seeds of Blood and Beauty: Scottish Plant Explorers* (Birlinn, 2005).

10. Brenda McLean, *George Forrest: Plant Hunter* (Royal Botanic Garden, 2004).

11. Nagoya Protocol, United Nations Treaty Collection, 2010. https://treaties.un.org/.

12. George Forrest, *Journeys and Plant Introductions,* ed. J. Macqueen Cowan (Oxford University Press, 1952).

13. *National Geographic* 105, no. 5 (May 1954).

14. "Dirt on Our Hands: Overcoming Botany's Hidden Legacy of Inequality, Unearthed—Journeys into the Future of Food, Featuring James Wong," Royal Botanic Gardens, Kew, *Unearthed* podcast, https://podcasts.apple.com/gb/podcast/dirt-on-our-hands-overcoming-botanys-hidden-legacy/id1524216431?i=1000512363938.

15. "Our Manifesto for Change 2021–2030," Royal Botanic Gardens, Kew, 2021.

16. Ibid.

17. Cornell Botanic Gardens, https://cornellbotanicgardens.org/.

18. Harvard Forest, https://harvardforest.fas.harvard.edu/.

19. Jamaica Kincaid, *My Garden (Book)* (Farrar, Strauss & Giroux, 2001).

20. Louise Bennett, "Linstead Market," *Jamaican Folk Songs,* 1954. https://folkways.si.edu/.

21. Alfred McFarland, *Mutiny in the "Bounty!" and Story of the Pitcairn Islanders* (1884).

22. Yuan-Chao Tung, *The Changing Chinese Ethnicity in French Polynesia* (UMI Dissertation Services, 1993).

23. "Flandreau Tribal Medical Pot Cards Leading to Arrests," *AP News,* February 19, 2022, https://apnews.com/article/health-arrests-marijuana-medical-marijuana-south-dakota-d567ee9bf1c7b1e9a7119a4e03127edd.

24. Chief Vincent Mann and Michaeline Picaro, Three Sisters Medicinal Farm, https://munseethreesisters.org/.

25. Ringwood Mines/Landfill Ringwood Borough, N.J. Superfund Site. United States Environmental Protection Agency. https://cumulis.epa.gov/.

26. Vincent Mann, *Our Land, Our Stories: Excavating Subterranean Histories of Ringwood Mines and the Ramapough Lunaape Nation* (Design for Public History, 2022).

27. Little Beach Harvest. https://littlebeachharvest.com/.

28. Dangerous Drugs (Amendment) Act. https://www.fid.gov.jm/www/wp-content/uploads/2017/09/The-Dangerous-Drugs-Amendment-Act-2015-Gazette-Fact-Sheet-Included.pdf.

29. Marcus Goffe. "The Rights of the Maroons in the Emerging Ganja Industry in Jamaica," *Social and Economic Studies* 67, no. 1 (March 2018): 85–115.

9. AFFECTIVE PLATE TECTONICS

1. Lina Bo Bardi, quoted in Isaac Julien's film *Lina Bo Bardi—A Marvellous Entanglement,* 2019, at the Yale Center of British Art.

2. Mia Mottley remarks at 2021 United Nations Climate Change Conference, COP 26, Glasgow.

3. Walter Benjamin, "On the Concept of History," 1940. https://www.marxists.org/reference/archive/benjamin/1940/history.htm.

4. Rosa Luxemburg, "Martinique," *Leipziger Volkszeitung*, 1902. https://www.marxists.org/archive/luxemburg/1902/05/15.htm.

5. Peter Morgan, *Fire Mountain: How One Man Survived the World's Worst Volcanic Disaster* (Bloomsbury, 2003).

6. Edmund Otis Hovey, The 1902–1903 Eruptions of Mont Pelée, Martinique, and the Soufrière: St Vincent, International Geological Congress, 1904.

7. Clyde Woods, *Development Arrested: The Blues and Plantation Power in the Mississippi Delta* (Verso, 1997).

8. Sriram Shamasunder, "What Native Americans Can Teach Rich Nations About Generosity in a Pandemic, *NPR*, May 10, 2021. https://www.npr.org/sections/goatsandsoda/2021/05/10/994254810/opinion-what-native-americans-can-teach-us-about-generosity-in-a-pandemic.

9. Howard A. Fergus, *Montserrat: History of a Caribbean Colony* (Macmillan Caribbean, 1994).

10. C. L. R. James, "Revolution and the Negro," "The Revolution and the Negro," New International, vol. V, December 1939, pp. 339–343. Published under the name J. R. Johnson.

11. The Beach Boys, "Kokomo," *Kokomo*, 1988.

12. Edward Brathwaite, "Calypso," in *Islands* (Oxford University Press, 1969).

13. Edward Kamau Brathwaite, "Carifesta '72: The Caribbean Cultural Revolution," *Advocate-News*, October 22, 1972.

14. Leslie Marmon Silko, *Almanac of the Dead: A Novel* (Simon & Schuster, 1991).

15. The term "unsinkable aircraft carrier" first appeared in World War II, used by the United States to describe Pacific atolls. It has been used more recently to describe islands in the South China Sea.

16. "Speech by His Excellency Mohamed Nasheed, at the Alliance of the Small Island States (AOSIS) Climate Change Summit," September 1, 2009.

17. Timothy Morton, "How I Learned to Stop Worrying and Love the Term Anthropocene," *Cambridge Journal of Postcolonial Literary Inquiry* 1, no. 2 (September 2014): 257–64, https://doi.org/10.1017/pli.2014.15. Dipesh Chakrabarty, "The Climate of History: Four Theses," *Critical Inquiry* 35, no. 2 (Winter 2009): 197–222, https://doi.org/10.1086/596640.

18. Kevin de Queiroz, "Ernst Mayr and the Modern Concept of Species," *Proceedings of the National Academy of Sciences of the United States of America* 102, supplement 1 (2005): 6600–6607. https://www.pnas.org/doi:10.1073/pnas.0502030102.

19. Malcolm X and Alex Haley, *The Autobiography of Malcolm X* (New York: Ballantine Books, 1989).

20. Psyche A. Williams-Forson, *Building Houses out of Chicken Legs: Black Women, Food, and Power* (The University of North Carolina Press, 2006).

21. X and Haley, *The Autobiography of Malcolm X*.

22. Ibid.

23. Jessica Russell, *The Life of Louise Norton Little: An Extraordinary Woman: Mother of Malcolm X and His Seven Siblings* (self-pub., Amazon Digital Services, 2021).

24. X and Haley, *The Autobiography of Malcolm X.*

CODA

1. In 1895, the LIRR was extended by Austin Corbin to Montauk after the Montauk land was expropriated in 1882.

2. Montauk Tribe of Indians. https://www.onemontauknation.org/.

3. Gerald Vizenor, *Survivance: Narratives of Native Presence* (University of Nebraska Press, 2008).

4. Jeannette Edwards Rattray, *East Hampton History: Including Genealogies of Early Families* (Country Life Press, 1953).

5. Jill Lepore, *The Name of War: King Philip's War and the Origins of American Identity* (Vintage Books, 1999).

6. Linford D. Fisher, "'Dangerous Designs': The 1676 Barbados Act to Prohibit New England Indian Slave Importation." *William and Mary Quarterly* 71, no. 1 (January 2014): 99–124. https://doi.org/10.5309/willmaryquar.71.1.0099. Native peoples from the United States were sold into slavery after the Tuscarora War of 1711. Others found themselves forcibly bound into indentureship in colonies such as Barbados and Bermuda.

7. *Conscience Point,* directed by Treva Wurmfeld (Conscience Point Film, LLC, 2019).

8. Shinnecock Indian Nation Climate Change Adaptation Plan, October 2013. https://www.epa.gov/sites/default/files/2016-09/documents/shinnecock_nation_ccadaptation_plan_9.27.13.pdf.

Bibliography

BOOKS AND ARTICLES

Baldwin, James. *Go Tell It on the Mountain*. Vintage International, 2013.

Bilby, Kenneth. *True-Born Maroons*. University Press of Florida, 2005.

Brown, Vincent. *Tacky's Revolt: The Story of an Atlantic Slave War*. Harvard University Press, 2020.

Browne, Patrick. *The Civil and Natural History of Jamaica*. 1756.

Cooper, Christian. *Better Living Through Birding: Notes from a Black Man in the Natural World*. Random House, 2023.

Creel, Margaret Washington. *A Peculiar People: Slave Religion and Community-Culture Among the Gullahs*. New York University Press, 1989.

Darwin, Charles. *The Descent of Man, and Selection in Relation to Sex*. 1871.

———. *The Structure and Distribution of Coral Reefs*. 1842.

Davenport, Charles Benedict, and Morris Steggerda. *Race Crossing in Jamaica*. Negro Universities Press, 1970.

Derrida, Jacques. *Archive Fever: A Freudian Impression*. Translated by Eric Prenowitz. University of Chicago Press, 1996.

Diaz, Natalie. *Postcolonial Love Poem*. Graywolf Press, 2020.

Dillon, Grace L., ed., *Walking the Clouds: An Anthology of Indigenous Science Fiction*. Tucson: University of Arizona Press, 2012.

Douglass, Frederick. "Cheap Labour." *New National Era*, August 17, 1871.

Du Bois, W. E. B. *The Souls of Black Folk*. A. C. McClurg & Co. 1903.

Emerson, Ralph Waldo. *The Conduct of Life*. 1860.

———. *English Traits*. 1856.

Fanon, Frantz. *The Wretched of the Earth*. Translated by Constance Farrington. Grove Press, 1965.

Ferrer, Ada. *Cuba: An American History*. Scribner, 2021.

Fisher, Linford D. " 'Dangerous Designs': The 1676 Barbados Act to Prohibit New England Indian Slave Importation." *William and Mary Quarterly* 71, no. 1 (January 2014): 99–124. https://doi.org/10.5309/willmaryquar.71.1.0099.

Fleming, Ian. *Dr. No.* Jonathan Cape, 1958.

———. *The Man with the Golden Typewriter: Ian Fleming's James Bond Letters.* Edited by Fergus Fleming. London: Bloomsbury, 2015.

Gaussoin, Eugene. *Memoir on the Island of Navassa, West Indies.* 1866.

Gilroy, Paul. *The Black Atlantic: Modernity and Double Consciousness.* Harvard University Press, 1993.

Glissant, Édouard. *Poetics of Relation.* Translated by Betsy Wing. Ann Arbor: University of Michigan Press, 1997.

———. *Caribbean Discourse.* Translated by J. Michael Dash. University Press of Virginia, 1989.

Gómez, Nicolás Wey. *The Tropics of Empire: Why Columbus Sailed South to the Indies.* MIT Press, 2008.

Gordon-Reed, Annette. *The Hemingses of Monticello.* W. W. Norton, 2008.

Hall, Stuart. *Familiar Stranger: A Life Between Two Islands.* Duke University Press, 2017.

Hamilton, Virginia. *The People Could Fly: American Black Folktales.* Alfred A. Knopf, 1985.

Honychurch, Lennox. *In the Forests of Freedom: The Fighting Maroons of Dominica.* Papillote Press, 2017.

hooks, bell. "Appalachian Elegy," in *Appalachian Elegy: Poetry and Place.* University Press of Kentucky, 2012.

Immerwahr, Daniel. *How to Hide an Empire: A History of the Greater United States.* Picador, 2019.

Jackson, George. *Soledad Brother: The Prison Letters of George Jackson.* Coward-McCann, 1970.

James, Marlon. *A Brief History of Seven Killings: A Novel.* Riverhead Books, 2014.

Kincaid, Jamaica. *My Garden (Book).* Farrar, Strauss & Giroux, 1999.

King Jr., Martin Luther. *Where Do We Go from Here: Chaos or Community?* Beacon Press, 1967.

Lee, Jessica J. *Dispersals: On Plants, Borders, and Belonging.* Catapult, 2024.

Lepore, Jill. *The Name of War: King Philip's War and the Origins of American Identity.* Vintage Books, 1999.

London Times. "The Coolie and Slave-Trade. Horrors of the Coolie-Trade—a Legalized System of Free Chinese Emigration—Important Dispatch from Lord John Russell." August 16, 1860.

Long, Stephen. *Thirty-Eight: The Hurricane that Transformed New England.* Yale University Press, 2016.

Lowe, Lisa. "The Intimacies of Four Continents." In *Haunted by Empire: Geographies of Intimacy in North American History,* edited by Ann Laura Stoler. Duke University Press, 2006.

Luxemburg, Rosa. *The Complete Works of Rosa Luxemburg. Volume 1, Economic Writings 1.* Edited by Peter Hudis. Translated by David Fernbach, Joseph Fracchia, and George Shriver. Verso, 2013.

Morgan, Peter. *Fire Mountain: How One Man Survived the World's Worst Volcanic Disaster.* Bloomsbury, 2003.

Morrison, Toni. *Playing in the Dark: Whiteness and the Literary Imagination*. Harvard University Press, 1992.

———. *Sula*. Alfred A. Knopf, 1973.

Nezhukumatathil, Aimee. *World of Wonders: In Praise of Fireflies, Whale Sharks, and Other Astonishments*. Milkweed Editions, 2020.

Parliamentary Papers. "Chinese Labourers and Emigration: Return to an Address of the Honourable House of Commons." March 19, 1846.

Ramey Berry, Daina. *The Price for Their Pound of Flesh: The Value of the Enslaved from Womb to Grave in the Building of a Nation*. Beacon Press, 2017.

Rhys, Jean. *Wide Sargasso Sea*. W. W. Norton, 1966.

Rosenthal, Gregory. *Beyond Hawai'i: Native Labor in the Pacific World*. University of California Press, 2018.

Sexton, Jared. *Amalgamation Schemes: Antiblackness and the Critique of Multiculturalism*. University of Minnesota Press, 2008.

Shakespeare, William. *The Tempest*, 1611.

Silko, Leslie Marmon. *Almanac of the Dead: A Novel*. Simon & Schuster, 1991.

Sloane, Hans. *A Voyage to the Islands Madera, Barbados, Nieves, S. Christophers and Jamaica*. 1707.

Stoler, Ann Laura. *Carnal Knowledge and Imperial Power: Race and the Intimate in Colonial Rule, with a New Preface*. University of California Press, 2010.

Swinton, Captain, and Mrs. Swinton. *Journal of a Voyage with Coolie Emigrants, from Calcutta to Trinidad*. Edited by James Carlile. 1859.

Teng, Emma Jinhua. *Eurasian: Mixed Identities in the United States, China, and Hong Kong 1842–1943*. University of California Press, 2013.

Thackeray, William Makepeace. *Vanity Fair*. 1848.

Trouillot, Michel-Rolph. *Silencing the Past: Power and the Production of History*. Beacon Press, 1995.

Vizenor, Gerald. *Survivance: Narratives of Native Presence*. University of Nebraska Press, 2008.

Wilder, Craig Steven. *Ebony and Ivy: Race, Slavery, and the Troubled History of America's Universities*. Bloomsbury, 2013.

Williams-Forson, Psyche A. *Building Houses out of Chicken Legs: Black Women, Food, and Power*. The University of North Carolina Press, 2006.

Woods, Clyde. *Development Arrested: The Blues and Plantation Power in the Mississippi Delta*. Verso, 1997.

X, Malcolm, and Alex Haley. *The Autobiography of Malcolm X*. Ballantine Books, 1989.

Yong, Ed. *An Immense World: How Animal Senses Reveal the Hidden Realms Around Us*. Random House, 2022.

FILMS

Cousteau, Jacques-Yves, and Louis Malle, dirs. *The Silent World*. 1953; Columbia Pictures.

Wurmfeld, Treva, dir. *Conscience Point*. 2019; Conscience Point Film, LLC.

Index

Page numbers in *italics* refer to illustrations.

Abinader, Luis, 193
Abraham, Gabrielle, 262
Academy of Natural Sciences (Philadelphia), 84
Accompong Cannabis Trust, 252
Accompong Maroons, 31–32, 34–36, 252
Acosta, José de, 150
Adam, 40, 59, 287
Africa, xxiv, 10, 22, 23, 41, 48, 62, 63, 78, 80, 97, 98, 108, 140, 161–62, 244, 261, 261. *See also specific groups and regions*
African Americans, xxi, xxxvii, 59, 83–84, 98, 133, 143, 161–65, 166, 171, 247
African-Asian heritage, 206, 292–93
Africans, xvi, xxx, 22, 62–63, 89, 95–98, 106, 108–9, 151, 159, 169, 190, 232, 244–47, 261, 269–71. *See also* Middle Passage; transatlantic slave trade; *and specific regions*
  enslaved, xvi, xxvii, 8–9, 12, 26, 59–63, 71, 76–79, 83, 89, 90, 100, 117, 119, 122–23, 142, 145, 151, 153, 170–71, 174–76, 182–85, 184, 189, 192, 205–6, 210–11, 221, 232, 244, 247, 267, 270–71, 287, 295
Afrofuturism, 95, 108–9
Agricultural Age, 60, 136
Agwé, 59–60
Ahmad, Asif, 35

AI conference of 2022 (Berlin), 87. *See also* Museum für Naturkunde (Berlin)
AIDS epidemic, 140
Ainslie, George Robert, 182
Air Quality Index (AQI), 17
AIR Studios, 271, 274
Aja, 246
Akan, 7, 246
Akeley, Carl, 80
Akwapim mountains, xxxiv
Alaska, 111
Alcan corporation, 33
Alexander, Monty, 227
Algonquian language, 126, 174, 290
Ali, Irfaan, 22
Alice's Adventures in Wonderland (Carroll), 77
Al Jazeera, 36
Almanac of the Dead (Silko), 277
Aloha 'Āina, xxxiv, xxxv
Alpine yodelers, xxvii–xxviii
Amazon rainforest, 57, 175–85
American Geological Union (AGU), 138, 144
American Museum of Natural History, 79–82, 294
American Ornithological Society, 188
American Thirteen Colonies, 172, 243
Amerindians, xvi, xxxiii, 8, 29, 43–44, 46, 55, 86, 123, 138, 148–50, 176, 177, 182, 233

Amsterdam, 176
Anansi's garden, 59
Anatolian plate, 97
Anatsui, El, 76
Andes, xxx, 147–51
Anglo-Saxons, 136–37, 143
Angola, 71
Anguilla, 212
Annual Maroon Treaty Celebration, 35
Antarctica, 238
Anthropocene, xxix, 134, 279
Antigua, 50, 244, 274
Antilles, 59, 264, 267
"Antilles, The" (Walcott), 226
Appalachia, xxi, xxvi, xxx, xxxi, 144
Appalachian Elegy (hooks), xxvii
Arabian Sea, 279
Arapaho, xxxv
Arawak, xxxiii, 8, 26, 30, 44, 56, 122, 138, 198,
    235, 252–53. *See also* Taíno
Arbery, Ahmaud, 165
Arcadia corporation, 88
Arctic, 238
Armit, Colonel, 214
Armstrong, Louis, 227
Arrow, 276
Arthur's Seat, xxvi, 16–17
Artibonite River, 142
Aruba, 181, 244
Aryans, 216
Asa Wright Nature Centre, 181
Ashanti, 246
Asia, xxiv, 41, 43, 58, 78
Asians, xxx, xxxvi, xxxvii, 62–63, 208, 232,
    236, 244–47, 249
    indentured, xvi, 62–73, 232, 156–58, 204,
    235. *See also specific groups and regions*
Atlanta, 129
Atlantic Ocean, 73, 95–96, 101–2, 109,
    294–96
Atlantis, 108–9, 126, 128
Attenborough, David, 167
Audubon, John James, xxxvii, 80, 84, 186–88,
    190–92, 222
Audubon Society of New York, 162, 187, 198
Australia, 220
Autobiography of Malcolm X, The, 281, 283,
    286
Azores, 238

Báez, Firelei, 109
Bahamas, *viii,* xxi, 41, 44, 56, 138, 228, 244,
    270, 278, 291
Bailey, Bridgette (Tutty Gran Rosie), 4

Bajan people, 25, 259
Baldwin, James, xxi
Baltimore, 192
bananas, 48–49, 265
Banga, Ajay, 259
Bangladesh, 61, 215
Barbados, *ix,* 20, 172–73, 189, 217–18, 222, 228,
    259–61, 290
Barnum & Bailey Circus, 266
Barrett, Aston "Family Man," 227
bats, 28–31, 147, 219
bauxite, xix, xxvi, xxxvi, 27, 29–30, 32–34, 36,
    157, 261, 273
Baxter, Dow V., 111
BBC, 275
Beach Boys, 271–72
Beagle (ship), 117–18, 122
Beaubois Estate, 181
Bedouin of al-Naqab, 19
Bedward, Alexander, 228
Beijing, 14
Belafonte, Harry, 227
Belgium, 69–70, 80, 197
Belize, xxii, 253
Belize Barrier Reef, 128–29
Benin, xxxiv
Benin Bronzes, 75
Benjamin, Walter, 264
Berea College, xxxi
Berlin, 81, 87
Bermuda, 102, 138, 290
Berry, Halle, 189
Better Common Names Project, 188
Better Living Through Birding (Cooper), 166
Betty (manumitted person), 123
Beyoncé, 129
Bhojpur, 236
Bhutan, 238
Bible, xxi, 40–41, 54, 59, 142, 237
Bihar, 232
biodiversity, xxii, xxx, xxxvii, 8, 37, 77, 86–87,
    101, 110, 119, 127, 146, 194, 204, 229, 239,
    242, 264, 296
Birchwood, Ms., 194
Bird Names for Birds, 188
birds and bird-watching, xxxvi–xxxviii, 57,
    86, 123, 161–68, 171–92, 179, 180, 196–200,
    207–8, 219, 226, 280–84
Birds of America (Audubon), xxxvii, 186, 188
Birds of the West Indies (Bond), 188–89
Bird Union, 187
Birmingham, Alabama, bombing, 186
Birmingham University, 68
Bishop, Maurice, 285

Black Atlantic, The (Gilroy), 109
Black Birders Week, 165, 187
Black Coalition of Greater New Haven, 171
Black Expo, 171
Black Feminist Fund, 259
Black feminist meteorology, 291
Black in Natural History Museums, 86–87
Black Like Water, 103
Black Lives Matter (BLM), xxxi, 46, 68, 78, 81, 162, 165, 169, 186, 194
Black nationalism, 283, 285
Black Panthers, xxviii, 171
Black People Will Swim, 103
Black Power movement, xv, xxviii, 62, 285
Black Seneca Village, 166
Black Skin, White Masks (Fanon), 29
Black Surf Week, 103–4
Blair, Tony, xxiv
Blight, William, 247–49
Bloom, Harold, 140
Bloomberg Philanthropies, 88
Blue Mountains, xxvi, 7, 57, 71, 144
Bo Bardi, Lina, 254
Bocas del Dragon, 197
Bogle, Paul, 34, 199
Bonaparte, Joséphine, 81
Bond, James (character), 30, 110–11, 155–57, 189, 274
Bond, James (ornithologist), 188–89
Bond Line of designation, 188–89
Books, F.L.S., 219
Boone, Daniel, xxxi, 80
Borderline (Portmore, Jamaica), 255–56
Borglum, Gutzon, xxxv
Bounty (ship), 247–48
Brando, Marlon, 129, 247
Branson, Richard, 129
Brathwaite, Kamau, 272–73
Brazil, 96, 100, 151, 197, 238
Brexit, 76, 242, 268
Brief Wondrous Life of Oscar Wao, The (Díaz), 224
Britain (U.K.), xxii, xxxiii–xxxiv, 3–4, 8, 11–23, 26, 29–31, 34–35, 37, 39, 42–49, 51, 58, 66–67, 72–73, 76, 81, 100, 102, 122, 152, 167, 195–97, 210–11, 214–15, 221–22, 235, 238–47, 259, 261–63, 269–70, 285, 289–90, 292, 295. *See also specific colonies and former colonies*
British citizenship, 194
British East India Company, 117
British Empire, xxiv, xxvii, xxix, xxxiii, 12–20, 58, 204, 205, 235, 249, 269
British Expeditionary Force, 222

British Ganja Law (1913), 237
British Guiana, 151, 184
British India, xxix, 63, 204, 213–15, 232, 235
British Jamaican Maroon Treaty (1739), 9, 24–25, 31, 34–36, 252–53
British Library, 88
British Museum, 26, 74–76, 80, 123–24, 124, 219
British North America, 100
British Overseas Territories, 268
British Parliament, 76, 169
British Pathé, 198, 199
British-Peruvian Treaty, 135
British West Indies, 18, 156, 172–73, 184, 192–93, 206–7
Brixton, xxxviii, 244, 246–47, 246
Bronx, 83, 146, 190, 195
Brooklyn, 37, 125
Brosnan, Pierce, 189
Browne, Danny, 28
Browne, Patrick, 26
Bryant, Carolyn, 163–64
Buchanan, Isat, 35
Buddhists, 41
Buff Bay plantation, 207, 217
Building Houses Out of Chicken Legs (Williams-Forson), 282
Bump Grave, 20
Bureau of Indian Affairs (BIA), 249
Burke's History of the Colonial Gentry, 222
Bush, George W., 150
Butler, Octavia, 168

Calhoun, John C., 174
California, 42, 151, 296
Callao, Peru, 148
Calusa, 126
Calypso, 223, 228, 276
"Calypso" (Brathwaite), 272
Canada, 10–12, 30, 33, 69, 125, 175, 238, 259–60
Canarsie, 126
Canboulay riots, xxvii, 270
Cantonese, 12, 14
Cape Verde, 108, 238
carbon emissions, xx, xxv, xxxi, 31, 42, 48, 55, 61, 242, 257, 268–69
Caribbean Climate Justice Project, 259
Caribbean Coastal Area Management Foundation, 210
Caribbean Quarterly, 86
Carib Reserve Act, Dominica (1978), 45
Caribs, 44, 47. *See also* Kalinago
Carlos III, King of Spain, 222
Carnival, 227, 270

Carriacouan, 25
Carroll, Lewis, 77
Carson, Rachel, 112
Castleton Botanical Garden, 241
Catharine (manumitted person), 123
Cathartes aura, 197
Catholics, 56, 59, 194
Catskill System reservoirs, 71
Cayenne, Guiana, 256
Cayman Islands, *viii*, 145
Cecillia (manumitted person), 123
Cedula of Population, 222
Center for Marine Conservation, 146
Centers for Disease Control (CDC), 279
Central America, 48–49, 56, 128, 277
Central Intelligence Agency (CIA)
    Cuban Project, 223
    World Factbook, 146
Central Park, 81
Central Park bird-watching incident, xxxviii,
    161–68, 185, 187–88
Central Park Five, 166
Césaire, Aimé, 58, 121, 256
Chad, Lake, 97
Chagos Islands, 119
Chaguanas, Trinidad, 226
Chakrabarty, Dipesh, 279
Charles I, King of England, 168–69
Charles II, King of England, 170
Charles III, King of England, 20, 269
Charleston, South Carolina, 100, 102
Charles Town, Jamaica, 34
Charles Town Maroons, 252
chattel slavery, xv–xvi, xxi, 64, 89, 96, 101,
    106, 114, 151–53, 158, 168, 171, 207, 271. *See
    also* Africans, enslaved; transatlantic
    slave trade
Chauvin, Derek, 83, 162
Chelsea Physic Garden, 76, 241
Chesapeake Bay, 100
Cheyenne, xxxv
Chicago Bird Alliance, 187
Chile, 238
China, xxxiii, 3, 10–14, *13*, 17–19, 48, 50, 61,
    72, 144, 157–58, 176, 238–39, 262, 277,
    284–85, 293
China Harbour Engineering Company, 158
Chinatown (Manhattan), 294
Chincha Islands, 153, 156, 158
Chinese, xvi, xvii, xxxii–xxxiv, 10–13, 30, 33,
    48, 155, 162, 169, 176–79, 197, 217
    debt or indentured laborers, 118, 133–35,
    150–59, 199, 205–7, 218, 220, 249
Chinese Exclusion Era, 13

Chirac, Jacques, 265
Choctaw, 270
Christian, Fletcher, 247–48
Christians, xxiv, 41–42
Christie's, 89
Chronixx, 66
Chung, Andrea, 108
CIBC, 260
Cimarrón, 8
Civil and Natural History of Jamaica
    (Browne), 26
Civil Rights Movement, 171
Civil War (US), 54, 153, 159, 190, 243, 248
Clara Lionel Foundation, 259
Clarendon, Jamaica, 16
Clark, William, 80
Clarke, Yvette, 35
Clinton, Bill, 48
Cockpit Country, 28–30, 32, 34
Cold Spring Harbor Laboratory, 37
Cold War, 33, 44, 284
Cole, Nat King, 173
Coleman, Captain, 137
Collect Pond, 294
Columbia University, 82
Columbus, Christopher, xvi, xxiv, xxxiii,
    xxxvi, 31, 33, 40–49, 52–56, 60, 66, 86,
    133, 137–38, 149, 205, 215, 233, 244, 254,
    278, 297
Columbus Day, 43
Columbus Tree (Santo Domingo), 54
Commonwealth, 20, 33–35, 260–61, 269, 281
Conduct of Life, The (Emerson), 140
Coney Island Reef, 126
Conference of Berlin (1885), 98
Congo, 69–71
Connecticut, 174, 234
Connecticut Freedom Trail, 171
conservationism, xxxvi–xxxvii, 42, 77, 80, 85,
    88, 137, 146–47, 163, 181, 214, 296
Cook, Captain, 247
Coolie Trade, 151
Cooper, Amy, 161, 163–68, 187
Cooper, Christian, 161–68, 187–88
corals, xix, xxii, xxx, xxxv, xxxvi, xxxvii,
    60, 98–106, *99*, 108, 110–31, *115*, 124, 132,
    146–47, 257, 272, 274, 278, 286, 294
Cornell University, 180, 193, 243, 275
Cosmos (von Humboldt), 81
Costa Rica, 237
Cotton Gardens (London), 196
Cousteau, Jacques, 124
Covid-19, 46, 52–53, 68, 129–31, 162, 165, 185,
    270, 274, 279

Crawle River, 16
Crenshaw, Kimberlé, 83
Crick, Francis, 37
crimson-hooded manakin, 179, *180*
Crockett, Davy, xxxi
Cromwell, Oliver, 172–73
Croton System reservoirs, 71
Crown Heights, 195
Cuba, *viii*, 138, 151, 212, 217–18, 228, 238, 244, 255
Cudjoe, Captain, 9, 26
Cummings, Scott, 194
Currie, Chief Richard, 31, 34–35, 252
Cushman, Gregory T., 134

Dakota Access Pipeline, xxvi
Daley, James "Scriber," 281
Dalit Panthers rebellion, xxviii
Dangerous Drugs (Amendment) Act, Jamaica (2015), 252
Dark Lab, xxiii, xxv, xxxv–xxxvi, 9, 46, 52, 67–73, 91, 103–4, 122, 125, 196, 293–95
Dark Princess (Du Bois), 59
Darwin, Charles, 38, 117–23, 178, 182–85, 205, 223, 231, 257
"Darwin's Negro Bird-Stuffer" (Freeman), 122
Davenport, Charles, 37–39, *38*, 85, 179
David Zwirner gallery, 185
Davis, Anthony, 218
Dawson, Kevin, 106
DDT, 265
Decolonial Glossary, 9
Decolonize This Place, 81
Defense Department, 25
Delaware System reservoirs, 71
Demerara, Guyana, 123, 184–85, *184*
Demerara River, 185
Deng Xiaoping, 17
Denmark, xxii, 8, 100
Derrida, Jacques, 49, 89
Descent of Man, The (Darwin), 119
Desert Island Discs, 278
Detroit, 109
Deverell, Richard, 242–43
Devonian extinction, 120
Diamond as Big as the Ritz, A (Fitzgerald), 145
Díaz, Junot, 224–25
DiCaprio, Leonardo, 129
Dickens, Charles, 168
Die Another Day (film), 189
Dillon, Lamorra "Hope," 35–36
Dilrosun, Sean, 177–81
"Dirt on Our Hands" (podcast), 242

Discourse on Colonialism (Césaire), 256
diversity, equity, and inclusion (DEI), 67, 84–85, 164, 167
Dixwell family, 170, *170*
Dolores Ugarte (ship) fire, 150–51
Dominica, *ix*, xxxvi, 27, 44–53, 57, 181, 228, 262
Dominica-China Friendship Hospital, 50
Dominican Republic, *viii*, 44–45, 130, 137–38, 145, 193, 224
Don, George, 238
Donald-Drexciya, Gerald, 107, 109
Donald-Hill, Willy, 30
Dotson, Katricia, 83
Douglas, David, 238
Douglass, Frederick, 116–17, 150–51
Dow Jones Industrial Average, 65
Down de Islands (DDI), 197
"Downpressor Man" (Tosh), 98
Drexciya, 95–99, 101–2, 108–9
Drexel University, 84
Dr. No (Everett), 157
Dr. No (film), 274
Dr. No (Fleming), 30, 110, 112, 155–59, 200
Droxler, André, 120
Du Bois, W. E. B., 59, 69, 164, 200
Duckenfield rebellion, 199
Duffield, A. J., 152–54
Duncan, Dr., 182–83
Duncan, Peter, 137–38
Dutch Antilles, 138
Dutch colonialism, xxii, 8, 21, 57, 88–89, 100, 102, 175–77, 190, 290, 295. *See also* Netherlands
Dutch Council for Culture, 88–89
Dutch East India Company, 21, 89, 112, 117
DX-listeners, 146

East African Herbarium Library (Kenya), 88
Eastern Caribbean, 103, 262
East Indies, xvi, 26
Ecology of Invasions by Animals and Plants, The (Elton), 229
Economist, 10
Ecuador, 119
Eden, xxv, xxxvi, 3, 40–54, 56, 58–66, 95, 110, 192, 215, 239–40, 247, 271, 278, 287
Edinburgh, 16
Edmonstone, Charles, 122–23, 183–84
Edmonstone, Helen, 123
Edmonstone, Jeanie, 123
Edmonstone, John, 120, 122–23, 182–85
Edmonstone, Princess Minda, 122, 183
Egypt, 97, 285
Eisenhower, Dwight D., 240

Ellis Island, 27
Elton, Charles, 229
Emerson, Ralph Waldo, 132–33, 136, 139–41, 143, 153, 159, 172, 190, 198, 214, 221
Endangered Archives Programme, 88
England, Kingdom of, 168
English language, 9
Enlightenment, xxx, 80, 264
Enriquillo–Plantain Garden fault zone, 255
environmentalism, xxv, xxxvii, 61, 82, 84–87, 110, 129, 134, 136, 139, 141, 143–44, 148, 209–10, 240, 259, 280
Environmental Protection Agency (EPA), xxxii, 139
environmental racism, xxxii, 61, 64, 73, 139, 193, 195, 282, 285, 297
environmental social and governance (ESG), 67
Eritreans, xxviii
Espeut, Caroline Marlton, 222
Espeut, Mrs. (née Armit), 205, 213–15
Espeut, Peter, elder, 222
Espeut, Peter, younger, 209–10
Espeut, W. B., 203–12, 208, 216–22, 224, 226
Espeut, William, 222
Ethiopia, xxviii, 73, 237
Ettu tradition, 22
eugenics, xvii, xxx, 37–38, 62, 79, 82, 120, 136, 176, 178–79, 182, 218, 279
Eugenics Records Office, 37
Eurasian plate, 97
Europe, xxiii–xxiv, xxix–xxx, xxxiv, 7–11, 40–43, 46, 54–55, 57, 64, 72, 97–98, 113–14, 169–76, 194, 255–56, 259, 287. *See also specific regions*
European ancestry, 161–62, 168–69
European Union, 48, 254, 265
Eve, 40, 51, 59, 271, 287
Everest, George, xxix
Everett, Percival, 157
Evers, Medgar, 281
Ewe, 246
Exodus, 142
Extraordinary Birder with Christian Cooper (TV show), 163
Exxon, 101

Facebook, 34
Fal, 198
Fanon, Frantz, 29, 57
Fante, 246
"Fate" (Emerson), 133, 136, 140
Federal Emergency Management Agency (FEMA), 150, 276

Ferdinand and Isabella, 54–55, 66
"Fight for Jamaica's Red Gold, The" (podcast), 36
Fiji, 158
Fillmore, Millard, 135
Filthy Riddim, 28
Finch, George, 196
First Nations, xvi
Fisk University, 115
Fitzgerald, Ella, 227
Fitzgerald, F. Scott, 145
Fitzroy, Captain, 122
Flandreau Santee Sioux, 249
Fleming, Ian, 30, 84, 110, 112, 153, 155–59, 188–89, 200
Florida, xxii, xxv, 126–27
Florida Keys, 100, 112
Floyd, George, 83, 161–67, 242
Fon, 7, 246
Ford Motor Company, xxxii, 250
"Foreigner's Home, The" (Morrison), 70
Forest Hills (Queens), 291–92
Forrest, George, 239
Fortune, Robert, 238
Foucault, Michel, 217–18
France, xxii, 8, 44, 46, 49, 100, 102, 137, 192, 215, 222, 240, 254–57, 265–71, 285
Franklin, Benjamin, 143
Franklin, Rosalind, 37
Fraser, John, 238
Frazier, Darnella, 165
Freeman, R. B., 122
Freeman Alley and the New Museum, 83
Freetown (East Hampton), 290
French Guiana, 8, 22, 54, 188, 254, 256–57
French Polynesia, 247
French Revolution, 270
Fugitive Slave Act, U.S. (1850), 133, 165
Fula, 246
Fu Manchu, Dr., 157
Fyfe, Alexander G., 24
Fyre Festival, 278

Gah-san tomb-sweeping ritual, 14. *See also* Qingming festival
Galapagos, 119, 183–84, 238
Gallagher, Ellen, 108
  Watery Ecstatic, 108
Galton, Francis, 37–38, 119–20
Gambia, 238
Ganga (goddess), 236
Ganga-Smith, Charles, 35
Ganges River, 232, 236
Gardner, Alexander, 153–55, 154

Garifuna, xxii
Garmin corporation, 262–63
Garvey, Marcus, 228, 283, 285
Gate of the Exonerated, 166
Gauguin, Paul, 72
Gaylads, 227
Gaza, 193, 255–56
General System of Gardening and Botany, A (Don), 238
Genesis, 41, 98
geology, xx, 33, 65, 78, 98–99, 113–21, 134, 144–45, 166, 255–57, 275, 280, 286, 293–94
George III, King of England, 44
George Washington University, 116
Georgia, 129, 283–84
German immigrants, 132, 136, 143
Germany, 52, 181, 248. *See also* Nazi Germany
Ghana, xxxiv, 18, 285
Ghanaian language, 23
Gilmore, Ruth Wilson, 268
Gilroy, Paul, 109
Glasgow, 259
Gleaner, 209
Glissant, Édouard, 47, 105
Global Positioning System (GPS), 25, 262–63
Gmelin, Johann Friedrich, 198
Goat Islands, 158
Goffe, Alfred Constantine, 189
Goffe, John Beecham, 189
Goffe, Dr. Marcus, 252–53
Goffe, Tao Leigh
    "Kill Them with the No!," 28
    Queen Nannies, 27–28, 27
Goffe, William, 168–69, 171–73
Goffe family tree, 168–71, 170, 189
GoFundMe, 34
Gold Coast, 7, 21, 96, 113, 117, 249, 293
GoldenEye estate, viii, 110, 189–90
Golden Hill Paugussett, 174
Gonaïves, Haiti, 138
Gonçalves-Ho Kang You, Lilian, 88
Goombeh tradition, 22
Go Tell It on the Mountain (Baldwin), xxi
"Go Tell It on the Mountain" (song), xxviii
Governors Island, 198
Grand Cayman, 145
Grandie Nanny, 20–22, 25–26, 27. *See also* Queen Nanny
Grant, Sir John Peter, 242
Great Blue Hole, 129
Great Experiment (Mauritius), 204
Greece, 97, 108, 193–94

Grenada, *ix*, 217, 223, 244, 278, 282–86
Grenadines, *ix*, 54
Grendel, 69
Guadeloupe, 27, 254–55, 265
Guam, 72
Guanahaní, *viii*, 41, 44, 55
Guangzhou, 293
guano, xix, xxxvi–xxxvii, 29–31, 131–60, 172–73, 190, 199, 219, 241, 271, 275
Guano and the Opening of the Pacific World (Cushman), 134
Guano Islands Act, U.S. (1856), 135
"Guano Song" (von Scheffel), 155
Guantanamo, 146
Guianan cock-of-the-rock, 177, 179–80, 179
Guinea, 238, 256
Guinea-Bissau, 138
Gullah, 102
Guyana, *ix*, 22, 54, 63, 101, 122, 182–85, 184, 188, 261
Guyanese, 259

Hadley, Massachusetts, 172
Haile Selassie, Emperor of Ethiopia, 237
Haiti, *viii*, xxi, 8, 130–31, 133, 136–47, 192–93, 199, 255, 261, 277
    earthquake of 2010, 142–43
    Revolution, xxxvii, 141, 192, 222, 271
Haitian immigrants, 140–41, 146, 193
Hakka Chinese, xxxiii, 14, 17–20, 217, 248
Hall, Stuart, 19, 68
Hāmākua Coast, 217
Hamilton, Lewis, 285
Hamilton, Virginia, 102
Hamptons (Long Island), 126, 251, 291, 294, 297
Hamza: Strictly Birds of Prey (TV show), 167
Han Chinese, 19
Harlem, 69, 156
Harlem African Burial Ground, 83
Harvard Forest, 243
Harvard University, 79, 243, 291
Haudenosaunee, 290
Hausa, 56
Hawai'i, xxvi, xxxiv–xxxv, 42, 71, 133, 156, 158–60, 199, 206, 216–19, 238, 276, 278–79
"Heads High" (song), 28
Heady Creek, 107
Hedonism resort, 271
Hellshire beach, 37
Helsinki, 69
Hemings family, 161
Hertfordshire, England, 222
Hilo, 69

Himalayas, xxx, 73
Hindi language, 59, 121, 213
Hinduism, xxxviii, 41, 59, 121, 143, 213–16, 235–37, 247
Hindustani, xvi, 177
Hispaniola, *viii*, 54, 130, 138, 145, 192–93, 212, 271. *See also* Dominican Republic; Haiti
Historia natural y moral de las Indias (Acosta), 150
Hitler, Adolf, 228
Hofstadter, Richard, 139–40
Home Away from Home Cave, 147
Homeland Security Department, 140
Hong Kong, xxvi, xxxiii, xxxvi, 3–4, 10–19, 13, 15, 39, 60–61, 72, 176, 150, 238, 292
  Fanling Lau, China, 16
Hong Kong Antiquities Advisory Board, 16
Honolulu, 188
Hope Botanical Gardens, 241
Hopi Nation, 270
Hortus Malabaricus (van Rheede), 21
"Hot, Hot, Hot" (song), 276
Houdini, Harry, 228
Houtouwan, 65
Howell, Leonard Percival, 236–37
Huawei corporation, 10–11
Hudson River, 71, 293, 294
Huggins, Nadia, 122
Hulu, 292
Humboldt, Alexander von, 81, 87, 121, 148–50
Hunga Tonga-Hunga Haʻapai, 61, 61
hurricanes, 18, 60, 129, 261–62, 264
  Beryl, 85
  Ida, 293
  Irene, 293
  Katrina, 4, 42, 61, 150
  Maria, 45–51, 53, 181, 262
  New England (1938), 291
  Sandy, 293

Ifa, 56
Ìgbò, 56, 95–96, 246
Iglesia de Nuestra Señora de los Remedios (Mexico), 72
Immense World, An (Yong), 28
Imperial College, 78
Imperial College of Tropical Agriculture, 209
Inca Empire, 148, 149, 150, 155
indentured labor, xvi, 77, 118–19, 121, 141, 150–60, 207, 218, 220, 236, 290–91
India, xxxviii, 58, 59, 203, 204, 214–16, 219, 227, 238

Indian, defined, xvi
Indiana, 186
Indian Ocean, 77, 118–21, 279
Indians, indentured, xxxviii, 59, 118–19, 121, 218, 230, 232, 235–37, 247
Indian Thanksgiving, 251
Indigenous Environmental Network, 259
indigenous peoples, xvi, xxii, xxiii, xxiv, xxix, xxx, xxxii, xxxvi, 22, 44, 48, 51, 53–54, 66, 69–70, 79, 86, 91, 106–7, 122, 126, 133, 135, 147–48, 151, 158–59, 165–66, 174, 182, 188, 233–34, 239, 242, 249–53, 262, 290. *See also* Native peoples; *and specific groups and regions*
Indigenous Peoples' Day, 43
Indo-Caribbeans, xvi, 227, 230
Indo-Jamaican labor, 237
Indonesia, 73, 89, 238
Industrial Revolution, 31, 40, 60, 136, 269
Institute of Jamaica (Kingston), 20, 26–27
International Labour Organization, 252
International Monetary Fund (IMF), 159
International Peacekeeping, 143
"Introduction to Political Economy" (Luxemburg), 149
invasive species, xxxviii, 204, 205, 208, 218, 222–23, 228–30
ʻIolani Palace (Honolulu), 188
Iran, 285
Ireland, 207, 270, 298
Irish, 132, 136–37, 143, 194, 270–71
Irish Potato Famine, 270
Iroquois longhouses, 290
Israel, xxiv
Italian immigrants, 137, 194
Italy, 97, 114, 193–94
Ivorians, 97

Jackson, Ayana V., 109
Jackson, George, 141–42, 159
Jaco, Chief, 182
Jagger, Mick, 271
Jamaica, *viii*, xxvi, xxxii–xxxiv, xxxvi–xxxviii, 4–39, 38, 48, 51, 56–61, 66, 71–72, 75–76, 85, 98, 110–11, 123, 138, 147, 151, 155–58, 172–78, 181, 188–223, 226–37, 241–56, 259, 262–63, 271–73, 280, 282, 285, 290–91, 294
Jamaica High Commission, 78
Jamaican Free Zones, 157–58
Jamaican National Environment and Planning Agency, 33
Jamaican National Heritage Trust, 35

Jamaican parliament, 158
Jamaican Revivalism, 228
Jamaicans, 76, 234–35
James, C. L. R., 58, 134, 270–71
James, Marlon, 123
JAM-NAV, 262–63
Japan, 15, 238
Javanese, 176–77
Jawaiian music, 71
Jay-Z, 129
Jesuit missionaries, 150
Jews, 41, 59, 241
Jim Crow, 54, 103
John, Elton, 271
John Crow Mountains, xxvi, 7, 196–97
"John Crow Skank" (song), 197
John Crow (vulture), 187–91, 196–97
Johnson, Andrew, 145
Jorry, Stéphan, 120
Journey of the Deep Sea Dweller IV (album), 109
Judges Cave, 172, 172
Judgment Day, 98
Juneteenth, 145
Jungle Book, The (Kipling), 203, 215–16
Junkanoo, 270

Kaiser corporation, 33
Kalina, 22
Kalinago, xxxvi, 44–52
Kanakas (South Sea Islanders), 158
Kamaʻehuakanaloa seamount (*previously* Lōʻihi), 278
Kannada language, 206
Kartel, Vybz, 255
Kauaʻi, 217
Kau Yi Chau, 16
Kennedy, John F., 281
Kentucky, xxvi, xxvii, 234
Kenya, 88
Kew Gardens, xxxviii, 63, 209, 240–43, 249
Kew Madagascar Conservation Centre, 242
Kick 'Em Jenny volcano, 278, 282
Kincaid, Jamaica, 58, 244
Kindred (Butler), 168
King, Martin Luther, Jr., xxi, 285
King Philip's War (1675–76), 290
Kingston, *viii*, 4–5, 10–11, 16, 18, 204, 241–42, 262, 292
Kingston Harbour, 18
Kingston Market, 198–200, 199
Kipling, Rudyard, 203, 214–16
Klein, Naomi, 276
"Kokomo" (song), 271–72

Koramantee, xxxiii, 9, 23, 246
Kowloon peninsula, 16
Kreyòl Ayisyen language, 137
Ku Klux Klan, xxxv, 283
Kumeyaay Territory, 103
Kumina tradition, 22
Kwinti, 22

Labeach, Alessandra, 35
Lagos, 293
La Jolla, California, 103
Lakota, xxxv
Lampedusa, 97
Lānaʻi, 217
Land Back, xxxi, 57, 66, 68, 107, 294
Lansing, Michigan, 282, 283
Lantau Island, 13, 16
Lantern Laws, 165
Latin America, xxi, 10, 12, 261, 271, 279
Laurelton (Queens), 195
Lebanese, 255, 298
Lee, Sabine, 143
Leeward Islands, 32, 285
Leeward Maroons, *viii*, 32, 35
Left Behind in Rosedale (Cummings), 194
Legba dances, 276
Leiden Garden, 243
Leiden University, 75
Lenape, 91, 166, 294
Lenapehoking, 294
Lenni-Lenape, xxxi
Leptospira, 217
Leung Ashew, 150
Levi, Paul, 149
Lewin, Olive, 21
Lewis, Sir Arthur, 17
LGBTQ, 163–64, 166, 192, 210
Library of Congress, 54
LIDAR, 262
Life, 284
Linnaeus, Carl, 38, 177–78, 187, 198, 244, 248, 279
Linnean Society, 67, 209, 219, 238
"Linstead Market" (song), 245
"Litany for Survival" (Lorde), 72
Little, Earl, 283
Little, Louise (Helen Louise Langdon), 283–85
Little Beach Harvest, 250–51
Locke, Hew, 76
Lokono, 22
London, 58, 71, 76–78, 123, 193, 196, 240, 244–46, 266, 274, 289
Long Celia, 145

Long Island, xxvi, 37, 107, 126, 167, 194, 250–51, 290–97
Long Island Aquarium, 126
Long Term Ecological Research Network, 243
Lorde, Audre, 24–25, 72
Lord Invader, 228
Lost (TV show), 278
Lo Ting (deity), 3–4, 14, 16
Louisiana, 61
Love Island (TV show), 278
Lucayans, 41, 44
Lukku-Cairi, 55–56
Lulu Bay, 146
Lumumba, Patrice, 281
Luohu, 17
Luxemburg, Rosa, 148–49, 265–66
Lyall, David, 238
Lydford Mining Company, 30

Madeira, 118, 238
Mātāwai, 22
Malcolm X, 168, 281–86
Maldives, 67, 121, 279
Mami Wata (River Mumma), 3, 14, 16, 59, 295
Mandela, Nelson, 268–69
Manhattan, 37, 83, 129, 156, 198, 294–95
Manifest Destiny, 135, 141, 143, 153, 296
"Manifesto for Change 2021–2030" (Kew Garden), 242
Manley, Michael, 33
Manley, Norman, 273
Mann, Vincent, 249
Manning, Sam, 227
Māori, xxviii
March on Washington, 54
marijuana, xxxv, 71, 231–37, *234*, 247–53, 278
Marijuana Regulation and Taxation Act, New York (2021), 250–51
Marley, Bob, xxviii, 71, 249, 259
Marley Natural Dispensary of Jamaica, 253
Marlton, Caroline, 222
Maroons, *viii*, xxv, xxxiii, xxxvi, 7–12, 18–36, *24*, 44–45, 47, 57, 107, 122, 177, 181–82, 188, 245, 252–53
Marshall, Kerry James, 185–87, 190
  Black and Part Black Birds in America, 185–86, 190
MARTA (rail network), 129
Martin, Sir George, 271
Martinique, *ix*, 58, 78, 81, 102, 121, 218, 222, 244, 254–56, 265–69, 275
Marxists, 147–48, 263, 266

Mashantucket Pequot, 174
Massachusetts Bay Colony, 172
Massachusetts Institute of Technology (MIT), 225
Massapequa people, 126
Masson, Francis, 238
Mauna a Wākea, xxvi
Mauritanians, 97
Mauritius, 76–77, 119, 121, 204
Maxam, Dr. Ava, 262
Maya, 233
Mbiri Creek, 184–85, *184*
McAuley, Louis Kirk, 204
McCartney, Paul, 271
McVeigh, Timothy, 54
Meadowlands, 125
Medical and Physical Society of Calcutta, 235
Mediterranean Sea, 97–99
Menafee, Corey, 174–75
Menehune (deities), 278–79
Menzies, Archibald, 238
Merchantman (ship), 204
Merwin, Hester, 47
Messy Mya, 4
Mexico, xxi, 52, 56, 103
Meye Tributary System, 142
Miami, 112
Michelangelo, 186
Michigan, 282–84
"Microstructural Changes in the Scleractinian Families" (Owens), 116
Middle East, xxiv, xxx, 255
Middle Passage, 4, 60, 95–97, 102, 106, 109, 117, 142, 152, 276. *See also* transatlantic slave trade
Midsummer Night's Dream, A (Shakespeare), 58
Mignolo, Walter, 104
Miller, George, 195
Milwaukee, 156, 282
mining, xix, xx, 8, 11, 19, 27, 29–30, 41, 64–65, 101, 132, 157, 261, 296
Minneapolis, 162, 165
Mississippi, 31, 208
Mississippi Valley, 233
Mohegan, 174
Mona Geoinformatics Institute, 262–63
mongoose, xxxv, xxxviii, 59, 203–8, *208*, 211–19, 222–30, 235, 278
"Mongoose and the Brahmin," 216
Mongoose Gang, 223
Monoprix, 254
Monroe Doctrine, 277
Montana, 145

Montauk, Long Island, 290–95
Montaukett (Montauk, Manitowoc), 126, 290, 294
Montego Bay, *viii,* 36
Montreal, 283
Montserrat, *ix,* xxxviii, 181, 255, 268–82, 286
Montserrat Springs Hotel, 273–74
Mo'orea island, 119–20
Moore Town, Jamaica, *viii,* 8, 20, 32, 34–36
Morant and Pedro Cays Act (1907), 200
Morant Bay Rebellion, 34, 199
Morant Cay, 198–200, *199*
Morgan, Derrick, 197
Morne Diablotin National Park, 181
Morrison, Toni, 57, 69–70, 85–86, 96
Morton, Timothy, 279
Moses, Robert, 290
Mottley, Mia, 259–61
Mount Everest, xxix
Mount Pelée eruption (1902), 265–69, *267*
Mount Rushmore, xxxv
MOVE bombing, 83
Moyers, Bill, 194–95
Mozambique, 266
Munsee bands, 295
Munsee Three Sisters Medicinal Farms, xxxii, 249–50
Museum für Naturkunde (Berlin), 87
Museum of Economic Botany, 242
museums, xxxvi–xxxvii, 74–84, 88–91
Muslims, 22, 41, 213
Mutiny on the Bounty (film), 247
My Garden (Book) (Kincaid), 244
Mystery of Dr. Fu Manchu, The (Rohmer), 157

Nagel, Thomas, 29
Nago tradition, 22
Nagoya Protocol, 239
Naipaul, V. S., 224–27
Nanny Town, Jamaica, 28, 252–53
NASA Thirty Meter Telescope, xxvi
Nasdaq, 65
Nasheed, Mohamed, 279
Nassau String Band, 228
National Anthropological Archives, 47
National Audubon Society, 187–88
National Geographic, 63, 127, 163, 240
National Guard, 171
National History Museum (London), 266
National Institutes of Health, 115
National Museums of Kenya, 88
National Oceanic and Atmospheric Administration (NOAA), 110
National Wildlife Refuge (Navassa), 137

Native American Graves Protection and Repatriation Act, U.S. (NAGPRA, 1990), 82–83
Native Americans, xvi, xxxi, 19, 43, 126–27, 79, 290–91. *See also* indigenous peoples; Native peoples; *and specific groups and regions*
Native Hawaiians, 276
Native Heritage Month, 251
Native peoples, xvii, xxi–xxii, xxv, xxvi, 25–26, 40–41, 44–46, 57, 60, 66, 68–69, 81, 86, 90, 103, 107, 118–19, 126, 143, 151, 173–74, 188, 199–200, 230, 233–34, 239, 249–50, 290–91, 294. *See also* indigenous peoples; *and specific groups and regions*
NATO, xxxviii
natural history, 62, 67, 75–91, 205, 209
Natural History Museum (London), 77–78
naturalists, 111, 124, 144, 146, 150, 156, 163, 165, 167, 183–87, 197, 219, 280
Nature Communications, 211–12
Navajo Nation, 270
Navassa, *viii,* 137–39, 145–46, 156–57, 219
Navassa, North Carolina, 139
Navassa Guano Factory, 139
Navassa Phosphate Company, 138
Nazi Germany, 37–38, 216
Ndyuka, 22
Nebraska, 282–83
Negril, Jamaica, *viii,* 36, 271
Negro World, The, 283
Nelson, David, 247
Nepal, xxviii–xxx, 73, 142, 239
Netherlands, 75, 88–89, 171, 176, 197, 215, 243. *See also* Dutch colonialism
New Caledonia, 158
New England, 170–73, 291, 296
New Guinea, 248
New Haven, Connecticut, 170–74, *170, 172*
New Jersey, xxxi, 91, 125, 249–51
New Jersey Botanical Garden, xxxii, 249
New Jersey Transit, xxxi
New Orleans, 4, 268
New Territories, *13,* 14–16, 19. *See also* Hong Kong
New York Botanical Garden, 146, 190–91, *191*
New York City, xxxviii, 37, 39, 42, 72–73, 81, 83, 104, 107, 125, 129–30, 165, 187, 190–98, 237, 244, 246, 249, 274, 289–95. *See also specific locations*
New York City Council, 166
New Yorker, 187, 225
New York State, 250–51, 290–91
New York Times, 10, 78, 145

New-York Tribune, 84
New Zealand, xxviii, 88, 220, 238
Nezhukumatathil, Aimee, 31
NGOs, 88, 97, 276
Nhu, Madam, 281
Niantic, 174
Nigeria, 35, 95, 285
Niger River, 293
Nina, 55
Nobel Prize, 17, 224–26
Nordic immigrants, 140
Norfolk, Virginia, 192
Norman Manley Airport, Jamaica, 18
North Carolina, 234
North Korea, 146
Norway, 136
Notting Hill Carnival, 76
Nova Scotia, xxi
Nubian Plate, 97
Nunnowa Shinnecock harvest holiday, 251
NYC Bird Alliance, 163, 187

Oakland, xxviii
Obeah tradition, 21, 56
Ober, Frederick Albion, 280
Ocean, Billy, 285
Oceania, 247
oceans and marine life, 51, 67, 95–108, 114–17,
    121, 124, 126–28, 130–31, 251, 278, 297
Oceans Back, 57, 107
Ocho Rios, Jamaica, *viii,* 189
Oklahoma, xxxi, 166
Oklahoma City bombing, 54
Olmsted, Frederick Law, 166
"On the Acclimatization of the Indian
    Mungoos" (Espreut), 218
Open University, 68
Operation Mongoose, 223
Opium Wars (1839–60), 18, 154, 236
Oracabessa, Jamaica, 189
Orientalism, 143
Orinoco River, 175
Orun, 56
O'Shaughnessy, William Brooke, 235
Otaheite apples, 71, 247, 247
Our Land, Our Stories, 250
"Out of Many, One People" (e pluribus
    unum), 85
Owens, Dr. Joan Murrell, 114–16, *115*
Oxford University, 76–77, 229, 241, 243
oysters and clams, 107, 126–27, 294–97

Pacific Islanders, xxx, xxxvii, 119, 135, 158–60
Pacific Ocean, 48, 104, 117–23, 132, 147

Painter, Nell Irvin, 133, 141
Pakistan, 215
Palestine, xxiv, xxxix, 70, 255–56, 298
Pan-Africanism, 261, 285
Panama, 138
Panama Canal, 237
Pan-Caribbean dream, 272
Panchatantra, 215–16
Pangaea, 244, 255, 286
Pansa, Fred, 177
Papa Legba, 276
Papua New Guinea, 158
Paramaribo, Suriname, 175
Parker, Charlie, 227
Parliamentary Papers, 205
Patterson, Ebony G., 190–92
    The Observation, 192
    ". . . things come to thrive . . . ," 190–91,
        191
Patwa, 190
Peace of Ryswick (1697), 137, 238
Pearl River, 293
Pearson, Karl, 38
Penn Museum, 83
People Could Fly, The (Hamilton), 102
Perry, Lee Scratch, 232
Perseus, 99
Peru, 133, 135, 147–55, 199, 219, 241
Philadelphia, 83, 138, 192
Philippines, 238, 266
"Photographic Foray" (Baxter), 111
Picaro, Michaeline, 249
Pinta, 55
Pizarro, Francisco, 148
Placencia reef system, 128–29
plantations, xx, xxi, xxxiii–xxxiv, xxxviii,
    8–9, 12, 22, 29, 39, 43, 51, 58–63, 66,
    71–72, 76, 85, 106–7, 118–19, 121, 126, 139,
    141, 145, 156–59, 169, 174, 179, 192, 204–10,
    213, 217–18, 223, 233, 235–37, 243, 276, 296
plants and gardens, 56, 191, 210, 230, 238–49,
    283–84, 286
plate tectonics, 97, 143, 255–58, 287
    affective, xxxviii, xxxix, 255, 257, 270, 279,
        283–84, 287
Plato, 108
Playing in the Dark (Morrison), 69
Play Whe, 197
Plomer, William, 156
Plymouth, Montserrat, 271, 273–74, 281
Poe, Edgar Allan, 198
Polynesians, 72, 89, 119, 121, 133
Portland, Jamaica, 20–21, 32
Portmore, Jamaica, 255

Portugal, 100, 102, 206, 215, 238
Power Fund Partners, 251
Pratt Institute, 125
Princeton University, xxiii, 69, 83, 174
Prinz Alexander (steamship), 220
Private Island Online, 129
*Proceedings of the Zoological Society of
    London,* 203
Protestantism, 59, 90
Puerto Rico, 212, 217, 229
Pulitzer Prize, 224
Punta Cana, 50
Putonghua language, 17

Qingming festival, 14
Quao, 26
Queen Nanny, 20–22, 25–28, 36–37. *See also*
    Grandie Nanny
Queens, 37, 73, 193–96, 291–92, 293
Queensland, Australia, *111,* 158
Quijano, Aníbal, 104
Quinnipiac, 174

Race Crossing in Jamaica (Davenport and
    Steggerda), 38–39, *38,* 85
Rachel Carson Center (Munich), 134
Raffles, Sir Stamford, 218–19
Ramapo Mountains, xxxii
Ramapough Lunaape Nation Turtle Clan,
    xxxi–xxxii, 250
Ramayana, 59, 215
Ranglin, Ernest, 227
Rastafari, xxxviii, 22, 66, 71–72, 98, 228,
    236–37, 247–48, 252
Rays of Sunlight from South America
    (Gardner), 153–55, *154*
Red Cross, 276
Reefapalooza, 125
Regicides Trail, 172–73
Residence in the West Indies and America, A
    (St. Clair), 184
Return Our American Rights (ROAR), 194
Revelations, 98
Reynolds corporation, 33
Rheede, Hendrik van, 21
Rhys, Jean, 51–52, 192
Richter scale, 266
Rigney, Clyde, 198
Rihanna, 259
Rijksmuseum, 89
"Rikki-Tikki-Tavi" (Kipling), 203, 214–15
Ringwood, New Jersey, xxxi–xxxii
Ringwood Mines, 250
Rio de Janeiro, 102

Riverhead, New York, 126
Riverside Park (New York), 198
Rodney, Walter, 62, 261, 285
Rolling Stones, 273
Roma, 19
Romans, 99
Roosevelt, Franklin D., 80
Roosevelt, Theodore, 49, 80–81, 84
Rose, Elizabeth Langley, 76
Roseau, Dominica, 49, 52
Rosedale (Queens), 193–96
Rosedale (documentary), 194
Rothschild, Dame Miriam, 30
Royal Academy, 209
Royal Bank of Canada, 12, 260
Royal Botanic Gardens, 209, 243
Royal Dutch Shell, 121
Royal Engineers, 214
Royal Garden of Edinburgh, 239
Royal Horticultural Show (Scotland), 240
Royal Horticultural Society, 238
Royal School of Mines, 78
Royal Society, 66–67, 76, 209
Royal Society Journal of the History of Science,
    122
Russia, 10, 30, 33, 238

Saamaka, 22
Sadhu aesthetics, 236
Sag Harbor, New York, 290
Saint Domingue, 137, 222. *See also* Haiti;
    Hispaniola; Dominican Republic
Saint Lucia, 48, 223
Saint-Pierre, Martinique, 265, *267,* 268
Salaam, Yusef, 166
Sandals Resorts, 36
San Diego, 103–4
San Francisco, 103
San Quentin, 142
Sanskrit texts, 215–16
Santa Maria, 55
Santo Domingo, Hispaniola, 54
Santuario de Mamíferos Marinos de la
    República Dominicana, 130
Saola, Super typhoon of 2023, 4
São Tomé, 71
Sarcoramphus papa, 197
Sardinia, 113–14, *113*
Sargasso Sea, 51–52, 291
Savannah, Georgia, 30
Say Her Name / Say Their Names, 83
Scandinavia, 102
Schaghticoke people, 174
Science, 211

Science Museum (London), 77
Scotiabank, 12, 260
Scotland, xxvi, 66, 73, 122–23, 167, 184, 207, 238–40
Scott, Winfield, 188
Scottish plant hunters, 238–40
Scott's Hall, Jamaica, 34
Scripps Institution of Oceanography, 103
Sea Around Us, The (Carson), 112
Sea Islands, 102
sea level rise, 16, 97–98, 110, 120–21, 279, 296
Seatucket people, 126
Sebeok, Thomas A., 28
Second International Eugenics Congress (1921), 82
Second Jungle Book, The (Kipling), 215
Second National Conference: Justice in Geoscience (2022), 138, 144
Secret Bay resort, 45
Selassie, Ras Moqapi, 237
Seminoles, xxii
Sewanhacky, 126, 297. *See also* Long Island
Shakespeare, William, 58–59, 69, 95
Shakira, 129
Shanghai, 14, 17
Sheba (Saba), 54
Shengshan Island, 65
Shenzhen, 13, 16–19, 293
Shenzhen River, 16–17
Shepherd, Verene, 237
Sherriff, George, 238
Shinnecock Hills Golf Club, xxvi
Shinnecock Nation, 107, 126, 250–51, 290, 294–97
Shiva, 236
Sierra Leone, 238
"Silence, The" (Díaz), 225
Silent World, The (Cousteau), 124
Silko, Leslie Marmon, 277
Simone, Nina, 98
Simpson, O. J., 268
"Sinnerman" (song), 98
"Slide Mongoose Slide" (song), 228
Sloane, Sir Hans, 26, 75–76, 77, 80–81, 84, 241
Slovenia, 140
"Sly Mongoose" (song), 227, 228
Small House Policy, Hong Kong (1972), 16
Smithsonian Institution, 47, 79, 116
Social Darwinism, 182
Soledad Brother (Jackson), 141–42
Solenoposis azteca (giant ants), 52
Sotheby's, 89
Soufrière, La (St. Vincent) eruption (2021), 275–76

Soufrière Hills (Montserrat) eruption (1995), 268–69, 273–74
#SoundDiAbeng, 35
sou sou (loan rotation), 48, 245
South Africa, 238, 266
South America, 55, 62, 149, 176, 182, 188
Southampton, New York, 107, 250–51, 290
South Asia, 78, 239
South Asians, xvi, 118, 121, 143, 213–14, 235–37, 285–86. *See also specific locations*
South Carolina, xxvii, 100, 102, 167, 174
South China Sea, 4, 13, 15, 277
Southeast Asia, 11, 26, 56, 240
Southern Europe, 97
South Jamaica (Queens), 194–95
South Pacific, 70–71, 119–20, 134, 251, 279
Spain, xxii, xxiv, xxxiii, 8, 41, 44, 54–55, 59, 63, 72, 100, 102, 137, 148, 150, 173, 238
Spanish language, 9
Springfield Gardens (Queens), 195–96
Spring Garden estate, 220
Sranan Tongo language, 181
Sri Lanka, 72, 89, 205, 209, 213, 214–15
Standing Rock, xxvi
St. Andrew's, Jamaica, 241
Starmer, Keir, 269
St. Catherine Parish, Jamaica, 76, 245
St. Clair, Thomas Staunton, 183, 184
St. Croix, 228
Steggerda, Morris, 38–39, 38, 85
St. Elizabeth parish, Jamaica, 12, 252
Sterling, Chief Wallace, 7, 20–25, 32–33, 35
Stewart, William, 248
Sting, 274
Stinson, James, 107–9
St. Kitts, 58
St. Mary parish, Jamaica, 34
Stoler, Ann, 217–18
St. Patrick's Day, 270–71
Structure and Distribution of Coral Reefs, The (Darwin), 117–18
St. Thomas parish, Jamaica, 199
Stuart Restoration, 172
St. Vincent and the Grenadines, ix, 54, 122, 228, 255, 257, 275–76, 282
Sudanese, 97, 167
Sugarbelly, 30
sugarcane, 31, 58, 60, 66, 71–72, 76, 107, 147, 153, 158, 170, 173, 185, 204–7, 217–18, 224, 228, 230, 235, 242, 270, 276, 296
Sula (Morrison), 96
Superfund sites, xxxii, 139, 250
Suriname, ix, 8, 22, 54, 57, 88, 175–78, 188, 197, 244, 253

Sweden, 100, 102
Switzerland, xxvii–xviii
Sylbaris, Ludger, 266–68
Syria, 255, 285

Tahiti, xxv, 70–72, 117–20, 124–25, 238, 247–49
Taíno, 44, 254. *See also* Arawak
Tai Ping Shan, xxvi, xxxiii. *See also*
    Victoria Peak
Taiwan, 285
Takyi revolt, 34
Tambo tradition, 22
Tanka people, 4
Taoism, 41, 59, 293
Tao Te Ching, 293
Tate, Greg, 95–96
taxidermy, 67, 84–85, 87, 122, 123, 176, 185, 187
taxonomy, 38, 111, 113–14, 120, 177–78, 182,
    185, 187, 217, 221–22, 224, 226–27
Taylor, Breonna, 165
Taylor, Zachary, 135
TEDxParamaribo Talk, 177
Telegu language, 206
Tempest, A (Césaire), 58
Tempest, The (Shakespeare), 58, 95
Temporary Protected Status (TPS), 140
Tenerife, 238
Tennessee, 234
Tetiaroa, 120
Texas, 145
Thackeray, William Makepeace, 58
Thakhek, 240
Thames River, 71
Thanksgiving, 251
Thiselton-Dyer, William Turner, 209
Tibet, 238
Tilbury Docks, 58
Till, Emmett, 164
TILT, 251
"Timbers of Jamaica, The" (Espeut),
    220
Tiwanaku people, 148
Tlachihualtepetl, 72
Tobago, 235
Togo, xxxiv
Tombs jail, 294
Tonga, 61, 257
Tongs, 157
Torabully, Khal, 121
Toronto Stock Exchange (TSX), 11
Tosh, Peter, xxviii, 98
"Towards the Sociogenic Principle"
    (Wynter), 29
Trafalgar Falls (Dominica), 52

transatlantic slave trade, xv, xxxvii, 8, 60, 62,
    78, 95–96, 100, 102, 106, 109, 117, 123, 127,
    141–42, 151, 173. *See also* Middle Passage
Trinidad and Tobago, ix, xxvii, 48, 56, 59,
    103, 121, 151, 181, 188, 196–97, 209, 212,
    217, 222, 226–28, 235, 244, 270, 276
Trinidad Experiment, 204–6, 218
Trio, 22
Trollope, Anthony, 206, 218, 221
Trump, Donald, 76, 132, 136, 139–40, 142, 162
Trump, Melania, 140
Turtle Island, 70, 290
Tutty Gran Rosie, 4–5
Tuvalu, 67, 279
23andMe, 179
Twitter (*later* X), 70, 140, 165

Uexküll, Jakob von, 28
Ukraine, 193
Undercurrents of Power (Dawson), 106
Underground Railroad, xxi, 106
UNESCO, 88, 128–29
United Nations, xxxviii, 35, 70, 142–43, 256,
    258
United Nations Climate Change
    Conferences, 261–62
  COP 26 (2021), 259
  COP 29 (2024), 42
  COP 30 (2025), xxxi
United States, xxi, xxv, 10–13, 30, 33, 39,
    43–44, 49, 69, 81, 85, 90, 100, 135–38, 141,
    144–47, 150, 158, 192, 204, 228, 233–35,
    238, 243, 247, 259, 261, 263, 277–78,
    284–85, 291
U.S. Congress, 35, 145, 195
U.S. military, 135, 139, 146
U.S. Miscellaneous Caribbean Islands, 146
U.S. Outlying Islands, 146
U.S. South, 54, 62, 102, 133, 186
Universal Negro Improvement Association
    (UNIA), 283, 285
University College London, 211
    Institute of Americas conference
      (2017), 76
University of Arizona, 134
University of Louisville, 194
University of Michigan, 86
University of Pennsylvania, 83
University of the West Indies, 86, 209, 278,
    263
Uttar Pradesh, 232

Valkyrie, 175
Van Cortlandt Park, 83

Vanity Fair, Jamaica, 245
Vanity Fair (Thackeray), 58
Vegas, Mr., 28
Vereenigde Oost-Indische Compagnie (VOC), 89, 112
Vermont, xxi
Vice News, 36
Victoria Harbor, 4, 18
Victorian era, 62, 123–23, 197, 205, 215, 257
Victoria Peak, xxxiii. *See also* Tai Ping Shan
Vietnam War, xxviii
Virginia, 8, 59, 234
Virgin Islands, 212
Vodou, 59
volcanoes, 78, 117, 118, 120, 122, 144, 255–58, 261, 265–69, 273–79, 286
Voyage to the Islands (Sloane), 26, 75, 241
vultures, 190–91, 196–97, 283

Wailers, xxviii, 227
Waitukubuli National Trail, 50
Walcott, Derek, 223–27
Wales, 207
Wanaque Reservoir, 250
Wari people, 148
Warren, Winni, 116
WASPs, 136
Waterton, Charles, 122, 182–83
Watson, G. Llewellyn, 227
Watson, James, 37
Wayana nation, 22
Web3, 263
Weber, Max, 90
WeChat, 17
West, Kanye, 150
West Africa, xxvii, xxxiv, 3, 7–8, 21, 27, 48, 56, 58, 60–61, 76, 96, 102–3, 106, 244, 246, 253, 291, 293. *See also* Gold Coast; *and specific groups and regions*
West Indies, 26, 58, 122, 225, 215, 220, 238, 246–47, 271–72
West Indies, The (Trollope), 221
West Indies Federation, 259, 273
Westminster, 71
Westphalian order, 61, 256
West Rock summit, 173
Westward Expansion, 143
whales, 108, 125, 129–31, 251, 294
Whalley, Edward, 170, 172–73
"What Is It Like to Be a Bat?" (Nagel), 29

White, William C., 228
white flight, 73, 192–96, 222
White Lotus, The (TV show), 277
white supremacy, xxx, xxxv, 54, 61, 85, 111, 114, 120, 136, 137, 186, 195, 216, 287
Wide Sargasso Sea (Rhys), 51–52, 192
Wildlife of the Caribbean, 86
Williams, Eric, 58
Williams-Forson, Psyche A., 282
Wilmington, North Carolina, 139
Wilson, Woodrow, 144–46
Windrush scandal, 76
Windrush (ship), 58
Windsor Caves, 28, 30, 219
Windward Channel, 138
Windward Maroons, *viii,* 7
Winfrey, Oprah, 269
Winti tradition, 59
Wisconsin, 290
Witness Tree, 54–55
Wonder, Stevie, 171, 271
Wong, James, 242
Woods, Clyde, 267
World Bank, 159, 259
World Trade Organization, 48
World War II, 30, 156, 228–29, 284
World Wildlife Fund, 30
Wormsley, Alisha, 109
Wreck Alley artificial reef, 104
Wutong Mountain, 16–17
Wynter, Sylvia, 29, 58

Xaymaca, *viii,* xxxii–xxxiii, 31, 72, 138, 198. *See also* Jamaica

Yaiullo, Joe, 126
Yale University, xxiii–xxiv, 170–75, 291
Yantian, 17
Yassin, Hamza, 167
Yellen, Janet, 259
Yellow Peril, 10, 157
Yellow Vests (Gilets Jaunes), 256
Yemayá, 3, 59
Yong, Ed, 28
Yorùbá, xxxiii, xxxiv, 56, 59, 246
Yunnan Province, 239

Zheng He, xxxvi, 11
Zoological Society of London, 203, 216, 218–19

Illustration Credits

Page

27 Institute of Jamaica, 2022. Courtesy of Romil Chouhan.

38 From Davenport and Steggerda, *Race Crossing in Jamaica*

61 Image courtesy of Nasa Earth Observatory

77 Sourced from the Wellcome Collection

99 Image from Album / Alamy Stock Photo

115 Photo courtesy of Joan Murrell Owens

124 Sourced from the Wellcome Collection

154 Image from Library of Congress's Rare Book and Special Collections Division

172 Image from the Library of Congress

199 Film still from British Pathé

All other images are in the public domain or have been provided courtesy of the author.

ABOUT THE AUTHOR

TAO LEIGH GOFFE is an award-winning historian, professor, and artist who grew up between London, where she was born, and New York City. Her research explores race, climate, and creative technology. She studied English literature at Princeton University before earning a PhD at Yale University. She lives and works in Manhattan, where she is an associate professor at Hunter College, City University of New York. She is also a fellow at Harvard Kennedy School.